Applied and Numerical Harmonic Analysis

Series Editor
John J. Benedetto
University of Maryland
College Park, MD, USA

Editorial Advisory Board

Akram Aldroubi
Vanderbilt University
Nashville, TN, USA

Andrea Bertozzi
University of California
Los Angeles, CA, USA

Douglas Cochran
Arizona State University
Phoenix, AZ, USA

Hans G. Feichtinger
University of Vienna
Vienna, Austria

Christopher Heil
Georgia Institute of Technology
Atlanta, GA, USA

Stéphane Jaffard
University of Paris XII
Paris, France

Jelena Kovačević
Carnegie Mellon University
Pittsburgh, PA, USA

Gitta Kutyniok
Technische Universität Berlin
Berlin, Germany

Mauro Maggioni
Duke University
Durham, NC, USA

Zuowei Shen
National University of Singapore
Singapore, Singapore

Thomas Strohmer
University of California
Davis, CA, USA

Yang Wang
Michigan State University
East Lansing, MI, USA

For further volumes:
http://www.springer.com/series/4968

Alexander I. Saichev • Wojbor A. Woyczyński

Distributions in the Physical and Engineering Sciences, Volume 2

Linear and Nonlinear Dynamics in Continuous Media

Alexander I. Saichev
Department of Management,
  Technology, and Economics
ETH Zürich
Zürich, Switzerland

Department of Radio Physics
University of Nizhniy Novgorod
Nizhniy Novgorod, Russia

Wojbor A. Woyczyński
Department of Mathematics,
  Applied Mathematics and Statistics, and Center
  for Stochastic and Chaotic Processes in Science
  and Technology
Case Western Reserve University
Cleveland, OH, USA

Additional material to this book can be downloaded from http://extras.springer.com

ISBN 978-0-8176-3942-6   ISBN 978-0-8176-4652-3 (eBook)
DOI 10.1007/978-0-8176-4652-3
Springer New York Heidelberg Dordrecht London

Library of Congress Control Number: 96039028

Mathematics Subject Classification (2010): 31-02, 31Axx, 31Bxx, 35-02, 35Dxx, 35Jxx, 35Kxx, 35Lxx, 35Qxx, 70-02, 76-02, 76Lxx, 76Nxx, 76Sxx

© Springer Science+Business Media New York 2013
This work is subject to copyright. All rights are reserved by the Publisher, whether the whole or part of the material is concerned, specifically the rights of translation, reprinting, reuse of illustrations, recitation, broadcasting, reproduction on microfilms or in any other physical way, and transmission or information storage and retrieval, electronic adaptation, computer software, or by similar or dissimilar methodology now known or hereafter developed. Exempted from this legal reservation are brief excerpts in connection with reviews or scholarly analysis or material supplied specifically for the purpose of being entered and executed on a computer system, for exclusive use by the purchaser of the work. Duplication of this publication or parts thereof is permitted only under the provisions of the Copyright Law of the Publisher's location, in its current version, and permission for use must always be obtained from Springer. Permissions for use may be obtained through RightsLink at the Copyright Clearance Center. Violations are liable to prosecution under the respective Copyright Law.
The use of general descriptive names, registered names, trademarks, service marks, etc. in this publication does not imply, even in the absence of a specific statement, that such names are exempt from the relevant protective laws and regulations and therefore free for general use.
While the advice and information in this book are believed to be true and accurate at the date of publication, neither the authors nor the editors nor the publisher can accept any legal responsibility for any errors or omissions that may be made. The publisher makes no warranty, express or implied, with respect to the material contained herein.

Printed on acid-free paper

Springer is part of Springer Science+Business Media (www.birkhauser-science.com)

ANHA Series Preface

The *Applied and Numerical Harmonic Analysis (ANHA)* book series aims to provide the engineering, mathematical, and scientific communities with significant developments in harmonic analysis, ranging from abstract harmonic analysis to basic applications. The title of the series reflects the importance of applications and numerical implementation, but richness and relevance of applications and implementation depend fundamentally on the structure and depth of theoretical underpinnings. Thus, from our point of view, the interleaving of theory and applications and their creative symbiotic evolution is axiomatic.

Harmonic analysis is a wellspring of ideas and applicability that has flourished, developed, and deepened over time within many disciplines and by means of creative cross-fertilization with diverse areas. The intricate and fundamental relationship between harmonic analysis and fields such as signal processing, partial differential equations (PDEs), and image processing is reflected in our state-of-theart *ANHA* series.

Our vision of modern harmonic analysis includes mathematical areas such as wavelet theory, Banach algebras, classical Fourier analysis, time–frequency analysis, and fractal geometry, as well as the diverse topics that impinge on them.

For example, wavelet theory can be considered an appropriate tool to deal with some basic problems in digital signal processing, speech and image processing, geophysics, pattern recognition, biomedical engineering, and turbulence. These areas implement the latest technology from sampling methods on surfaces to fast algorithms and computer vision methods. The underlying mathematics of wavelet theory depends not only on classical Fourier analysis, but also on ideas from abstract harmonic analysis, including von Neumann algebras and the affine group. This leads to a study of the Heisenberg group and its relationship to Gabor systems, and of the metaplectic group for a meaningful interaction of signal decomposition methods. The unifying influence of wavelet theory in the aforementioned topics illustrates the justification for providing a means for centralizing and disseminating information from the broader, but still focused, area of harmonic analysis. This will be a key role of *ANHA*. We intend to publish the scope and interaction that such a host of issues demands.

Along with our commitment to publish mathematically significant works at the frontiers of harmonic analysis, we have a comparably strong commitment to publish major advances in the following applicable topics in which harmonic analysis plays a substantial role:

Biomedical signal processing	*Numerical partial differential equations*
Compressive sensing	*Prediction theory*
Communications applications	*Radar applications*
Data mining/machine learning	*Sampling theory*
Digital signal processing	*Spectral estimation*
Fast algorithms	*Speech processing*
Gabor theory and applications	*Time–frequency and time-scale analysis*
Image processing	*Wavelet theory*

The above point of view for the *ANHA* book series is inspired by the history of Fourier analysis itself, whose tentacles reach into so many fields.

In the last two centuries, Fourier analysis has had a major impact on the development of mathematics, on the understanding of many engineering and scientific phenomena, and on the solution of some of the most important problems in mathematics and the sciences. Historically, Fourier series were developed in the analysis of some of the classical PDEs of mathematical physics; these series were used to solve such equations. In order to understand Fourier series and the kinds of solutions they could represent, some of the most basic notions of analysis were defined, e.g., the concept of "function". Since the coefficients of Fourier series are integrals, it is no surprise that Riemann integrals were conceived to deal with uniqueness properties of trigonometric series. Cantor's set theory was also developed because of such uniqueness questions.

A basic problem in Fourier analysis is to show how complicated phenomena, such as sound waves, can be described in terms of elementary harmonics. There are two aspects of this problem: first, to find, or even define properly, the harmonics or spectrum of a given phenomenon, e.g., the spectroscopy problem in optics; second, to determine which phenomena can be constructed from given classes of harmonics, as done, e.g., by the mechanical synthesizers in tidal analysis.

Fourier analysis is also the natural setting for many other problems in engineering, mathematics, and the sciences. For example, Wiener's Tauberian theorem in Fourier analysis not only characterizes the behavior of the prime numbers, but also provides the proper notion of spectrum for phenomena such as white light; this latter process leads to the Fourier analysis associated with correlation functions in filtering and prediction problems, and these problems, in turn, deal naturally with Hardy spaces in the theory of complex variables.

ANHA Series Preface

Nowadays, some of the theory of PDEs has given way to the study of Fourier integral operators. Problems in antenna theory are studied in terms of unimodular trigonometric polynomials. Applications of Fourier analysis abound in signal processing, whether with the fast Fourier transform (FFT), or filter design, or the adaptive modeling inherent in time–frequency-scale methods such as wavelet theory. The coherent states of mathematical physics are translated and modulated Fourier transforms, and these are used, in conjunction with the uncertainty principle, for dealing with signal reconstruction in communications theory. We are back to the raison d'être of the *ANHA* series!

University of Maryland
College Park

John J. Benedetto

To Tanya and Liz—
  with love and respect

Introduction to Volume 2

This book continues our multivolume project that endeavors to show how the theory of distributions, also often called the theory of generalized functions, can be used by a theoretical researcher or graduate student working in the physical and engineering sciences or applied mathematics as well as by advanced undergraduate students. Our general goals, the intended audience, and the philosophy we are pursuing here are already described in detail in the introduction to Volume 1, which covers the distributional and fractal (fractional) calculus, the integral transform, and wavelets. However, given the long time that has elapsed since publication of the first volume, for the benefit of the reader of the present volume, we are repeating the main points below.

Goals and Audience: The usual calculus/differential equations sequence taken by the physical sciences and engineering majors is too crowded to include an in-depth study of many widely applicable mathematical tools that should be a part of the intellectual arsenal of any well-educated scientist or engineer. So it is common for the calculus sequence to be followed by elective undergraduate courses in linear algebra, probability, and statistics, and by a graduate course that is often labeled *Advanced Mathematics for Engineers and Scientists*. Traditionally, it contains such core topics as equations of mathematical physics, special functions, and integral transforms. This series of books is designed as a text for a modern version of such a graduate course and as a reference for theoretical researchers in the physical sciences and engineering. Nevertheless, inasmuch as it contains basic definitions and detailed explanations of a number of traditional and modern mathematical notions, it can be comfortably and profitably taken by advanced undergraduate students.

It is written from the unifying viewpoint of distribution theory, and the aim is to give readers a major modern analytic tool in their research.

Students will be able to independently attack problems in which distribution theory is of importance.

Prerequisites include a typical science or engineering three- or four-semester calculus sequence (including elementary differential equations, Fourier series, complex variables, and linear algebra—we review the basic definitions and facts as needed) and, for Volume 2, familiarity with the basic concepts and techniques developed in Volume 1. In solving some problems, familiarity with basic computer programming methods is necessary, although knowledge of a symbolic manipulation language such as Mathematica, MATLAB, or Maple would suffice. Such skills are usually acquired by students during their freshman and sophomore years.

Whereas in the first volume we delved into some foundational topics, this book concentrates on the body of mathematical techniques that are often called equations of mathematical physics and that deal with modeling issues for the linear and nonlinear dynamics of continuous media. However, our approach is different from that of most books in this area, and it consistently relies on the distributional tools and paradigm.

The book can form the basis of a special one- or two-semester course on the equations of mathematical physics and partial differential equations. Typically, a course based on this text would be taught in a mathematics or applied mathematics department or in a department of physics. However, in many schools, some other department (such as electrical, systems, mechanical, or chemical engineering) could assume responsibility for it.

Finally, we should make it clear that the book is not addressed to pure mathematicians who plan to pursue research in distributions theory. They have many other excellent sources some of which are listed in the bibliographical notes.

Philosophy: This book employs distribution theory from the applied viewpoint; abstract functional-analytic constructions are reduced to a minimum. There is no mention of the framework of nuclear locally convex topological vector spaces. The unifying theme is a study of the Dirac delta and related one- and multidimensional distributions, a rich enough family given the variety of curves and surfaces on which they can be concentrated. To be sure, these are the distributions that appear in the vast majority of problems encountered in practice. Indeed, at some point we were toying with the idea of entitling the series *The Dirac Delta in the Physical and Engineering Sciences*, but decided it was too corny and backed off.

Our choice was based on long experience in teaching graduate mathematics courses to physical scientists and engineers, which indicated to us that

Introduction to Volume 2

distributions, although commonly used in their faculty's professional work, are very seldom learned by students in a systematic fashion; there is simply not enough room in the engineering curricula. This induced us to weave distributions into an exposition of integral transforms (including wavelets and fractal calculus), equations of mathematical physics, and random fields and signals, where they enhance the presentation and allow the student to achieve both additional insight into the subject matter and a degree of computational efficiency. In some sense, their use replaces the formal limit procedures very few scientists and engineers have stomach for by reliable and efficient algorithmic operations. Dirac deltas are used as a practical computational tool (for engineers and scientists it is a long-established shortcut) and permit the reader to avoid (fairly safely) the delicate issues of convergence in the weak and vague (etc. etc.) topologies that normally petrify practitioners in the applied sciences.

Also, we should mention that we were careful not to be too religiously orthodox about using distributional tools in applied problems to the exclusion of other approaches. When a simpler and more commonsense solution was available, we just went for it, as any applied scientist would.

We have made an effort to be reasonably rigorous and general in our exposition: results are proved and assumptions are formulated explicitly, and in such a way that the resulting proofs are as simple as possible. Since in realistic situations, similar sophisticated assumptions may not be valid, we often discuss ways to expand the area of applicability of the results under discussion. Throughout we endeavor to favor constructive methods and to derive concrete relations that permit us to arrive at numerical solutions. Ultimately, this is the essence of most of problems in the applied sciences, and we treat the job of illuminating each problem from both the mathematical and physical perspectives as essential to our success.

As a byproduct, the book should help in improving communication between applied scientists and mathematicians. The first group is often only vaguely aware of the variety of modern mathematical tools that can be applied to physical problems, while the second is often innocent of how physicists and engineers reason about their problems and how they adapt purely mathematical theories to turn them into effective tools. Experts in one narrow area often do not see the vast chasm between the mathematical and physical mentalities. For instance, a mathematician rigorously proves that

$$\lim_{x \to \infty} \bigl(\log(\log x)\bigr) = \infty,$$

while a physicist, not generally disposed to follow the same logic, might say:

> Wait a second, let's check the number 10^{100}, which is bigger than most physical quantities—I know that the number of atoms in our galaxy is less than 10^{70}. The iterated logarithm (in base 10) of 10^{100} is only 2, and this seems to be pretty far from infinity.

This little story illustrates the sort of psychological difficulties that one encounters in writing a book such as this one.

Finally, it is worth mentioning that some portions of the material, especially the parts dealing with the basic distributional formalism, can be treated within the context of symbolic manipulation languages such as Maple or Mathematica, where the package `DiracDelta.m` is available. Their use in student projects can enhance the exposition of the material contained in this book, both in terms of symbolic computation and visualization. We have used them successfully with our students.

Organization: Major topics included in the book are divided among three volumes: Volume 1, *Distributions in the Physical and Engineering Sciences: Distributional and Fractal Calculus, Integral transforms and Wavelets*, contained the following:

Part I. *Distributions and Their Basic Physical Applications*, containing the basic formalism and generic examples;

Part II. *Integral Transforms and Divergent Series*, which contains chapters on Fourier, Hilbert, and wavelet transforms and an analysis of the uncertainty principle, divergent series, and singular integrals.

The present Volume 2, *Distributions in the Physical and Engineering Sciences: Linear and Nonlinear Dynamics of Continuous Media*, is also divided into two parts:

Part III. *Potentials, Diffusions and Waves* contains an analysis of the three basic types of linear partial differential equations: elliptic, parabolic, and hyperbolic;

Part IV. *Nonlinear Partial Differential Equations* contains chapters on one- and multidimensional first-order nonlinear partial differential equations and conservation laws, generalized solutions of first-order nonlinear partial differential equations, Kardar–Parisi–Zhang (KPZ) and Burgers' equations, Korteweg-de Vries (KdV) equations, the equations of gas dynamics, and flows in porous media.

Finally, the third and last volume of this series, *Distributions in the Physical and Engineering Sciences: Random and Fractal Signals and Fields* (in preparation), will contain the following:

Part V. *Random Signals and Fields* will include chapters on probability theory, stationary signals and white noise, stochastic differential equations and generalized random fields, statistics of turbulent fluids, and branching processes;

Part VI. *Fractal Random Dynamics in Continuous Media* will contain an exposition of anomalous super- and subdiffusions, both linear and nonlinear, and will employ the tools of fractional calculus developed in Volume 1 in the context of applied problems.

The needs of the applied sciences audience are addressed by a careful and rich selection of examples arising in real-life industrial and scientific labs and a thorough discussion of their physical significance. They form the background for our discussions as we proceed through the material. Numerous illustrations help the reader attain a better understanding of the core concepts discussed in the text.

The many exercises at the end of each chapter expand on themes developed in the main text. Perhaps the name "exercises" is something of a euphemism here, since many of them are really mini research projects. For that reason, we have deliberately included solutions to all of them in the appendix. This was not a lighthearted decision. Some of the problems may be challenging, but they provide an essential complement to the material in the book. We ourselves teach this type of material in graduate and advanced undergraduate classes that proceed through a lectures–take-home-project–lectures sequence rather than through the usual sequence of lectures followed by an in-class exam. But the projects we assign are usually some variations of the exercises in our books, so the question of copying the solutions verbatim does not arise. On the other hand, having a complete set of solutions gives both teachers and students the confidence to attempt the more difficult problems in the knowledge that even if they do not succeed independently, they can learn why they failed. This resource also serves as a replacement for what in calculus courses are recitation sections. Including all the solutions involved considerable extra work, but following publication of the first volume, we received quite a bit of feedback from industrial and government researchers thanking us for them. So we have done this again in Volume 2.

A word about notation and the numbering system for formulas. A list of notation follows this introduction. Formulas are numbered separately in each section to reduce clutter, but outside the section in which they appear, referred to by three numbers. For example, formula (4) in Sect. 3 of Chap. 1 will be referred to as formula (1.3.4) outside Sect. 1.3, but only as formula (4) within Sect. 1.3. Sections and chapters can be easily located via the running heads.

Finally, a Springer Extras *Appendix C: Distributions, Fourier Transform, and Divergent Series*, freely available at http://extras.springer.com, provides a compact version of the foundational material on distributions (generalized functions) contained in Volume 1 of this book series. It explains the basic concepts and applications needed for the development of the theory of linear and nonlinear dynamics in continuous media and its diverse applications in physics, engineering, biology, and economics as presented in Volume 2. The goal is to make the present book more self-contained if the reader does not have easy access to the first volume. We aimed at a compression level of about 30 %, a compromise that permits the reader to obtain sufficient (for our purposes) operational acquaintance with the distributional techniques while skipping more involved theoretical arguments. Obviously, browsing through it is no replacement for a thorough study of Volume 1. The structure of Appendix C roughly mimics that of Volume 1, with chapters replaced by sections, sections by subsections, etc., while in some cases several units were merged into one.

Acknowledgments: The authors would like to thank David Gurarie, of the Mathematics Department, Case Western Reserve University, Cleveland, Ohio; Valery I. Klyatskin, of the Institute for Atmospheric Physics, Russian Academy of Sciences, Moscow, Russia; and Gennady Utkin, of the Radiophysics Faculty of the Nizhny Novgorod University, Nizhny Novgorod, Russia, who read parts of the book and offered their valuable comments. Martin W. Woyczynski, of Humedica, Inc., Boston, Massachusetts, kindly read through several chapters of the book in great detail and offered valuable advice on the style and clarity of our exposition. We appreciate his assistance. Finally, the anonymous referees issued reports on the original version of the book that we found extremely helpful. They led to several revisions of our initial text. The original Birkhäuser editors, Ann Kostant and Tom Grasso, took the series under their wing, and we are grateful to them for their encouragement and help in producing the final copy. Their successors, Allen Mann, Mitch Moulton, and Brian Halm, ably guided the present volume to its successful completion.

Introduction to Volume 2

About the Authors: *Alexander I. Saichev* received his B.S. in the Radio Physics Faculty at Gorky State University, Gorky, Russia, in 1969, a Ph.D. from the same faculty in 1975 for a thesis on kinetic equations of nonlinear random waves, and his D.Sc. from the Gorky Radiophysical Research Institute in 1983 for a thesis on propagation and backscattering of waves in nonlinear and random media. Since 1980 he has held a number of faculty positions at Gorky State University (now Nizhny Novgorod University) including senior lecturer in statistical radio physics and professor of mathematics and chairman of the mathematics department. Since 1990 he has visited a number of universities in the West, including Case Western Reserve University, the University of Minnesota, New York University, and the University of California, Los Angeles. He is coauthor of the monograph *Nonlinear Random Waves and Turbulence in Nondispersive Media: Waves, Rays and Particles* and has served on the editorial boards of *Waves in Random Media* and *Radiophysics and Quantum Electronics*. His research interests include mathematical physics, applied mathematics, waves in random media, nonlinear random waves, and the theory of turbulence. In 1997 he was awarded the Russian Federation's State Prize and Gold Medal for research in the area of nonlinear and random fields. He is currently professor of mathematics at the Radio Physics Faculty of the Nizhny Novgorod University and a Professor in the Department of Management, Technology, and Economics at the Swiss Federal Institute of Technology (ETH) in Zurich, Switzerland.

Wojbor A. Woyczyński received his B.S./M.Sc. in electrical and computer engineering from Wrocław Polytechnic in 1966 and a Ph.D. in mathematics in 1968 from Wrocław University, Poland. He moved to the United States in 1970, and since 1982 has been professor of mathematics and statistics at Case Western Reserve University, in Cleveland, and served as chairman of the department there from 1982 to 1991. He has held tenured faculty positions at Wrocław University, Poland, and at Cleveland State University, and visiting appointments at Carnegie Mellon University, Northwestern University, the University of North Carolina, the University of South Carolina, the University of Paris, Göttingen University, Aarhus University, Nagoya University, the University of Tokyo, the University of Minnesota, the National University of Taiwan, Taipei, and the University of New South Wales, in Sydney. He is also (co-)author and/or editor of twelve books on probability theory, harmonic and functional analysis, and applied mathematics, and serves as a member of the editorial board of the *Applicationes Mathematicae* and as a managing editor of *Probability and Mathematical Statistics*. His research interests include probability theory, stochastic models, functional analysis, and partial differential equations and their applications in statistics, statistical physics, surface chemistry, hydrodynamics, and biomedicine. He is currently a professor in the Department of Mathematics, Applied Mathematics and Statistics, and Director of the Case Center for Stochastic and Chaotic Processes in Science and Technology.

Zürich, Switzerland
Cleveland, OH, USA

Alexander I. Saichev
Wojbor A. Woyczyński

In memoriam notice: On June 8, 2013, my friend, collaborator and co-author Alexander I. Saichev unexpectedly passed away as a result of a short illness. He has been widely appreciated as a mathematician, teacher, and colleague. The book would hardly have been written without his energy, knowledge and enthusiasm for the project. He will be missed by all of us.

W.A.W.

Contents

Introduction to Volume 2   xi

Notation   xxiii

III   Potentials, Diffusions, and Waves   1

9 Potential Theory and Fundamental Solutions of Elliptic Equations   3
- 9.1 Poisson Equation 3
- 9.2 1-D Helmholtz Equation 9
- 9.3 3-D Helmholtz Equation 12
- 9.4 2-D Helmholtz Equation 13
- 9.5 Diffraction of a Monochromatic Wave 17
- 9.6 Helmholtz Equation in Inhomogeneous Media 29
- 9.7 Waves in Waveguides 33
- 9.8 Sturm–Liouville Problem 37
- 9.9 Waves in Waveguides Revisited 53
- 9.10 Exercises 57

10 Diffusions and Parabolic Evolution Equations   59
- 10.1 Diffusion Equation and Its Green's Function 59
- 10.2 Self-Similar Solutions 60
- 10.3 Well-Posedness of Initial Value Problems with Periodic Data 64
- 10.4 Complex Parabolic Equations 65
- 10.5 Fresnel Zones 68
- 10.6 Multidimensional Parabolic Equations 72
- 10.7 The Reflection Method 75
- 10.8 Moving Boundary: The Detonating Fuse Problem . . . 81

10.9 Particle Motion in a Potential Well	87
10.10 Exercises	89

11 Waves and Hyperbolic Equations 93
11.1 Dispersive Media 94
11.2 Examples of Dispersive Media 97
11.3 Integral Laws of Motion for Wave Packets 100
11.4 Asymptotics of Waves in Dispersive Media 102
11.5 Energy Conservation Law in the Stationary Phase Method . . 104
11.6 Wave as Quasiparticle 106
11.7 Wave Packets with Narrow-Band Spectrum 110
11.8 Optical Wave Behind a Phase Screen 111
11.9 One-Dimensional Phase Screen 114
11.10 Caustics 115
11.11 Telegrapher's Equation 119
11.12 Exercises 141

IV Nonlinear Partial Differential Equations 143

12 First-Order Nonlinear PDEs and Conservation Laws 145
12.1 Riemann Equation 145
12.2 Continuity Equation 150
12.3 Interface Growth Equation 158
12.4 Exercises 167

13 Generalized Solutions of First-Order Nonlinear PDEs 171
13.1 Master Equations 172
13.2 Multistream Solutions 174
13.3 Summing over Streams in Multistream Regimes 180
13.4 Weak Solutions of First-Order Nonlinear Equations 186
13.5 E–Rykov–Sinai Principle 198
13.6 Multidimensional Nonlinear PDEs 214
13.7 Exercises 223

14 Nonlinear Waves and Growing Interfaces: 1-D Burgers–KPZ Models 229
14.1 Regularization of First-Order Nonlinear PDEs: Burgers, KPZ, and KdV Equations 229
14.2 Basic Symmetries of the Burgers and KPZ Equations and Related Solutions 236

	14.3 General Solutions of the Burgers Equation	254
	14.4 Evolution and Characteristic Regimes of Solutions of the Burgers Equation	258
	14.5 Exercises	278

15 Other Standard Nonlinear Models of Higher Order   281
    15.1 Model Equations of Gas Dynamics 281
    15.2 Multidimensional Nonlinear Equations 289
    15.3 KdV Equation and Solitons 294
    15.4 Flows in Porous Media 320
    15.5 Exercises 324

A Appendix A: Answers and Solutions   327
    Chapter 9 327
    Chapter 10 330
    Chapter 11 340
    Chapter 12 352
    Chapter 13 362
    Chapter 14 381
    Chapter 15 392

B Appendix B: Bibliographical Notes   397

Index   405

Notation

$\lceil \alpha \rceil$	Least integer greater than or equal to α
$\lfloor \alpha \rfloor$	Greatest integer less than or equal to α
C	Concentration
$\mathbf{C}$	Complex numbers
$C(x)$	$= \int_0^x \cos(\pi t^2/2)\, dt$, Fresnel cosine integral
C^∞	Space of smooth (infinitely differentiable) functions
$\mathcal{D}$	$= C_0^\infty$, Space of smooth functions with compact support
$\mathcal{D}'$	Dual space to $\mathcal{D}$, space of distributions
$\bar{D}$	The closure of domain D
D/Dt	$= \partial/\partial t + \boldsymbol{v} \cdot \boldsymbol{\nabla}$, substantial derivative
$\delta(x)$	Dirac delta centered at 0
$\delta(x-a)$	Dirac delta centered at a
Δ	Laplace operator
$\mathcal{E}$	$= C^\infty$-space of smooth functions
$\mathcal{E}'$	Dual to $\mathcal{E}$, space of distributions with compact support
$\operatorname{erf}(x)$	$= (2/\sqrt{\pi}) \int_0^x \exp(-s^2)\, ds$, The error function
$\tilde{f}(\omega)$	Fourier transform of $f(t)$
$\{f(x)\}$	Smooth part of function f
$\lfloor f(x) \rceil$	Jump of function f at x
ϕ, ψ	Test functions
$\gamma(x)$	Canonical Gaussian density
$\gamma_\epsilon(x)$	Gaussian density with variance ϵ
$\Gamma(s)$	$= \int_0^\infty e^{-t} t^{s-1} dt$, gamma function
(h, g)	$= \int h(x) g(x) dx$, the Hilbert space inner product
$\chi(x)$	Canonical Heaviside function, unit step function
$\hat{H}$	The Hilbert transform operator
j, J	Jacobians
$I_A(x)$	The indicator function of set A ($= 1$ on A, $= 0$ off A)
$\operatorname{Im} z$	The imaginary part of z
$\lambda_\epsilon(x)$	$= \pi^{-1} \epsilon (x^2 + \epsilon^2)^{-1}$, Cauchy density

$L^p(A)$	Lebesgue space of functions f with $\int_A |f(x)|^p\,dx < \infty$
$\mathbf{N}$	Nonnegative integers
$\phi = O(\psi)$	ϕ is of order not greater than ψ
$\phi = o(\psi)$	ϕ is of order smaller than ψ
$\mathcal{PV}$	Principal value of the integral
$\mathbf{R}$	Real numbers
$\mathbf{R}^d$	d-dimensional Euclidean space
Re z	The real part of z
ρ	Density
sign (x)	$= 1$ if $x > 0$, -1 if $x < 0$, and 0 if $x = 0$
sinc ω	$= \sin \pi\omega / \pi\omega$
$\mathcal{S}$	Space of rapidly decreasing smooth functions
$\mathcal{S}'$	Dual to $\mathcal{S}$, space of tempered distributions
$S(x)$	$= \int_0^x \sin(\pi t^2/2)\,dt$, Fresnel sine integral
T, S	Distributions
$T[\phi]$	Action of T on test function ϕ
T_f	Distribution generated by function f
$\tilde{T}$	Generalized Fourier transform of T
z^*	Complex conjugate of z
$\mathbf{Z}$	Integers
∇	Gradient operator
$\mapsto$	Fourier map
$\to$	Converges to
$\Rightarrow$	Uniformly converges to
$*$	Convolution
$[\![\,\cdot\,]\!]$	Physical dimensionality of a quantity
$\emptyset$	Empty set
∎	End of proof, example
$\boldsymbol{x} \cdot \boldsymbol{y}$	Dot (also called inner) product of $\boldsymbol{x}$ and $\boldsymbol{y}$
$|\boldsymbol{x}|$	$= \sqrt{\boldsymbol{x} \cdot \boldsymbol{x}}$, the norm of vector $\boldsymbol{x}$
$\int$	The integral over the whole space

Part III

Potentials, Diffusions, and Waves

Chapter 9

Potential Theory and Fundamental Solutions of Elliptic Equations

This chapter is devoted to the theory of linear elliptic partial differential equations and the related problems of potential theory. The basic concept of the Green's function and the source solution are introduced and explored. This is followed by a detailed analysis of the Helmholtz equation in one, two, and three dimensions with applications to the diffraction problem for monochromatic waves. The inhomogeneous media case sets the stage for the Helmholtz equation with a variable coefficient and an analysis of waves in waveguides. The latter can be reduced to the celebrated Sturm–Liouville problem, and we study properties of its eigenvalues and eigenfunctions.

9.1 Poisson Equation

The analytic properties of distributions, such as their infinite differentiability, make them a handy tool in mathematical physics. They allow us to construct generalized (and in particular, fundamental) solutions of partial differential equations.

A *fundamental solution* $G(\boldsymbol{x}, \boldsymbol{y})$, $\boldsymbol{x}, \boldsymbol{y} \in \mathbf{R}^n$ (also called a *Green's function*) corresponding to a given linear partial differential operator $L(\boldsymbol{x}, \partial/\partial \boldsymbol{x})$ is defined as a solution of the equation

$$LG(\boldsymbol{x}, \boldsymbol{y}) = \delta(\boldsymbol{x} - \boldsymbol{y}). \tag{1}$$

A Green's function depends on two parameters: the coordinates $\boldsymbol{x}$ of the current point and the source coordinates $\boldsymbol{y}$. The above definition is somewhat incomplete, since the boundary conditions are missing. These necessary details will be discussed separately for each type of equation to be considered in this chapter.

Once the fundamental solution is found, every solution of the corresponding inhomogeneous problem

$$Lu(\boldsymbol{x}) = g(\boldsymbol{x}), \qquad \boldsymbol{x} \in V \subset \mathbf{R}^n, \tag{2}$$

with an arbitrary function $g(\boldsymbol{x})$ on the right-hand side, can be found by a simple integration. Indeed, the function

$$u(\boldsymbol{x}) = \iint G(\boldsymbol{x}, \boldsymbol{y}) g(\boldsymbol{y}) d^n y \tag{3}$$

satisfies (2), as is easily verified by an application of the operator L to both sides of (3), taking into account (1), and then applying the probing property of the Dirac delta.

We begin with fundamental solutions of several simple elliptic equations that are frequently encountered in applications. All of them involve the *Laplace operator* (or *Laplacian*) Δ, which in the n-dimensional Cartesian coordinate system has the form

$$\Delta = \frac{\partial^2}{\partial x_1^2} + \cdots + \frac{\partial^2}{\partial x_n^2}.$$

For this reason, we will try to find first the so-called principal singularity of a fundamental solution corresponding to the Laplacian, i.e., a distribution $G_0(\boldsymbol{x})$ satisfying the *Poisson equation*

$$\Delta G_0(\boldsymbol{x}) = \delta(\boldsymbol{x}), \tag{4}$$

in spaces of dimension $n = 1, 2, 3$.

1-D case. Consider the 1-D case ($n = 1$), that is, the equation

$$\frac{d^2}{dx^2} G_0(x) = \delta(x). \tag{5}$$

We shall assume symmetry of the solution in both directions on the x-axis and look for an even Green's function. A solution

$$G_0(x) = |x|/2 \tag{6}$$

9.1. Poisson Equation

of (5) has already been encountered in Volume 1. It is clear, however, that any function that differs from the above function $G_0(x)$ by a solution of the corresponding homogeneous equation

$$\frac{d^2}{dx^2} G_0(x) = 0$$

is also a fundamental solution, and that a difference of two fundamental solutions is a solution of the homogeneous problem. Hence, the general solution of (5) is of the form

$$G_0(x) = |x|/2 + Ax + B,$$

where A and B are arbitrary constants. A similar comment is applicable to the 2-D and 3-D cases considered below.

2-D case. We shall show that the fundamental solution of a two-dimensional Poisson equation

$$\Delta G_0 = \left(\frac{\partial^2}{\partial x_1^2} + \frac{\partial^2}{\partial x_2^2}\right) G_0(\boldsymbol{x}) = \delta(\boldsymbol{x}) \tag{7}$$

is a function of the form

$$G_0(\boldsymbol{x}) = \frac{1}{2\pi} \ln \rho, \quad \rho = |\boldsymbol{x}| = \sqrt{x_1^2 + x_2^2}. \tag{8}$$

The verification depends on the *second Green's identity* of classical calculus,

$$\iiint_V (f\Delta\phi - \phi\Delta f)\, d^n x = \oint_S \Big(f(\boldsymbol{n} \cdot \nabla\phi) - \phi(\boldsymbol{n} \cdot \nabla f)\Big)\, dS, \tag{9}$$

which is here written for a function f that is twice continuously differentiable in a closed bounded domain V with a smooth-boundary hypersurface S and an external normal $\boldsymbol{n}$.[1]

First, let us check that the function (8) satisfies the 2-D Laplace equation

$$\Delta G_0 \equiv 0 \tag{10}$$

in the whole space with the exception of the point $\boldsymbol{x} = 0$. We shall do so by introducing the polar coordinate system

---

[1] As in Volume 1, ϕ stands, usually, for an infinitely differentiable test function with compact support contained in V (in this case).

Chapter 9. Potential Theory and Elliptic Equations

$$x_1 = \rho \cos\varphi, \quad x_2 = \rho \sin\varphi.$$

The Laplacian in the polar coordinate system is expressed by the formula

$$\Delta = \frac{1}{\rho}\frac{\partial}{\partial \rho}\left(\rho \frac{\partial}{\partial \rho}\right) + \frac{1}{\rho^2}\frac{\partial^2}{\partial \varphi^2}.$$

Thanks to the radial symmetry of the function G_0, we have $\partial^2 G_0/\partial \varphi^2 = 0$, so that to prove (10), it suffices to show that

$$\frac{1}{\rho}\frac{d}{d\rho}\left(\rho \frac{d}{d\rho}\right) G_0 \equiv 0.$$

The latter can be checked, for any $\rho > 0$, by direct substitution.

Now we shall prove that in a generalized sense, the function G_0 appearing in (8) satisfies the Poisson equation (7). In other words, for any test function $\phi(\boldsymbol{x}) \in \mathcal{D}$,

$$\phi(0) = \iint \phi \Delta G_0 \, d^2x = \iint G_0 \Delta \phi \, d^2x, \tag{11}$$

the last equality being justified by the definition of the distributional derivative. Indeed, let us cut out from the whole plane of integration $\mathbf{R}^2$ a disk of radius ε, and write

$$\iint G_0 \Delta \phi \, d^2 x = \iint_{\rho < \varepsilon} G_0 \Delta \phi \, d^2 x + \iint_{\rho \geq \varepsilon} G_0 \Delta \phi \, d^2 x. \tag{12}$$

Furthermore, note that the contribution from the first integral on the right-hand side tends to zero as $\varepsilon \to 0$. This follows from the boundedness of all the derivatives of any test function (implied by each derivative's continuity and compact support via the Weierstrass theorem), so that $|\Delta \phi| < K < \infty$, and as a result,

$$\left|\iint_{\rho<\varepsilon} G_0 \Delta\phi \, d^2x\right| < K \left|\iint_0^\varepsilon \rho \ln(\rho) d\rho\right| \sim -K \frac{1}{2}\varepsilon^2 \ln \varepsilon \to 0.$$

Therefore, asymptotically,

$$\iint G_0 \Delta \phi \, d^2 x \sim \iint_{\rho \geq \varepsilon} G_0 \Delta \phi \, d^2 x, \quad \varepsilon \to 0. \tag{13}$$

9.1. Poisson Equation

Since $G_0 \in C^\infty$ in the integration region $\rho \geq \varepsilon$, one can transform the latter integral with the help of Green's formula (9), taking as f the anticipated fundamental solution G_0 from (8). Also note that the boundedness of the region V automatically guarantees compactness of the support of any test function ϕ. So, since the identity (10) is valid for every $\rho > 0$, Green's formula (9) implies that

$$\iint_{\rho \geq \varepsilon} G_0 \Delta \phi \, d^2x = \oint_{\rho = \varepsilon} \Big(G_0 (\boldsymbol{n} \cdot \nabla \phi) - \phi (\boldsymbol{n} \cdot \nabla G_0) \Big) dl,$$

where, in the 2-D case being considered, the integral on the right-hand side is a line integral along the circle $\rho = \varepsilon$. The integral of the first term, like the first integral on the right-hand side of (12), vanishes as $\varepsilon \to 0$. Since

$$\boldsymbol{n} \cdot \nabla G_0 = -\frac{d}{d\rho} G_0 = -\frac{1}{2\pi \rho}, \qquad (14)$$

the integral of the second term,

$$-\oint_{\rho = \varepsilon} \phi (\boldsymbol{n} \cdot \nabla G_0) dl = \frac{1}{2\pi \varepsilon} \oint_{\rho = \varepsilon} \phi \, dl,$$

equals the mean value of function $\phi(\boldsymbol{x})$ on the circle $\rho = \varepsilon$. In view of the continuity of ϕ, this mean value tends to $\phi(0)$ as $\varepsilon \to 0$. This completes the proof of formula (11).

3-D case. In a similar fashion, one can prove that the fundamental solution of the Poisson equation (4) in the 3-D case ($n = 3$) has the form

$$G_0(\boldsymbol{x}) = -\frac{1}{4\pi |\boldsymbol{x}|}. \qquad (15)$$

We would now like to make the reader aware that compared to the 1-D and 2-D problems, the above 3-D problem has additional interesting aspects. Observe that the Laplace operator, as well as other operators expressing fundamental laws of physics, enjoys various symmetry properties. The Laplace operator is invariant under translations, reflections, and rotations of space. This means, in particular, that if the origin of the Cartesian coordinate system is shifted and if that operation is followed by a rotation of the whole space, then in the new coordinate system, the expression for the Laplace operator will not change. The fundamental solution of the Poisson equation

$$\Delta G(\boldsymbol{x}, \boldsymbol{y}) = \delta(\boldsymbol{x} - \boldsymbol{y})$$

enjoys the same symmetry properties. In unbounded space it is of the form

$$G(x,y) = G(|x-y|) = -\frac{1}{4\pi|x-y|},$$

thus depending only on the distance between the "observation" point x and the "source" point y. This gives rise to the representation

$$\delta(x-y) = -\frac{1}{4\pi}\Delta\frac{1}{|x-y|} \qquad (16)$$

of the shifted Dirac delta on 3-D space, which is often encountered in the physics literature.

The presence of boundary surfaces, or as mathematicians say, boundary conditions, can destroy—completely or partially—the above symmetry of fundamental solutions, and therefore make the search for them much more difficult. However, sometimes the boundaries themselves have symmetries that makes it possible to use the knowledge of fundamental solutions in the whole space.

Example 1. Consider the Green's function for the half-space $x_3 > 0$, with the extra condition that it vanishes on the "reflecting" boundary $x_3 = 0$. It is easy to see that this Green's function can be obtained by subtracting the Green's function for the whole space from its "mirror" reflection in the plane $x_3 = 0$:

$$G_0(x,y) = \frac{1}{4\pi}\left[\frac{1}{\sqrt{(x_1-y_1)^2+(x_2-y_2)^2+(x_3+y_3)^2}} - \frac{1}{\sqrt{(x_1-y_1)^2+(x_2-y_2)^2+(x_3-y_3)^2}}\right],$$
$$(x_3, y_3 > 0).$$

The above *reflection method* can also be used in the case of regions more complex than the half-space, such as a wedge-shaped region with reflecting boundaries at the angle $\alpha = \pi/n$. The corresponding Green's function can be easily written analytically with the help of a schematic diagram, which, in the case of $n = 3$, is shown in Fig. 9.1.1.

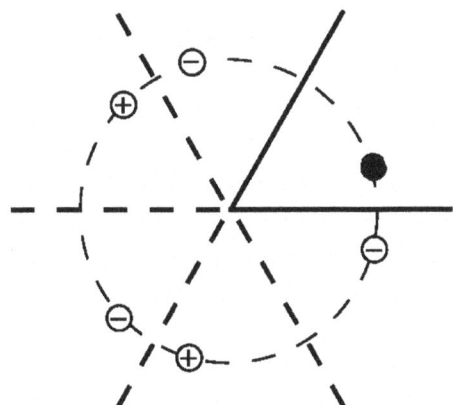

FIGURE 9.1.1
Illustration of the reflection method in the case of a wedge region. A *solid dot* marks the position of the real source, while *hollow dots* indicate positions of imaginary sources. Signs + and − signify whether the sign of the imaginary source is the same as that of the real source or the opposite.

9.2 1-D Helmholtz Equation

In this section we construct the fundamental solution of the somewhat more complicated problem described by the 1-D *Helmholtz equation*

$$\frac{d^2}{dx^2}G(x) + k^2 G(x) = \delta(x). \tag{1}$$

We begin by outlining a useful universal asymptotic relationship between the fundamental solution G of an arbitrary elliptic equation (e.g., (1), and the Green's function G_0 of the corresponding Poisson equation (9.1.5). Roughly speaking, in any elliptic equation, the highest-order derivatives determine the nature of the principal singularity of the fundamental solutions. Since the Laplacian of the Poisson equation contains all the highest-order derivatives of the Helmholtz equation, we have the asymptotic relation[2]

$$G \sim G_0, \qquad (x \to 0). \tag{2}$$

---

[2] Recall that G_0 appears here as a function of the Euclidean distance, so it is also a function of two variables.

Now, let us return to the 1-D Helmholtz equation (1). Initially, we shall solve it in the region $x > 0$, where $\delta(x) \equiv 0$, and the equation becomes the homogeneous equation

$$\frac{d^2}{dx^2}G(x) + k^2 G(x) = 0.$$

Its general solution has the form

$$G(x) = Ae^{ikx} + Be^{-ikx}. \tag{3}$$

In order to select the unique solution of interest to us, we must determine the values of the constants appearing in (3). They can be specified if two conditions are imposed on the solution. The first is dictated by the asymptotic relation (2),

$$G(x) \sim G_0(x) = |x|/2, \quad (x \to 0). \tag{4}$$

The second condition, at $x \to \infty$, is chosen on physical grounds and is dictated by the celebrated *radiation condition*, which we shall formulate a little later.

At this point it is worthwhile to recall that the Helmholtz equation (1) is closely related to the *wave equation*,

$$\frac{\partial^2}{\partial x^2} E(x,t) = \frac{1}{c^2} \frac{\partial^2}{\partial t^2} E(x,t) + D(x,t), \tag{5}$$

where the function $D(x,t)$ describes the wave source. Indeed, if the source is *monochromatic* with frequency ω, then it is analytically convenient to solve the wave equation with the help of a complex wave function $U(x,t)$ such that the real wave solution is given by

$$E(x,t) = \operatorname{Re} U(x,t).$$

The corresponding complex source function

$$W(x,t) = w(x)e^{i\omega t} \tag{6}$$

is related to the real source through the equality

$$D(x,t) = \operatorname{Re} W(x,t).$$

The monochromatic source radiates a monochromatic wave of the same frequency. This means that the complex solution

$$U(x,t) = u(x)e^{i\omega t} \tag{7}$$

9.3. 3-D Helmholtz Equation

has a form identical to (6). Substituting $U(x,t)$ (7) for $E(x,t)$, and $W(x,t)$ (6) for $D(x,t)$ in (5), and then canceling the common factor $\exp(i\omega t)$, we arrive at the Helmholtz equation

$$\frac{d^2}{dx^2}u(x) + k^2 u(x) = w(x), \tag{8}$$

which describes the complex amplitude $u(x)$ of a monochromatic wave. The constant

$$k = \frac{\omega}{c},$$

where ω is the wave frequency and c its speed, has a clearcut physical meaning and is called the *wavenumber*. Its dimension is reciprocal length. Alternatively,

$$k = 2\pi/\lambda,$$

where λ is the *wavelength*.

We are now ready to formulate the radiation condition. It postulates that the only wave present in space is the wave radiated by the source. Physically, it is obvious that any source is of finite spatial extent. As a mathematical condition, this translates to compactness of the support of the function $D(x)$. Without loss of generality we can assume that the support is contained in the negative half-line. This implies, in particular, that to the right of the source, where $D(x) \equiv 0$, the solution has a traveling waveform $E(x - ct)$, and the corresponding complex monochromatic wave is proportional to

$$\exp(i\omega t - ikx).$$

Comparing this radiating monochromatic wave with the general solution (3), we see that the radiation condition implies $A = 0$.

The assumed evenness of Green's function (or equivalently, the radiation condition on both sides of the source) gives

$$G(x) = B \exp(-ik|x|).$$

To separate the principal singularity (for $x \to 0$) of this expression, we shall use Euler's formula

$$G(x) = B \cos(k|x|) - iB \sin(k|x|).$$

The first term on the right-hand side is infinitely differentiable and does not contribute to the singularity. However, as $x \to 0$,

$$iB \sin(k|x|) \sim iBk|x|,$$

so that the second term behaves like the Green's function of the 1-D Poisson equation (9.1.6). Taking into account the asymptotic relation (4), we obtain $B = i/2k$. As a result, the Green's function of the 1-D Helmholtz equation has the form

$$G(x) = -\frac{1}{2ik} \exp(-ik|x|). \tag{9}$$

9.3 3-D Helmholtz Equation

For the 3-D Helmholtz equation

$$\Delta G + k^2 G = \delta(\boldsymbol{x}), \tag{1}$$

considered in the whole space, a Green's function is spherically symmetric, depends only on $r = |\boldsymbol{x}|$, and is governed by the homogeneous equation

$$\frac{1}{r^2} \frac{d}{dr}\left(r^2 \frac{d}{dr}\right) G + k^2 G = 0, \qquad r > 0. \tag{2}$$

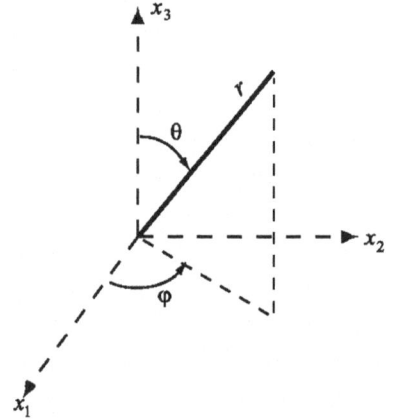

FIGURE 9.3.1
The spherical coordinate system.

It is obtained from (1) by passing from the Cartesian coordinates (x_1, x_2, x_3) to the spherical coordinates (r, θ, φ) (Fig. 9.3.1). The latter are connected with the former by a system of equations

$$x_1 = r \sin\theta \cos\varphi, \quad x_2 = r \sin\theta \sin\varphi, \quad x_3 = r \cos\theta.$$

The Laplacian in spherical coordinates is given by the expression

$$\Delta G = \frac{1}{r^2}\frac{\partial}{\partial r}\left(r^2\frac{\partial}{\partial r}\right)G + \frac{1}{r^2 \sin\theta}\frac{\partial}{\partial \theta}\left(\sin\theta\frac{\partial}{\partial \theta}\right)G + \frac{1}{r^2 \sin^2\theta}\frac{\partial^2}{\partial^2\varphi}G,$$

and the Helmholtz equation (1) for a spherically symmetric (i.e., independent of angles θ and φ) function G becomes the ordinary differential equation (2).

A substitution $v(r) = G(r)r$ reduces (2) to the equation

$$v'' + k^2 v = 0,$$

with constant coefficients, for an unknown function v. Thus, a solution of (2) satisfying the radiation condition has the form

$$G(r) = B r^{-1} e^{-ikr}.$$

A comparison of the singular (nondifferentiable at $r = 0$) component

$$\operatorname{Re} G(r) = B r^{-1} \cos kr$$

of that solution to the Green's function of the corresponding Poisson equation (9.1.15) yields $B = -1/4\pi$. Consequently, in three dimensions, the Green's function of the Helmholtz equation is of the form

$$G(r) = -\frac{1}{4\pi} r^{-1} e^{-ikr}. \tag{3}$$

9.4 2-D Helmholtz Equation

9.4.1 Reduction of Dimension

Since we already have a solution in 3-D space, there is no need to construct the 2-D solution from scratch. The desired 2-D Green's function can be found by the *method of dimension reduction*, commonly utilized in mathematical physics: a solution in a smaller-dimensional space can be calculated from the solution in a larger-dimensional space by averaging out "extra" coordinates.

In particular, the Green's function of the 2-D Helmholtz equation is equal to the integral of the 3-D Green's function over the whole x_3-axis (Fig. 9.4.1). This corresponds to solving the three-dimensional equation with the source function a line Dirac delta concentrated on the x_3-axis (rather than at the origin).

Passing to the cylindrical coordinates whose z-axis coincides with the x_3-axis and whose radial coordinate ρ is located in the (x_1, x_2)-plane, we arrive at an expression for the 2-D Green's function:

$$G(\rho) = -\frac{1}{2\pi} \int_0^\infty \frac{\exp(-ik\sqrt{z^2 + \rho^2})}{\sqrt{z^2 + \rho^2}} dz.$$

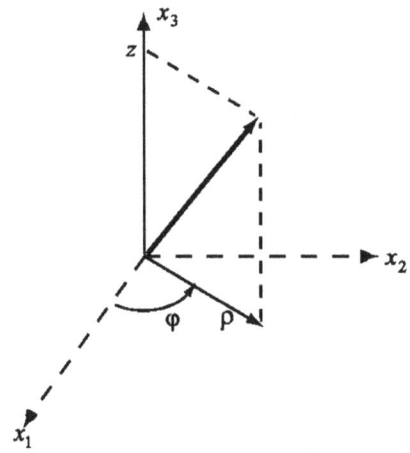

FIGURE 9.4.1
Cylindrical coordinate system.

Adopting μ as the new variable of integration, related to z through the formula $z = \rho \sinh \mu$, we obtain that

$$G(\rho) = -\frac{1}{2\pi} \int_0^\infty \exp(-ik\rho \cosh \mu) \, d\mu, \tag{1}$$

which, up to a constant factor, coincides with one of the integral representations of $H_0^{(2)}(k\rho)$, where

$$H_0^{(2)}(z) = \frac{2i}{\pi} \int_0^\infty \exp(-iz \cosh \mu) \, d\mu$$

is the *zero-order Hankel function of the second kind*. Thus, the Green's function of the 2-D Helmholtz equation can be written in the form

$$G(\rho) = \frac{i}{4} H_0^{(2)}(k\rho). \tag{2}$$

9.4. 2-D Helmholtz Equation

As $\rho \to 0$ ($k\rho \ll 1$), $G(\rho)$ asymptotically approaches the fundamental solution (9.1.8) of the 2-D Poisson equation

$$G(\rho) \sim \frac{1}{2\pi} \ln \rho, \qquad (\rho \to 0),$$

and for large ρ ($k\rho \gg 1$), its asymptotic behavior is described by the formula

$$G(\rho) \sim \sqrt{\frac{1}{8\pi k \rho}} \exp\left(-i\left(k\rho - \frac{3\pi}{4}\right)\right), \qquad (\rho \to \infty). \tag{3}$$

The minus sign in front of $ik\rho$ reflects the fact that the above fundamental solution of the Helmholtz equation in the plane satisfies the radiation condition that postulates that the only wave present in space is the wave radiated by the source.

9.4.2 Bessel Functions

To complete our description of the 2-D case, let us recall the differential equations for the Hankel function and its relationship to other special functions. Note that the Helmholtz equation in the polar coordinate system takes the form

$$\frac{1}{\rho} \frac{\partial}{\partial \rho}\left(\rho \frac{\partial}{\partial \rho}\right) G + \frac{1}{\rho^2} \frac{\partial^2}{\partial \varphi^2} G + k^2 G = \delta(\boldsymbol{x}).$$

For $\rho > 0$, where the Dirac delta on the right-hand side is identically equal to zero, and for a Green's function independent of the angular variable φ, the Helmholtz equation reduces to the equation

$$\rho^2 G'' + \rho G' + k^2 \rho^2 G = 0.$$

Here, the primes denote derivatives with respect to ρ. Introducing a new dimensionless variable $z = k\rho$ and function $u(z) = G(z/k)$, we obtain the equation

$$z^2 u'' + z u' + z^2 u = 0,$$

where the primes denote derivatives with respect to z. This last equation is a special case of the *Bessel equation*

$$z^2 u'' + z u' + (z^2 - \nu^2) u = 0,$$

which has numerous physical applications. The two linearly independent real solutions of the Bessel equation, which can be found in mathematical tables or generated by computer packages, are traditionally denoted by $J_\nu(z)$ and $N_\nu(z)$, and are called the *Bessel functions* and *Neumann functions of order* ν, respectively.

In our problem, they arise from the effort to find solutions of the Helmholtz equation that are not radially symmetric and that are of the form

$$G_n = g_n(\rho)e^{in\varphi}.$$

Substituting this expression into the Helmholtz equation, we note that it becomes an identity (for every φ and every $\rho > 0$) if the function $u_n(z) = g_n(z/k)$ satisfies the Bessel equation of order $\nu = n$. Since the Bessel equation is linear, its general solution $U(\rho, \varphi)$ is, for $\rho > 0$, a superposition of the above particular solutions, i.e.,

$$U(\rho, \varphi) = \sum_{n=-\infty}^{\infty} \left[A_n J_n(k\rho) + B_n N_n(k\rho)\right] e^{in\varphi}.$$

The reader should not be disconcerted by the presence of an infinite number of arbitrary coefficients A_n, B_n in a solution of a second-order differential equation. It is perfectly normal, since an arbitrary equation in partial derivatives, even of the first order, is equivalent to an infinite system of ordinary differential equations.

Without attempting an in-depth exposition of the theory of cylindrical functions (to which the Bessel, Neumann, and Hankel functions are related), we shall only remark that the Bessel and Neumann functions are analogous to the trigonometric solutions of the equation

$$u'' + u = 0.$$

The complex Hankel functions that satisfy the radiation condition are related to the Bessel and Neumann functions through the equality

$$H_\nu^{(2)}(z) = J_\nu(z) - iN_\nu(z),$$

which is reminiscent of the familiar Euler's formula for trigonometric functions:

$$e^{-iz} = \cos z - i \sin z.$$

9.5. Diffraction of a Monochromatic Wave

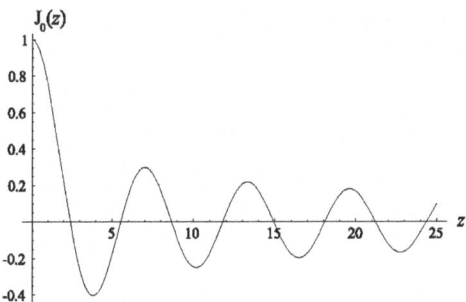

FIGURE 9.4.2
Graph of the Bessel function $J_0(z)$.

In conclusion, we also would like to mention the following integral representation of the Bessel functions of integer order:

$$J_n(z) = \frac{1}{2\pi} \int_{-\pi}^{\pi} \exp\left(-in\theta + iz \sin \theta\right) d\theta$$
$$= \frac{1}{\pi} \int_0^{\pi} \cos\left(n\theta - z \sin \theta\right) d\theta. \qquad (4)$$

The graph of the zero-order Bessel function is plotted in Fig. 9.4.2. For large z, the graph of the function $J_0(z)$ becomes more and more like the graph of the sine or cosine function. From the physical viewpoint this is not very surprising, since a propagating cylindrical wave becomes locally planar far away from the source, and also because the monochromatic harmonic planar waves are characterized by their harmonic dependence on both spatial coordinates and time.

9.5 Diffraction of a Monochromatic Wave

Many applied wave problems can be reduced to the following *wave diffraction problem*: for a given scalar field $u_0(x)$ on a surface S, find the corresponding wave away from S such that the radiation condition is satisfied. Let us solve this problem in the special case in which S is the plane $x_3 = 0$, and the wave propagates onto the half-space $x_3 > 0$. To simplify our notation, we will call the axis x_3 (along which the wave propagates) the longitudinal axis and denote it by z. The transverse coordinates (x_1, x_2) will be denoted by the vector $\mathbf{y} = (y_1, y_2)$; accordingly, the desired solution of the diffraction problem will be denoted by $u(\mathbf{y}, z)$.

The wave diffraction problem, or the *exterior Dirichlet problem* in the parlance of mathematical physics, can be solved with the help of the reflection method of Sect. 9.1. However, in this section we will take a different route, which once again illustrates the general effectiveness of our distributional tools, and the particular effectiveness of the Dirac delta distribution.

A little aside is here in order. Mathematics textbooks often present material in a deductive, linear fashion. That approach makes exposition of formal proofs easier but ignores the fact that the human brain usually arrives at solutions following a complex path of trial and errors, often relying initially on vague analogies and guesses rather than on rigid logic. Approaching the diffraction problems, we shall follow that more intuitive path and begin by looking around the mathematical "kitchen" and searching for potential "ingredients" in a "recipe" for the solution.

Let us begin by writing an equation for the Green's function $G(\boldsymbol{y}, z)$ (9.3.3) of the 3-D Helmholtz equation in the form

$$\frac{\partial^2 G}{\partial z^2} + \Delta_\perp G + k^2 G = \delta(z)\delta(\boldsymbol{y}), \quad \boldsymbol{x} = (\boldsymbol{y}, z) \in \mathbf{R}^3, \tag{1}$$

which separates the longitudinal and the transverse parts of the Laplacian; $\Delta_\perp$ stands for the Laplacian in coordinates $\boldsymbol{y}$.

Consider first some of the properties of the Green's function $G(\boldsymbol{y}, z)$ (9.3.3), which we can rewrite as follows:

$$G(\boldsymbol{y}, z) = \frac{1}{4\pi\sqrt{\rho^2 + z^2}} e^{-i\kappa\sqrt{\rho^2+z^2}} = G(\rho, z),$$

with $\rho = |\boldsymbol{y}| = \sqrt{y_1^2 + y_2^2}$. So $G(\boldsymbol{y}, z)$ is an even function of the variable z. The situation is shown in Fig. 9.5.1.

Pursuing our nonrigorous hunt for a hint of what the solution of the diffraction problem is like, let us integrate (1) with respect to z in the vicinity of zero (ignoring, for now, the contribution due to the last two terms on the left-hand side; granted, it is not obvious that this is justified!). We then get a heuristic equality,

$$\int_{0_-}^{0_+} \frac{\partial^2 G}{\partial z^2} dz = \left.\frac{\partial G}{\partial z}\right|_{z=+0} - \left.\frac{\partial G}{\partial z}\right|_{z=-0} = \delta(\boldsymbol{y}).$$

Indeed, the integral of the right-hand side of (1) is given by $\int_{0_-}^{0_+} \delta(z)\delta(\boldsymbol{y})\,dz = \delta(\boldsymbol{y})$.

This suggests that $G(\boldsymbol{y}, z)$ is an infinitely differentiable function for $z > 0$, and with $\partial G(\boldsymbol{y}, z)/\partial z$ an odd function of z (because $G(\boldsymbol{y}, z)$ is even), its

9.5. Diffraction of a Monochromatic Wave

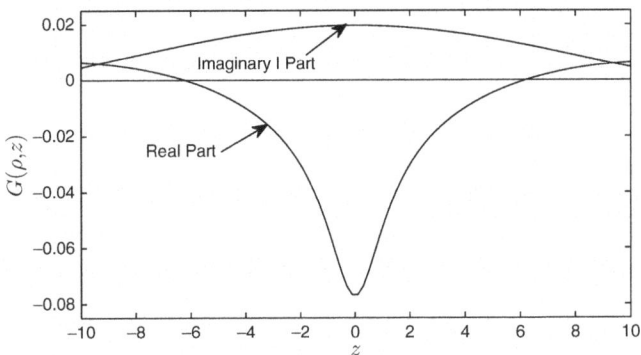

FIGURE 9.5.1
Plots of the real and imaginary parts of the Green's function $G(y, z)$ illustrating their evenness with respect to the variable z. Here, $\rho = 1$, and $\kappa = 1/4$.

derivative with respect to z weakly converges, as $z \to 0+$, to one-half of the Dirac delta:
$$\frac{\partial}{\partial z} G(y, z) \to \frac{1}{2} \delta(y), \qquad z \to 0+. \tag{$*$}$$

The above intuitive arguments make it plausible that the solution $u(y, z)$ of the above wave diffraction problem is given by the integral

$$u(y, z) = 2 \iint u_0(p) \frac{\partial}{\partial z} G(y - p, z) d^2 p. \tag{2}$$

And indeed, the right-hand side satisfies (1) for $z > 0$ (and any y), simply because the Green's function $G(y, z)$ (9.3.3) satisfies the homogeneous (for $z > 0$) equation (1), and thus the Helmholtz equation

$$\Delta u + k^2 u = 0. \tag{3}$$

Furthermore, as a result of the probing property of the Direc delta $\delta(y)$ and ($*$), the right-hand side of (2) satisfies the additional boundary condition

$$u(y, z) \big|_{z \to 0_+} = u_0(y), \tag{4}$$

where $u_0(y)$ is a known wave field in the plane $z = 0$.

A rigorous implementation of the above arguments requires careful evaluation of the derivative of the Green's function in formula (2). This gives

$$2 \frac{\partial}{\partial z} G(y, z) = G_z(y)(1 + ikr) e^{-ikr}, \tag{5}$$

with $r = |\boldsymbol{x}| = \sqrt{y^2 + z^2}$, where the factor

$$G_z(\boldsymbol{y}) = \frac{z}{2\pi r^3} = \frac{z}{2\pi(y^2 + z^2)^{3/2}} \tag{6}$$

is the derivative with respect to z of the Green's function (9.1.15) of the Poisson equation (multiplied by 2). Observe that $G_z(\boldsymbol{y})$ becomes singular as $z \to 0$, and it is easy to check that it weakly converges to the 2-D Dirac delta $\delta(\boldsymbol{y})$. Indeed, for every $\boldsymbol{y} \neq 0$,

$$\lim_{z \to 0} G_z(\boldsymbol{y}) = 0,$$

and the integral $\iint G_z(\boldsymbol{y}) d^2y$ is evaluated thus:

$$\iint G_z(\boldsymbol{y}) d^2y = z \int_0^\infty \frac{\rho \, d\rho}{(\rho^2 + z^2)^{3/2}} = 1, \qquad z > 0.$$

Here, we used the fact that the function $G_z(\boldsymbol{y})$ is radially symmetric. Hence, the scalar field (2), which satisfies the radiation condition in the upper half-plane $z > 0$, gives a rigorous solution to the diffraction problem.

9.5.1 Antenna Radiation

Let us take a closer look at the solution (2) of the Helmholtz equation (3). We will first rewrite it in the form

$$u(y, z) = \iint u_0(\boldsymbol{p}) g(\boldsymbol{y} - \boldsymbol{p}, z) \, d^2y, \tag{8}$$

where

$$g(\boldsymbol{y} - \boldsymbol{p}, z) = 2 \frac{\partial}{\partial z} G(\boldsymbol{y} - \boldsymbol{p}, z) = \frac{z}{2\pi R^3}(1 + ikR)e^{-ikR} \tag{9}$$

and

$$R = \sqrt{z^2 + (\boldsymbol{y} - \boldsymbol{p})^2} = \sqrt{r^2 + p^2 - 2(\boldsymbol{y} \cdot \boldsymbol{p})}, \qquad r = \sqrt{z^2 + y^2}. \tag{10}$$

It will be easier to realize the meaning of the expression (8) if we recall its physical interpretation. Indeed, it represents the complex amplitude of a monochromatic, say acoustic, wave radiated onto the half-space $z > 0$ by an antenna. In this case, $u_0(\boldsymbol{y})$ is the complex amplitude of the wave at the antenna itself, which is located in the plane $z = 0$. Multiplying (8) by the function $\exp(i\omega t)$, which expresses the dependence of the wave on time, and

9.5. Diffraction of a Monochromatic Wave

separating the real part, we will find the scalar field radiated by the antenna which in the acoustic case, represents pressure.

To analyze this field, or the complex amplitude (8), physicists distinguish a number of zones wherein the complex amplitude is described by qualitatively different asymptotics. In the first zone, $kz < 2\pi$, called the *near-field zone*, $g(\boldsymbol{y}, z)$, see (9), practically coincides with the expression (6), and weakly converges, for $z \to 0$, to the Dirac delta $\delta(\boldsymbol{y})$. In physical terms, the first zone is determined by the inequality $z < \lambda$, which explicitly includes the length $\lambda = 2\pi/k$ of the radiated wave.

The domain in which the opposite inequality $z \gg \lambda$ is satisfied is called the *wave zone*. Here we can neglect the first term on the right-hand side of (9), and write

$$g(\boldsymbol{y} - \boldsymbol{p}, z) \approx \frac{ikz}{2\pi R^2} e^{-ikR}. \tag{11}$$

In turn, the wave zone can be decomposed into several further zones. We shall just mention the one that turns out to be important for the purpose of understanding the antenna's functioning. Assume that the antenna has a finite size, say $2a$. In other words, $u_0(\boldsymbol{y}) \equiv 0$, for $|\boldsymbol{y}| > a$. Moreover, the characteristic scale of the function $u_0(\boldsymbol{y})$ is usually equal to a and much larger than the wavelength: $a \gg \lambda$. In addition, as will become clear from the arguments provided below, the field radiated by the antenna is negligibly small far from the z-axis, and in the analysis of the radiated wave one can assume that the condition

$$|\boldsymbol{y}| \ll z, \qquad (a \ll z), \tag{12}$$

is satisfied. Inside this domain one can, with high accuracy, replace R in the denominator of (11) by r, and rewrite it in the form

$$g(\boldsymbol{y} - \boldsymbol{p}, z) \approx \frac{ikz}{2\pi r^2} \exp\left(-ikr\sqrt{1 + \frac{p^2}{r^2} - \frac{2(\boldsymbol{y} \cdot \boldsymbol{p})}{r^2}}\right). \tag{13}$$

Condition (13) also permits a simplification of the exponent in (13). However, this simplification requires the following more precise arguments, because in the wave zone, the multiplier kr is much larger than 1. Indeed, let us expand the square root in the Taylor series

$$kr\sqrt{1 + \frac{p^2}{r^2} - \frac{2(\boldsymbol{y} \cdot \boldsymbol{p})}{r^2}} = kr + \frac{k\,p^2}{2\,r} - \frac{k}{r}(\boldsymbol{y} \cdot \boldsymbol{p}) + \ldots. \tag{14}$$

Depending on the mutual relationships among the various quantities $k, r, \boldsymbol{y}, \boldsymbol{p}$, one needs to retain an appropriate number of terms in the above Taylor series. Since $|\boldsymbol{p}| \leq a$, the second term on the right-hand side of (14) can be neglected, provided that

$$\frac{ka^2}{2r} \ll 1. \tag{15}$$

Moreover, it is easy to show (do it yourself!) that the higher-order terms that do not appear explicitly in (14) can indeed be neglected when the inequality

$$\frac{k}{8}\frac{|\boldsymbol{y}|^4}{r^3} \ll 1 \tag{16}$$

is satisfied. Thus, assuming that (15) and (16) are valid, one can retain in (14) only the first two terms and rewrite (13) in the form

$$g(\boldsymbol{y} - \boldsymbol{p}, z) = \frac{ikz}{2\pi r^2} e^{-ikr} \exp\left(i\frac{k}{r}(\boldsymbol{y}\cdot\boldsymbol{p})\right). \tag{17}$$

Physicists call the zone determined by the inequalities (15) and (16) the *Fraunhofer zone*.

Note that we have replaced in (17) the symbol $\approx$ by the equality $=$. The reason is that physicists working with the Fraunhofer zone take the expression (17) as the definition of the Green's function. The corresponding complex amplitude of the radiated wave is, as can be seen from (8),

$$u(\boldsymbol{y}, z) = 2\pi \frac{ikz}{r^2} e^{-ikr} \tilde{u}\left(-\frac{k}{r}\boldsymbol{y}\right). \tag{18}$$

We use here the following definition of the 2-dimensional Fourier image that was introduced in Volume 1:

$$\tilde{f}(\boldsymbol{s}) = \frac{1}{4\pi^2} \iint f(\boldsymbol{p}) \exp(-i(\boldsymbol{p}\cdot\boldsymbol{s})) d^2p. \tag{19}$$

The complex amplitude (18) of the radiated wave can be analyzed by rewriting (18) in the spherical coordinate system (see Fig. 9.3.1), where the role of the x_3-coordinate is played by the antenna axis z, and the coordinates (x_1, x_2) are coordinates $\boldsymbol{y}$ of a point on the antenna. Thus

$$y_1 = r\sin\theta\cos\phi, \qquad y_2 = r\sin\theta\sin\phi, \qquad z = r\cos\theta. \tag{20}$$

Substituting these equalities into (18) and introducing the unit vector $\boldsymbol{m}$ with components $m_1 = \cos\phi, m_2 = \sin\phi$, one can rewrite the complex amplitude in the form

$$u(\boldsymbol{y}, z) = 2\pi ik \frac{e^{ikr}}{r} \cos\theta\, \tilde{u}(-\boldsymbol{m}k\sin\theta). \tag{21}$$

9.5. Diffraction of a Monochromatic Wave

Observe that the right-hand side of the formula is now split into two factors. The first depends only on the radial coordinate r, while the second depends only on the angular coordinates. The radial part is proportional to the Green's function (9.3.3) of the point source located at the origin. This means that the expression (21) satisfies the radiation condition and decays at the rate $1/r$, as $r \to \infty$. Sometimes, the above-mentioned conditions are written in the form of the asymptotic relationships

$$\begin{cases} \partial u/\partial r + iku = o\,(1/r), \\ u = O\,(1/r) \end{cases} \quad (r \to \infty). \tag{22}$$

Mathematical physics arguments show that these conditions guarantee the uniqueness of the solutions of the external boundary problem for the Helmholtz equation. Recall that the radiation condition means that the complex monochromatic wave

$$u(\boldsymbol{y}, z)e^{i\omega t} \sim \exp\left(i\omega\left(t - \frac{r}{c}\right)\right)$$

propagates away from the antenna (and not toward it).

The dependence of $u(\boldsymbol{y}, z)$ on the angles θ and ϕ expresses the mathematical fact that the antenna is directional. This property of the antenna can be described via a *directional diagram* (*radiation pattern*). Often, in a directional diagram one shows the square of the modulus of the complex amplitude of the radiated wave as a function of the angles:

$$D(\theta, \phi) = \cos^2\theta \,|\tilde{u}(-\boldsymbol{m}k\sin\theta)|^2. \tag{23}$$

9.5.2 Bessel Transform

In this subsection we concentrate on the case of an antenna with radial symmetry, an important application. The directional diagram depends here only on the angle θ between the direction to the point of observation and the antenna's axis. This setup gives us an opportunity to introduce the *Bessel transform* (also called *Hankel transform*), a useful addition to the family of integral transforms introduced in Volume 1. For this purpose consider (19) in polar coordinates,

$$s_1 = \rho\cos\phi, \qquad s_2 = \rho\sin\phi,$$

and change variables in the integral (19) as follows:

$$p_1 = \gamma\cos\psi, \qquad p_2 = \gamma\sin\psi. \tag{24}$$

Then the inner product appearing in the integral (19) can be written as $(\boldsymbol{p}\cdot\boldsymbol{s}) = \rho\gamma\cos(\phi-\psi)$, and the 2-dimensional Fourier transform of (19) takes the form

$$\tilde{f}(\rho,\phi) = \frac{1}{4\pi^2}\int_{-\pi}^{\pi} d\psi \int_0^{\infty} f(\boldsymbol{p})\exp(-i\rho\gamma\cos(\phi-\psi))\gamma\,d\gamma. \qquad (25)$$

In what follows we shall also assume that the integrand $f(\boldsymbol{p})$ is radially symmetric, i.e. $f(\boldsymbol{p}) = f(\gamma)$. In this case the right-hand side of (25) splits into the product of two integrals, and the integral with respect to ψ can be calculated via the formula (9.4.4), which yields

$$\int_{-\pi}^{\pi} \exp\left(-i\rho\gamma\cos(\phi-\psi)\right) d\psi = 2\pi J_0(\rho\gamma). \qquad (26)$$

In the radially symmetric case considered here, the result of the integration in (25) is independent of the angle ϕ, and (25) takes the form

$$\tilde{f}(\rho) = \frac{1}{2\pi}\int_0^{\infty} f(\gamma) J_0(\rho\gamma)\gamma\,d\gamma. \qquad (27)$$

For convenience, let us introduce a new function

$$F(\rho) = 2\pi\tilde{f}(\rho). \qquad (28)$$

In view of (27), it can be expressed in terms of the original function $f(\gamma)$ as follows:

$$F(\rho) = \int_0^{\infty} f(\gamma) J_0(\rho\gamma)\gamma\,d\gamma. \qquad (29)$$

This is the *integral Bessel transform*, mapping an original function $f(\gamma)$ to its *Bessel image* $F(\rho)$. Similarly, calculating the inverse 2-dimensional Fourier transform of the radially symmetric function $\tilde{f}(\rho)$, one arrives at the formula for the *inverse Bessel transform*

$$f(\gamma) = \int_0^{\infty} F(\rho) J_0(\rho\gamma)\rho\,d\rho. \qquad (30)$$

Calculation of the Bessel image of particular functions requires knowledge of the analytic properties of the Bessel functions; the theory thereof is available in numerous monographs devoted to the subject.[3] Here, we just restrict ourselves to quoting a couple of useful relationships.

---

[3] See, e.g., the classic monograph *Higher Transcendental Functions* by H. Bateman and A. Erdélyi, Mc Graw-Hill, Inc., New York 1953.

9.5. Diffraction of a Monochromatic Wave

Substituting (29) into (30), we arrive at a new integral representation for the Dirac delta:

$$2\delta(\rho^2 - \mu^2) = \int_0^\infty J_0(\gamma\rho) J_0(\gamma\mu) \gamma \, d\gamma. \tag{31}$$

On the other hand, Bessel functions of different orders are connected by the following *recurrence relations*:

$$z^{n+1} J_n(z) = \frac{d}{dz} z^{n+1} J_{n+1}(z), \qquad \frac{J_{n+1}(z)}{z^n} = -\frac{d}{dz}\left(\frac{J_n(z)}{z^n}\right). \tag{32}$$

To conclude this subsection, we provide three examples of calculations of directional diagrams for radially symmetric antennas. For this purpose, let us rewrite the expression (23) in the form

$$D(\theta) = \frac{1}{4\pi^2} \cos^2\theta |U_0(k\sin\theta)|^2, \tag{33}$$

where

$$U_0(\gamma) = \int_0^\infty u_0(\rho) J_0(\gamma\rho) \rho \, d\rho \tag{34}$$

is the Bessel image of the complex amplitude of the wave in the antenna plane $z = 0$.

Example 1. Circular antenna. Consider a circular antenna of radius a with

$$u_0(\rho) = \delta(\rho - a). \tag{35}$$

In this case,

$$U_0(\gamma) = J_0(\gamma a), \tag{36}$$

and the directional diagram has the form

$$D(\theta) = D\cos^2\theta J_0^2(ka\sin\theta). \tag{37}$$

The constant D is just the value of the directional diagram at $\theta = 0$.

Example 2. Disk antenna, uniform field. Consider an antenna in the shape of a circular disk of radius a, and assume that the field on the antenna is uniform: $u_0(\mathbf{y}) = 1$. Then the function (34) takes the form

$$U_0(\gamma) = \frac{1}{\gamma^2} \int_0^{\gamma a} z J_1(z) \, dz. \tag{38}$$

The above integral can be evaluated using the first recurrence relation in (32). Indeed, for $n = 0$, this relation is of the form

$$zJ_0(z) = \frac{d}{dz}zJ_1(z). \tag{39}$$

Substituting the right-hand side into the integral (38), one obtains

$$U_0(\gamma) = \frac{a}{\gamma}J_1(\gamma a),$$

and the corresponding directional diagram (33) is described by the function

$$D(\theta) = D\frac{4}{k^2a^2}\cot^2\theta J_1^2(ka\sin\theta). \tag{40}$$

Here, as above, the constant D denotes the value of the directional diagram at $\theta = 0$, and we have taken into account the fact that

$$J_n(z) \sim \frac{1}{n!}\left(\frac{z}{2}\right)^n, \qquad (z \to 0). \tag{41}$$

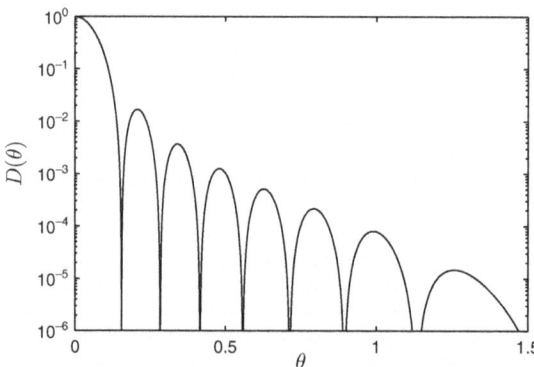

FIGURE 9.5.2
Directional diagram $D(\theta)$ of a disk antenna for which the field $u_0(y)$ is constant at all points. The value $D = D(\theta = 0)$ is assumed to be equal to 1.

The above property of the directionality of the antenna radiation can be observed only if the antenna's size is larger than the wavelength, that is, if the inequality

$$ka \gg 1 \tag{42}$$

9.5. Diffraction of a Monochromatic Wave

is satisfied. In this case, even for a small angle θ, the argument $ka\sin\theta$ can assume large values for which the factor $U_0(ka\sin\theta)$ is negligibly small. This is the essence of antenna directionality: the wave radiated by the antenna propagates primarily along the z-axis. The shape of the directional diagram is illustrated in Fig. 9.5.2. The graph shows the function (40), for $ka = 8\pi$, in which case the antenna's radius is equal to four wavelengths.

Observe that the directional diagram (40) decays very slowly as the angle θ increases. This phenomenon is related to the *slow decay of the Bessel function* as its argument increases, and can be demonstrated via the method of stationary phase discussed in Chap. 5 of Volume 1. Indeed, this method permits us to pass from the integral (9.4.4) to the asymptotic formula

$$J_n(z) = \sqrt{\frac{2}{\pi z}} \cos\left(z - \frac{\pi}{2}n - \frac{\pi}{4}\right) + O\left(\frac{1}{z\sqrt{z}}\right) \qquad (z \to \infty).$$

In practical terms, the slow decay observed above signifies that the antenna's directionality is not very sharp. An analysis of the asymptotics of the Fourier images provided in Chap. 4 of Volume 1 gives a deeper understanding of the reason for this poor directionality: it is due to the jump of the field $u_0(\boldsymbol{y})$ at the antenna's edge. The next example shows that the directionality would improve if such jumps were absent.

Example 3. Disk antenna, field vanishing on the boundary. This time, consider the field of a radiated wave of the form

$$u_0(\boldsymbol{y}) = \begin{cases} a^2 - |\boldsymbol{y}|^2, & for \ |\boldsymbol{y}| < a, \\ 0, & for \ |\boldsymbol{y}| > a. \end{cases} \qquad (44)$$

In this case, the auxiliary function (34) is

$$U_0(\gamma) = \frac{a^2}{\gamma^2}\int_0^{\gamma a} z J_0(z)\,dz - \frac{1}{\gamma^4}\int_0^{\gamma a} z^3 J_0(z)\,dz. \qquad (45)$$

The first integral on the right-hand side has already been calculated in the preceding example, and we can write

$$U_0(\gamma) = \frac{a^3}{\gamma} J_1(\gamma a) - \frac{1}{\gamma^4}\int_0^{\gamma a} z^3 J_0(z)\,dz.$$

To calculate the second integral, we will also take advantage of the first recurrence relation in (32) to obtain

$$\int_0^{\gamma a} z^3 J_0(z)\,dz = \int_0^{\gamma a} z^3 \frac{d}{dz} z J_1(z)\,dz.$$

Integration by parts gives

$$\int_0^{\gamma a} z^3 J_0(z)\, dz = (\gamma a)^3 J_1(\gamma a) - 2 \int_0^{\gamma a} z^2 J_1(z)\, dz,$$

and another application of the recurrence relation (32) yields

$$\int_0^{\gamma a} z^3 J_0(z)\, dz = (\gamma a)^3 J_1(\gamma a) - 2(\gamma a)^2 J_2(\gamma a).$$

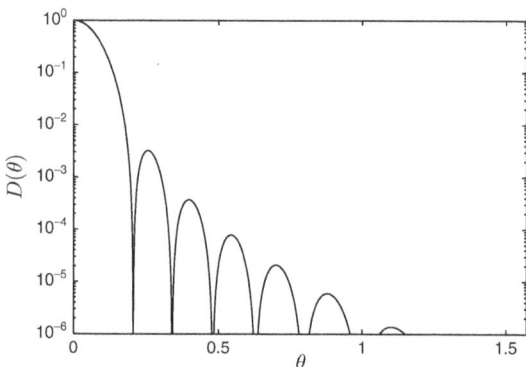

FIGURE 9.5.3
Directional diagram $D(\theta)$ of a disk antenna with surface field $u_0(y)$ vanishing toward the antenna's boundary according to (44).

Substituting this identity into (45), we finally obtain

$$U_0(\gamma) = 2 \frac{a^2}{\gamma^2} J_2(\gamma a). \tag{46}$$

Thus, in view of (33), the directional diagram has the form

$$D(\theta) = D \frac{64}{k^4 a^4} \frac{\cos^2 \theta}{\sin^4 \theta} J_2^2(ka \sin \theta). \tag{47}$$

Its graph is shown in Fig. 9.5.3.

A comparison of Figs. 9.5.2 and 9.5.3 shows that making the surface field decay toward the edge of the antenna produces an improvement in its directional quality.

9.6 Helmholtz Equation in Inhomogeneous Media

The elliptic problems discussed in the preceding sections were formulated for homogeneous media in unbounded domains. In the present section, we consider elliptic equations describing phenomena in media in bounded domains in which inhomogeneities are of fundamental importance.

Consider the Green's function G satisfying the 3-dimensional Helmholtz equation

$$\Delta G + k^2(\boldsymbol{x})G = \delta(\boldsymbol{x} - \boldsymbol{x}_1). \tag{1}$$

In contrast to (9.3.1), the coefficient $k^2(\boldsymbol{x})$ depends now on the coordinate $\boldsymbol{x}$, reflecting the inhomogeneity of the medium. In addition, we shall assume that the domain V in which the problem is being considered has a boundary S that is a smooth, closed surface on which the solution vanishes:

$$G(\boldsymbol{x}, \boldsymbol{x}_1)|_{x \in S} = 0, \qquad \boldsymbol{x}_1 \in V. \tag{2}$$

9.6.1 The Reciprocity Theorem

Inhomogeneity and the presence of a boundary destroy several symmetry properties enjoyed by the Green's function (9.3.3) in the case of homogeneous media and unbounded domains. Nevertheless, the solution of (1) preserves the general symmetry expressed by the *reciprocity theorem*, which can be formulated as follows. Rewrite (1) in the form

$$\Delta G + k^2(\boldsymbol{x})G = \delta(\boldsymbol{x} - \boldsymbol{x}_2), \tag{3}$$

which differs from (1) only by the location of the source, $\boldsymbol{x}_2 \neq \boldsymbol{x}_1$. Let us multiply (1) by $G(\boldsymbol{x}, \boldsymbol{x}_2)$, and (3) by $G(\boldsymbol{x}, \boldsymbol{x}_1)$, and then subtract the second resulting equation from the first to obtain

$$\begin{aligned} G(\boldsymbol{x}, \boldsymbol{x}_2)\Delta G(\boldsymbol{x}, \boldsymbol{x}_1) - G(\boldsymbol{x}, \boldsymbol{x}_1)\Delta G(\boldsymbol{x}, \boldsymbol{x}_2) \\ = G(\boldsymbol{x}, \boldsymbol{x}_2)\delta(\boldsymbol{x} - \boldsymbol{x}_1) - G(\boldsymbol{x}, \boldsymbol{x}_1)\delta(\boldsymbol{x} - \boldsymbol{x}_2). \end{aligned} \tag{4}$$

Integrating both sides of the above equality over the domain V, applying the second Green's formula (9.1.9), and taking into account the probing property of the Dirac delta, we get

$$\oint_S \left(G(\boldsymbol{x}, \boldsymbol{x}_2)(\boldsymbol{n} \cdot \nabla G(\boldsymbol{x}, \boldsymbol{x}_1)) - G(\boldsymbol{x}, \boldsymbol{x}_1)(\boldsymbol{n} \cdot \nabla G(\boldsymbol{x}, \boldsymbol{x}_2)) \right) dS$$

$$= G(\boldsymbol{x}_1, \boldsymbol{x}_2) - G(\boldsymbol{x}_2, \boldsymbol{x}_1), \tag{5}$$

where $\boldsymbol{n}$ denotes the external normal to the surface S. Also note that in view of the boundary condition (2), the surface integral is equal to zero, which gives us the *reciprocity theorem* in the following form:

$$G(\boldsymbol{x}_2, \boldsymbol{x}_1) = G(\boldsymbol{x}_1, \boldsymbol{x}_2). \tag{6}$$

This fundamental symmetry of the Green's function plays an important role in the study of properties of fields and waves of various types, and we complement the above derivation by several comments.

Remark 1. The reciprocity property means that the field will not change if the source point $\boldsymbol{x}_1$ is interchanged with the observation point $\boldsymbol{x}$. If two observers look at each other through a complex system of mirrors and optical guides, then if the first can see the eye of the second, the second can see the eye of the first.

Remark 2. In the proof of the reciprocity theorem we have used the *homogeneous boundary condition of the first kind*, sometimes called the *homogeneous Dirichlet condition* (2). However, the surface integral in (5) vanishes, and the reciprocity theorem remains valid if the *homogeneous boundary condition of the second kind*, sometimes called the *homogeneous Neumann condition*,

$$\left. (\boldsymbol{n} \cdot \nabla G(\boldsymbol{x}, \boldsymbol{y})) \right|_{x \in S} = 0, \tag{7}$$

is satisfied. Moreover, even a more general *homogeneous boundary condition of the third kind*,

$$\left. \alpha G(\boldsymbol{x}, \boldsymbol{y}) + \beta \left(\boldsymbol{n} \cdot \nabla G(\boldsymbol{x}, \boldsymbol{y}) \right) \right|_{x \in S} = 0 \quad (\alpha \neq 0, \beta \neq 0), \tag{8}$$

guarantees the validity of the reciprocity theorem.

Remark 3. The Green's function inside the closed surface S provides the solution of the *interior boundary problem*. In a physical context it describes, for example, the acoustic or electromagnetic fields in resonators. However, the reciprocity theorem is also valid for the analogous *exterior boundary problem*, in which one has to find the Green's function outside the surface S.

9.6.2 Inhomogeneous Boundary Conditions

Recall that once the Green's function has been found, one can write the solution of the inhomogeneous Helmholtz equation

$$\Delta u + k^2(\boldsymbol{x})u = f(\boldsymbol{x}), \qquad (\boldsymbol{x} \in V) \tag{9}$$

in the form of the integral (9.1.3) involving a given $f(\boldsymbol{x})$ and $G(\boldsymbol{x}, \boldsymbol{x}')$. Such a solution satisfies the homogeneous boundary conditions such as the Dirichlet condition

$$u(\boldsymbol{x})|_S = 0. \tag{10}$$

In a situation such that (9) is augmented by an *inhomogeneous* boundary condition

$$u(\boldsymbol{x})|_S = g(\boldsymbol{x}), \tag{11}$$

where $g(\boldsymbol{x})$ is a given function on a surface S, the sought solution can be found by splitting u into a sum of two functions

$$u(\boldsymbol{x}) = v(\boldsymbol{x}) + w(\boldsymbol{x}), \tag{12}$$

where $v(\boldsymbol{x})$ satisfies the inhomogeneous equation augmented by the homogeneous condition

$$\Delta v + k^2(\boldsymbol{x})v = f(\boldsymbol{x}), \qquad v(\boldsymbol{x})|_S = 0, \tag{13}$$

and $w(\boldsymbol{x})$ satisfies the homogeneous equation complemented by the inhomogeneous condition

$$\Delta w + k^2(\boldsymbol{x})w = 0, \qquad w(\boldsymbol{x})|_S = g(\boldsymbol{x}). \tag{14}$$

The auxiliary problem (13) has the solution (9.1.3),

$$v(\boldsymbol{x}) = \iint_V G(\boldsymbol{x}, \boldsymbol{x}') f(\boldsymbol{x}') d^3 x'. \tag{15}$$

The solution of the auxiliary boundary value problem (14) is possible via the previously used Green's formula (9.1.9), which we will rewrite in a form more suitable for our purposes:

$$\iint_V (\psi L \phi - \phi L \psi) \, d^3 x = \oint_S \left(\psi (\boldsymbol{n} \cdot \nabla \phi) - \phi (\boldsymbol{n} \cdot \nabla \psi) \right) dS, \tag{16}$$

where

$$L = \Delta + k^2 \tag{17}$$

is the Helmholtz operator. Let us apply (16) to

$$\psi(\boldsymbol{x}) = w(\boldsymbol{x}), \quad \text{and} \quad \phi(\boldsymbol{x}) = G(\boldsymbol{x}, \boldsymbol{x}_0), \quad \boldsymbol{x}_0 \in V. \quad (18)$$

In this case, in view of (14), the integral of the second term on the left-hand side of (16) vanishes,

$$\iint_V \phi L \psi \, d^3 x = \iint_V G(\boldsymbol{x}, \boldsymbol{x}')(\Delta + k^2) w(\boldsymbol{x}) \, d^3 x = 0,$$

and the integral of the first term can be calculated as follows:

$$\iint_V \psi L \phi \, d^3 x = \iint_V w(\boldsymbol{x})(\Delta + k^2) G(\boldsymbol{x}, \boldsymbol{x}_0) \, d^3 x \\ = \iint_V w(\boldsymbol{x}) \delta(\boldsymbol{x} - \boldsymbol{x}_0) \, d^3 x = w(\boldsymbol{x}_0). \quad (19)$$

Consequently, the Green's formula (16) takes the form

$$w(\boldsymbol{x}_0) = \oint_S \Big(w(\boldsymbol{x})(\boldsymbol{n} \cdot \nabla G(\boldsymbol{x}, \boldsymbol{x}_0)) - G(\boldsymbol{x}, \boldsymbol{x}_0)(\boldsymbol{n} \cdot \nabla w(\boldsymbol{x})) \Big) \, dS. \quad (20)$$

Taking into account the boundary condition (14) and the corresponding homogeneous boundary condition (2) for the Green's function, we arrive at the sought solution of the boundary value problem (14):

$$w(\boldsymbol{x}) = \oint_S g(\boldsymbol{x}') \Big(\boldsymbol{n} \cdot \nabla G(\boldsymbol{x}, \boldsymbol{x}') \Big) \, dS. \quad (21)$$

Here we have dropped the subscript in $\boldsymbol{x}_0$, denoted the variable of integration over S by $\boldsymbol{x}'$, and applied the reciprocity theorem. In (21), ∇ denotes the gradient with respect to $\boldsymbol{x}'$, and $\boldsymbol{n}$ is the external normal to the surface S at the point $\boldsymbol{x}'$.

Putting together (15) and (21), we obtain the final form of the solution of the boundary value problem (9),(11):

$$u(\boldsymbol{x}) = \iint f(\boldsymbol{x}_0) G(\boldsymbol{x}, \boldsymbol{x}_0) \, d^3 x_0 + \oint_S g(\boldsymbol{x}') \Big(\boldsymbol{n} \cdot \nabla G(\boldsymbol{x}, \boldsymbol{x}') \Big) \, dS. \quad (22)$$

Remark 1. Formula (21) generalizes formula (9.5.2)—thoroughly discussed in the previous section—to the case of an arbitrary smooth surface S, and is valid for the interior as well as the exterior boundary value problem. In the latter case, to find a physically meaningful solution, one should take conditions at infinity into account in (22), such as the radiation conditions.

Remark 2. We would like to emphasize once more that formula (22) provides the solution of the inhomogeneous problem (9) with the *inhomogeneous boundary condition* (11) via the Green's function satisfying (1) with the *homogeneous boundary condition (2)*. In general, an analytic or numerical calculation of the Green's function can be a very difficult endeavor. In the remainder of this chapter we will discuss a few methods that are useful in such calculations.

9.7 Waves in Waveguides

In this section we will study scalar fields in *waveguides*, cylindrical domains V bounded in some directions but unbounded in others. Often the symmetry of the waveguide boundary coincides with the symmetry of the inhomogeneous medium inside the waveguide. For instance, the ocean can be considered to be a layered waveguide because of stratification of salinity, density, temperature, etc. In terms of (9.6.1), this means that the coefficient k^2 depends only on the distance of the point of observation from the ocean surface. In such cases we can apply the method of *separation of variables*, which simplifies the problem considerably.

9.7.1 Method of Separation of Variables

For the sake of concreteness we shall speak about the hydroacoustic waveguide created by the flat ocean floor and the ocean surface. Denote the distance of the point of observation from the ocean surface by z, and the 2-D coordinates of its projection on the ocean surface by $\boldsymbol{y}$. Our assumption of a stratified ocean mass means that the coefficient k^2 in (9.6.1) depends only on the vertical coordinate z, i.e., $k^2 = k^2(z)$. In this case, the equation for the Green's function $G(\boldsymbol{y}, z, z_0)$ can be written in the form

$$\Delta_\perp G + \frac{\partial^2 G}{\partial z^2} + k^2(z)G = \delta(\boldsymbol{y})\delta(z - z_0). \tag{1}$$

Here $\Delta_\perp$ denotes the 2-D Laplacian in the horizontal plane $\boldsymbol{y}$, and the source is located at the point with coordinates $z = z_0, \boldsymbol{y} = 0$ ($0 < z_0 < h$, where h denotes the ocean's depth).

Equation (1) can now be augmented by various boundary conditions. They could be, e.g., the conditions of an "ideally soft ocean surface and ideally hard ocean floor":

$$\frac{\partial}{\partial z}G(\boldsymbol{y}, z, z_0)|_{z=0} = 0, \qquad G(\boldsymbol{y}, z = h, z_0) = 0. \tag{2}$$

We shall solve the boundary value problem (1)–(2) via the method of separation of variables. Its essence will be better understood if we begin by explaining the intuitions underlying it; (1) is too complex to be solved in one big swoop. So, as a first step, we shall try to find a solution $u(\boldsymbol{y},z)$ of a simpler homogeneous equation,

$$\Delta_\perp u + \frac{\partial^2 u}{\partial z^2} + k^2(z)u = 0. \tag{3}$$

Although this equation is still quite complex, we can find its particular solutions by separating the variables, yielding an even simpler equation. This is accomplished by assuming that the solution of (3) can be written in the product form:

$$y(\boldsymbol{y},z) = Y(\boldsymbol{y})Z(z), \tag{4}$$

with the first function, $Y(\boldsymbol{y})$, depending only on the variable $\boldsymbol{y}$, and the second function, $Z(z)$, depending only on the variable z. Substituting (4) into (3) and dividing the result by YZ gives

$$\frac{\Delta_\perp Y}{Y} = -\frac{Z'' + k^2(z)Z}{Z}, \tag{5}$$

where the primes denote ordinary differentiation with respect to z.

Observe that the two sides of the above equation depend on different variables: the left-hand side depends only on $\boldsymbol{y}$, and the right-hand side on z. The conclusion is that the two sides of (5) must be equal to an identical constant, independent of the variables $\boldsymbol{y}$ and z. Denote this constant by $-\mu^2$. Then (5) splits into two equations:

$$\Delta_\perp Y + \mu^2 Y = 0 \tag{6}$$

and

$$Z'' + [k^2(z) - \mu^2]Z = 0. \tag{7}$$

Let us denote their solutions by $Y_\mu(\boldsymbol{y})$ and $Z_\mu(z)$, respectively. By finding these, we will find a particular solution of (3). We emphasize once more that instead of a complex partial differential equation with variable coefficients, we are now working with two simpler equations: (6) is a partial differential equation but with constant coefficients, with a solution in the whole unbounded 2-D plane, while (7) is an ordinary differential equation, the theory of which is better understood than that of partial differential equations.

Starting with (6) we recall that in Sect. 9.4, we found a solution of the equation

$$\Delta_\perp G + \mu^2 G = \delta(\boldsymbol{y}), \tag{8}$$

9.7. Waves in Waveguides

which coincides with (6) for arbitrary $y \neq 0$, where $\delta(y) \equiv 0$. Thus, in particular, for a positive μ^2,

$$Y_\mu(y) = -\frac{1}{2\pi}\int_0^\infty \exp(-i|\mu|\rho\cosh\nu)\, d\nu, \qquad (\mu^2 > 0). \tag{9}$$

This is a complex-valued function satisfying the radiation condition, and has asymptotic behavior for large ρ described by the relation (9.4.3):

$$Y_\mu(y) \sim \sqrt{\frac{1}{8\pi|\mu|\rho}}\exp\left(-i\left(|\mu|\rho - \frac{3\pi}{4}\right)\right), \qquad (\rho\to\infty), \tag{10}$$

where $\rho = \sqrt{y_1^2 + y_2^2}$ is the radial coordinate.

For a negative μ^2, the formula (9.4.3) gives

$$Y_\mu(y) = \frac{1}{2\pi}\int_0^\infty \exp(-|\mu|\rho\cosh\nu)\, d\nu, \qquad (\mu^2 < 0). \tag{11}$$

The asymptotic behavior of this integral, for large ρ, can be found by the method of steepest descent described in Sect. 5.5 of Volume 1. It calls for the replacement of $\cosh\mu$ in (11) by the first two terms of its Taylor expansion in μ, and yields

$$Y_\mu(y) \sim \sqrt{\frac{1}{8\pi|\mu|\rho}}\exp(-|\mu|\rho), \qquad (\rho\to\infty). \tag{12}$$

This asymptotic formula, in contrast to (10), implies a very fast exponential decay of Y as ρ goes to infinity.

Now let us turn to (7). Although the analytic and numerical methods for this type of equation are well developed, here we will just provide an explicit formula for its solution in the case of a constant coefficient $k^2(z) = k^2$ when (7) is of the form

$$Z'' + (k^2 - \mu^2)Z = 0. \tag{13}$$

As will become clear from what follows, only solutions corresponding to the case $k^2 > \mu^2$ are of importance. Then

$$Z_\mu(z) = C_1\cos\left(z\sqrt{k^2-\mu^2}\right) + C_2\sin\left(z\sqrt{k^2-\mu^2}\right), \tag{14}$$

where C_1 and C_2 are arbitrary constants.

Having found an arbitrary partial solution of (3) via the method of separation of variables (4), we find ourselves in possession of infinitely many solutions corresponding to different values of the separation constant μ:

$$u_\mu(\boldsymbol{y}, z) = Y_\mu(\boldsymbol{y}) Z_\mu(z).$$

Since (4) is linear, the composition

$$u(\boldsymbol{y}, z) = \sum_\mu A(\mu) Y_\mu(\boldsymbol{y}) Z_\mu(z) \tag{15}$$

is also a solution of (3), representing, we hope, a general solution. The notation $\sum_\mu$ indicates summation (or integration) of particular solutions over the index μ, and $A(\mu)$ is an arbitrary function that is an analogue of arbitrary constants entering in the general solutions of the ordinary differential equations.

The solution (15) is sufficiently rich for the purpose of constructing the Green's function for the boundary value problem (1)–(2). In the next step, we will require not only that the solution (15) satisfy (3) but also that it satisfy the boundary conditions (2). It is clear that a sufficient condition here is that similar conditions be satisfied by all functions $Z_\mu(z)$ appearing in (15). Thus all $Z_\mu(z)$ need to satisfy the following boundary value problem:

$$Z'' + [k^2(z) - \mu^2]Z = 0, \qquad Z'(0) = 0, \ Z(h) = 0. \tag{16}$$

Now suppose that we have found all solutions of this boundary value problem and substituted them into (15). The functions $Y_\mu(\boldsymbol{y})$ entering into (15) now satisfy not (6) but an inhomogeneous equation (8). The function $u(\boldsymbol{y}, z)$ thus produced will not yield zero on the right-hand side after the substitution in the left-hand side of (3); its value at the point $\boldsymbol{y} = 0$ is now significant. Indeed,

$$\Delta_\perp u + \frac{\partial^2 u}{\partial z^2} + k^2(z) u = \sum_\mu A(\mu) \Big(Z_\mu \Delta_\perp Y_\mu + Y_\mu \{Z''_\mu + k^2(z) Z_\mu\} \Big).$$

Substituting

$$\Delta_\perp Y_\mu = -\mu^2 Y_\mu + \delta(\boldsymbol{y})$$

gives the equation

$$\Delta_\perp u + \frac{\partial^2 u}{\partial z^2} + k^2(z) u =$$

$$\sum_\mu A(\mu) Y_\mu (Z''_\mu + [k^2(z) - \mu^2] Z_\mu) + \delta(\boldsymbol{y}) \sum_\mu A(\mu) Z_\mu(z).$$

9.8. Sturm–Liouville Problem

The first sum on the right-hand side vanishes, in view of (16), and the second sum becomes the right-hand side of the original equation (1), provided that $A(\mu)$ can be selected in such a way that

$$\sum_\mu A(\mu) Z_\mu(z) = \delta(z - z_0). \tag{17}$$

This issue can be resolved via a detailed analysis of the so-called *Sturm–Liouville problem*, which will be the main topic in the next section. We shall return to the solution of the boundary value problem (1)–(2) in Sect. 9.9.

9.8 Sturm–Liouville Problem

The boundary value problem (9.7.16), which arose during our study of waves in waveguides, can be reduced to the celebrated *Sturm–Liouville problem*, which plays a key role in several areas of mathematical physics.

9.8.1 Basic Definitions

Consider a boundary value problem consisting of the equation

$$LZ = \lambda \rho(z) Z, \tag{1}$$

where $\rho(z)$ is a positive continuous function on the interval $[0, h]$, augmented by the *boundary conditions of the third kind*

$$a_1 Z(0) - b_1 Z'(0) = 0, \qquad a_2 Z(h) + b_2 Z'(h) = 0, \tag{2}$$

where a_1, a_2, b_1, b_2 are constants such that

$$a_1 \geq 0, \quad a_2 \geq 0, \quad b_1 \geq 0, \quad b_2 \geq 0, \quad a_1 + b_1 > 0, \quad a_2 + b_2 > 0. \tag{3}$$

The operator L in (1) is assumed to be the so-called *Sturm–Liouville operator*

$$L = -\frac{d}{dz} p(z) \frac{d}{dz} + q(z). \tag{4}$$

We require that its coefficients satisfy the following conditions:

$$p(z) > 0, \quad q(z) \geq 0, \quad p(z) \in C^1[0, h], \quad q(z) \in C[0, h]. \tag{5}$$

In plain language, the last two conditions mean that for z contained in the interval $[0, h]$, the function $q(z)$ is continuous and that $p(z)$ and its first derivative are continuous as well.

The previously studied boundary value problem (9.7.16) clearly can be reduced to the problem (1)–(2). Indeed, if we introduce the new notation

$$k^2 = \max_{x \in [0,h]} k^2(z), \qquad (6)$$

assume that $k^2(z)$ is continuous on the interval $[0, h]$, so that in particular, $k^2 < \infty$, and define

$$q(z) = k^2 - k^2(z), \qquad p = 1, \qquad \lambda = k^2 - \mu^2, \qquad (7)$$

then (9.7.16) becomes (1), and all the conditions in (5) are satisfied. The boundary conditions (9.7.16) are a particular case of the conditions (2) with $a_1 = 0, a_2 = 1, b_1 = 1, b_2 = 0$.

Excluding the trivial solution $Z_m(z) \equiv 0$, it turns out that solutions of the boundary value problem (1)–(2) exist only for certain special values of λ, which are called the *eigenvalues* of the Sturm–Liouville problem. The corresponding solutions are called the *eigenfunctions* of the Sturm–Liouville problem.

9.8.2   Auxiliary Facts from Functional Analysis

Our study of the Sturm–Liouville problem will utilize a few concepts from functional analysis, which are introduced below.

The *scalar product* of functions f and g defined on the interval $[0, h]$ is defined via the formula

$$(f, g) = \int_0^h f(z) g(z) \, dz. \qquad (8)$$

We shall say that a function $f(z)$ *belongs to the class M* (in brief, $f \in M$) if $f(z)$ is twice continuously differentiable on the interval $[0, h]$ (in brief, $f \in C^2[0, h]$) and if it satisfies conditions (2)–(3). An operator L is said to be *self-adjoint* if for all functions $f, g \in M$,

$$(Lf, g) = (f, Lg). \qquad (9)$$

The first important observation is that the *Sturm–Liouville operator* (4) *is self-adjoint*. Indeed,

$$(Lf, g) - (f, Lg) = \int_0^h g(z) \frac{d}{dz}\left(p(z) f'(z)\right) dz - \int_0^h f(z) \frac{d}{dz}\left(p(z) g'(z)\right) dz. \qquad (10)$$

9.8. Sturm–Liouville Problem

Integrating the last integral by parts twice removes the derivatives of the function $g(z)$, and we obtain

$$\int_0^h f(z)\tfrac{d}{dz}\Big(p(z)g'(z)\Big)\,dz = \int_0^h g(z)\tfrac{d}{dz}\Big(p(z)f'(z)\Big)\,dz$$
$$-p(z)\Big[g(z)f'(z) - f(z)g'(z)\Big]_0^h.$$

A substitution of this equality into (10) yields

$$(Lf,g) - (f,Lg) = p(0)\Big[f(0)g'(0) - g(0)f'(0)\Big] \\ +p(h)\Big[g(h)f'(h) - f(h)g'(h)\Big]. \tag{11}$$

We will show that if $f,g \in M$, then the expressions in the brackets vanish. For this purpose, consider a system of homogeneous algebraic equations

$$\begin{aligned} a_1 f(0) - b_1 f'(0) &= 0, \\ a_1 g(0) - b_1 g'(0) &= 0. \end{aligned} \tag{12}$$

According to (3), the system has nontrivial solutions such that $a_1 + b_1 > 0$. Consequently, the determinant of this system vanishes:

$$\begin{vmatrix} f(0) & -f'(0) \\ g(0) & -g'(0) \end{vmatrix} = g(0)f'(0) - f(0)g'(0) = 0. \tag{13}$$

This means that the coefficient of $p(0)$ on the right-hand side of (11) must vanish as well. The coefficient of $p(h)$ can be shown to vanish in a similar fashion.

The functions $f(z)$ and $g(z)$ are called *orthogonal* on the interval $[0,h]$ with respect to a nonnegative *weight function* $\rho(z)$ if

$$(\rho f, g) = (f, \rho g) = \int_0^h f(z)g(z)\rho(z)\,dz = 0. \tag{14}$$

Given a nonnegative weight function $\rho(z)$, the norm $\|f\|_\rho$ of a function $f(z)$ is defined by the equality

$$\|f\|_\rho^2 = (\rho f, f) = \int_0^h f^2(z)\rho(z)\,dz. \tag{15}$$

Finally, functions $Z_1(z),\ldots,Z_n(z)$ are said to be *linearly dependent* if one can find constants $C_1,\ldots,C_n$, not all equal to zero, such that

$$Z(z) := \sum_{k=1}^n C_k Z_k(z) \equiv 0. \tag{16}$$

Otherwise, the functions $Z_1(z), \ldots, Z_n(z)$ are called *linearly independent*.

The next result is important for what follows.

Theorem. *For a system, $Z_1(z), \ldots, Z_n(z)$ of particular solutions of a linear homogeneous ordinary differential equation*

$$Z^{(n)} + a_1(z)Z^{(n-1)} + \cdots + a_n(z)Z = 0, \tag{17}$$

linear dependence is equivalent to the condition that the determinant

$$W(z) := \begin{vmatrix} Z_1(z) & \cdots & Z_n(z) \\ Z_1'(z) & \cdots & Z_n'(z) \\ \cdots & \cdots & \cdots \\ Z_1^{(n-1)}(z) & \cdots & Z_n^{(n-1)}(z) \end{vmatrix}, \tag{18}$$

called the Wronskian of the system, vanishes at some point $z = z_1$.

The condition (18) is obviously easier to verify than the original linear dependence condition (16).

To prove the above observation, note that the function $Z(z)$ defined in (16), differentiated repeatedly, yields the following system of n equalities:

$$\begin{array}{ll} \sum_{k=1}^n C_k Z_k(z) & = Z(z), \\ \sum_{k=1}^n C_k Z_k'(z) & = Z'(z); \\ \cdots & = \cdots \\ \sum_{k=1}^n C_k Z_k^{(n-1)}(z) & = Z^{(n-1)}(z). \end{array} \tag{19}$$

If the Wronskian $W(z)$ vanishes at a certain point $z = z_1$, then the system of equations

$$\begin{array}{ll} \sum_{k=1}^n C_k Z_k(z_1) & = 0; \\ \sum_{k=1}^n C_k Z_k'(z_1) & = 0; \\ \cdots & = \cdots \\ \sum_{k=1}^n C_k Z_k^{(n-1)}(z_1) & = 0; \end{array} \tag{20}$$

with $C_1, C_2, \ldots, C_n$ as unknowns has a nonzero solution.

On the other hand, if $Z_1(z), Z_2(z), \ldots, Z_n(z)$ are solutions of (17), then so is $Z(z)$, and (20) means that the latter satisfies the initial conditions

$$Z(z_1) = Z'(z_1) = \cdots = Z^{(n-1)}(z_1) = 0.$$

9.8. Sturm–Liouville Problem

But such a solution must be identically equal to zero, and this means that the condition (16) is fulfilled and the functions $Z_1(z), \ldots, Z_n(z)$ are linearly dependent. The reverse implication is obvious. ∎

9.8.3 Properties of Eigenvalues and Eigenfunctions

Now we are ready to formulate some basic properties of solutions of the Sturm–Liouville problem.

Property 1. Eigenfunctions corresponding to different eigenvalues are orthogonal with respect to the weight function ρ.

To prove this, let $f(z)$ and $g(z)$ be solutions of the boundary value problem (1)–(2) corresponding to λ_1 and λ_2, respectively, $\lambda_1 \neq \lambda_2$. This means that

$$Lf \equiv \lambda_1 \rho f \quad \text{and} \quad Lg \equiv \lambda_2 \rho g.$$

Multiplying the first equation by $g(z)$, the second by $f(z)$, subtracting the second from the first, and integrating over the interval $[0, h]$ we get

$$(g, Lf) - (f, Lg) \equiv (\lambda_1 - \lambda_2)(\rho f, g).$$

The left-hand side of the above identity is zero in view of the self-adjointness of the Sturm–Liouville operator. Since $\lambda_1 \neq \lambda_2$, we obtain that $(\rho f, g) = 0$. ∎

At this point we would like to introduce the concept of *simple eigenvalue*. The linearity of the problem (1)–(2) implies that there are infinitely many eigenfunctions corresponding to each eigenvalue. For example, if $Z_1(z)$ is such an eigenfunction, then for every $C \neq 0$, so is $Z_2(z) = CZ_1(z)$. If all eigenfunctions corresponding to an eigenvalue λ are of this form, then λ is called a *simple eigenvalue*. Otherwise, such an eigenvalue is called *multiple*. If every eigenfunction $Z(z)$ corresponding to λ can be written as a linear combination (18) of linearly independent eigenfunctions $Z_1(z), \ldots, Z_n(z)$ corresponding to λ, then the eigenvalue λ is said to have *multiplicity n*.

Property 2. All eigenvalues of the Sturm–Liouville problem are simple.

Indeed, let $f(z)$ and $g(z)$ be two eigenfunctions of the boundary value problem (1)–(2) corresponding to the eigenvalue λ, which means that both of them satisfy (1). We verify directly that their Wronskian,

$$W(z) = \begin{vmatrix} f(z) & g(z) \\ f'(z) & g'(z) \end{vmatrix} = f(z)g'(z) - g(z)f'(z), \tag{22}$$

satisfies the equation

$$\frac{d}{dz}\bigl(p(z)W(z)\bigr) = 0, \tag{23}$$

and consequently,

$$W(z) = W(0)\frac{p(0)}{p(z)}. \tag{24}$$

Furthermore, note that $f(z)$ and $g(z)$ must satisfy algebraic equations (12), which have nontrivial solutions. Therefore, in view of (13), we have $W(0) = 0$. Hence, by (24), $W(z) = 0$, $f(z)$, and $g(z)$ are linearly dependent, so that the eigenvalue λ is necessarily simple. ∎

Property 3. The Sturm–Liouville operator is positive definite, i.e., for every real function $f(z) \in M$,

$$(Lf, f) \geq 0. \tag{25}$$

Indeed,

$$(Lf, f) = (q, f^2) - \int_0^h f(z)\frac{d}{dz}(p(z)f'(z))\, dz. \tag{26}$$

Integrating by parts yields

$$-\int_0^h f(z)\frac{d}{dz}(p(z)f'(z))\, dz = (p, f'^2) + p(0)f(0)f'(0) - p(h)f(h)f'(h).$$

Substituting the result into (26) and utilizing condition (2), which is satisfied for every function from the set M, we finally obtain

$$(Lf, f) = (q, f^2) + (p, f'^2) + p(0)\frac{a_1}{b_1}f^2(0) + p(h)\frac{a_2}{b_2}f^2(h). \tag{27}$$

Observe that for $b_1 > 0, b_2 > 0$, all the terms on the right-hand side are nonnegative in view of (3) and (5), and for $b_1 = 0, b_2 = 0$, the last two terms vanish. This concludes the proof of Property 3. ∎

Property 4. Eigenvalues of the Sturm–Liouville problem (1)–(2) are real and nonnegative.

First, we shall show that complex eigenvalues would lead to a contradiction. Let $\lambda = \alpha + i\beta$ be a complex eigenvalue ($\beta \neq 0$), and let $Z(z)$ be the corresponding eigenfunction. Since the Sturm–Liouville operator and the weight function $\rho(z)$ are real-valued, the complex conjugate $Z^*(z)$ is an eigenfunction corresponding to the eigenvalue $\lambda^* = \alpha - i\beta$. Since eigenfunctions

9.8. Sturm–Liouville Problem

corresponding to different eigenvalues are orthogonal with respect to ρ, we would have
$$(\rho Z, Z^*) = \int_0^h |Z(z)|^2 \rho(z)\, dz = 0,$$
which is impossible. Thus the eigenvalue λ must be real-valued.

Now we can prove that it also has to be nonnegative. Multiply (1) by the eigenfunction $Z(z)$ and integrate the resulting equality over the interval $[0, h]$ to obtain
$$\lambda = \frac{(Z, LZ)}{\|Z\|^2}. \tag{28}$$

Since both the numerator and denominator of the above fraction are nonnegative, the Property 4 follows. ∎

Example. We shall provide a simple example illustrating the above properties in the case in which the coefficients of the Sturm–Liouville operator (4) and the weight coefficient in (1) are constant, i.e., independent of z. Now (1) takes the form
$$pZ'' + (\lambda\rho - q)Z = 0. \tag{29}$$

Let us complement this equation with the boundary conditions of the first kind
$$Z(0) = Z(h) = 0. \tag{30}$$
Equation (29) can be written in the simplified form
$$Z'' + \mu Z = 0, \tag{31}$$
where
$$\mu = \frac{\lambda\rho - q}{p}, \quad \text{or, equivalently,} \quad \lambda = \frac{\mu p + q}{\rho}. \tag{32}$$

First, consider the case $\mu < 0$. Then the general solution of (31) is of the form
$$Z = C_1 e^{-\sqrt{|\mu|}z} + C_2 e^{\sqrt{|\mu|}z}, \tag{33}$$
where C_1 and C_2 are arbitrary constants. Substituting this expression into the two boundary conditions (30), we arrive at the following system of equations for C_1 and C_2:
$$\begin{aligned} C_1 + C_2 &= 0, \\ C_1 e^{-\sqrt{|\mu|}h} + C_2 e^{\sqrt{|\mu|}h} &= 0. \end{aligned} \tag{34}$$

The determinant of this homogeneous linear algebraic system of equations,

$$\Delta = \begin{vmatrix} 1 & 1 \\ e^{-\sqrt{|\mu|}h} & e^{\sqrt{|\mu|}h} \end{vmatrix} = 2\sinh\left(\sqrt{|\mu|}h\right), \tag{35}$$

is nonzero for every $h > 0$ and $\mu < 0$. This means that the system (34) has only the trivial solutions $C_1 = C_2 = 0$. Thus $\mu < 0$ cannot be an eigenvalue, because it corresponds to a trivial, identically vanishing, solution of (31).

It is not difficult to show that in the case of the boundary conditions (30), the same argument remains valid for $\mu = 0$ when the general solution of (31) is of the form

$$Z = C_1 + C_2 z. \tag{36}$$

Now let us consider the case $\mu > 0$. Here the general solution of (31) is of the form

$$Z(z) = C_1 \cos\sqrt{\mu}z + C_2 \sin\sqrt{\mu}z. \tag{37}$$

The boundary conditions (30) imply that

$$\begin{aligned} C_1 &= 0; \\ C_1 \cos\sqrt{\mu}h + C_2 \sin\sqrt{\mu}h &= 0. \end{aligned} \tag{38}$$

The determinant of this system,

$$\Delta = \begin{vmatrix} 1 & 0 \\ \cos\sqrt{\mu}h & \sin\sqrt{\mu}h \end{vmatrix} = \sin\sqrt{\mu}h, \tag{39}$$

vanishes for

$$\mu_n = \left(\frac{\pi n}{h}\right)^2, \quad n = 1, 2, \ldots, \tag{40}$$

so that in view of (32), the eigenvalues of the Sturm–Liouville problem (29)–(30) are

$$\lambda_n = \frac{p}{\rho}\left(\frac{\pi n}{h}\right)^2 + \frac{q}{\rho}. \tag{41}$$

The corresponding nontrivial eigenfunctions of the Sturm–Liouville problem (29)–(30) are

$$Z_n(z) = A_n \sin\sqrt{\mu_n}z = A_n \sin\left(\frac{\pi n}{h}z\right), \tag{42}$$

where the constants A_n are selected so that each of the eigenfunctions has norm one, i.e.,

$$\|Z_n\|^2 = (Z_n, Z_n) = A_n^2 \int_0^h \sin^2\left(\frac{\pi n}{h}z\right) dz = 1. \tag{43}$$

9.8. Sturm–Liouville Problem

Here, for the sake of simplicity, we set $\rho = 1$. Evaluation of the integral in (43) leads to

$$A_n = \sqrt{\frac{2}{h}}. \tag{44}$$

Since in view of Property 1, the eigenvalues (42) are mutually orthogonal,

$$(Z_n, Z_m) = 0. \tag{45}$$

The normalized eigenfunctions

$$Z_n(z) = \sqrt{\frac{2}{h}} \sin\left(\frac{\pi n}{h} z\right), \qquad n = 1, 2, \ldots, \tag{46}$$

form an *orthonormal* system of functions. ∎

9.8.4 Extremal Properties of Eigenvalues

Consider the quadratic functional (Lf, f) (27), which is well defined for all $f \in M$. It follows from (27) that $(Lf, f) \geq 0$, which implies that there exists a lower bound,

$$\mu := \inf_{f \in M} \frac{(Lf, f)}{\|f\|^2} \geq 0. \tag{47}$$

It turns out that there exists an intimate connection between the eigenvalues and eigenfunctions of the Sturm–Liouville problem and relation (47). More precisely, if $f = Z_1(z)$ is a function realizing the minimum in (47), then it is also an eigenfunction of the Sturm–Liouville problem corresponding to the eigenvalue $\lambda_1 = \mu$. Moreover, μ is necessarily the smallest eigenvalue.

Indeed, consider the functional

$$J[f] = (Lf, f) - \mu \|f\|^2 \geq 0.$$

Since this functional attains its smallest value at the function $Z_1(z)$, we have

$$J[Z_1] = (LZ_1, Z_1) - \mu \|Z_1\|^2 = 0.$$

The auxiliary function

$$\varphi(\epsilon) := J[Z_1(z) + \epsilon \eta(z)],$$

where $\eta(z)$ is an arbitrary function from the class M, is a differentiable function of the variable ϵ that attains its minimal value at $\epsilon = 0$, where $\varphi'(0) = 0$. On the other hand,

$$\varphi'(0) = \frac{d}{d\epsilon}\left\{\Big(L(Z_1 + \epsilon \eta), Z_1 + \epsilon \eta\Big) - \mu\Big(\rho(Z_1 + \epsilon \eta), Z_1 + \epsilon \eta\Big)\right\}\bigg|_{\epsilon=0}.$$

In view of the linearity of the operator L, we obtain
$$\varphi'(0) = (L\eta, Z_1) + (LZ_1, \eta h) - \mu[(\rho\eta, Z_1) + (\rho Z_1, \eta)].$$

The self-adjointness of the operator L, (9), and the symmetry of the scalar product with weight ρ,
$$(L\eta, Z_1) = (\eta, LZ_1), \qquad (\rho\eta, Z_1) = (\rho Z_1, \eta),$$

imply

$$\varphi'(0) = 2(LZ_1 - \mu\rho Z_1, \eta) = 2\int_0^h [LZ_1(z) - \mu\rho(z)Z_1(z)]\,\eta(z)\,dz = 0.$$

Since the function $\eta \in M$ is arbitrary, it follows that the above equality is satisfied if and only if
$$LZ_1 - \mu\rho Z_1 \equiv 0.$$

This means that $Z_1(z)$ is an eigenfunction of the Sturm–Liouville operator, and $\mu = \lambda_1$ is the corresponding (smallest) eigenvalue.

Now let us narrow the class of functions $f(z)$ to the subspace of M defined as follows:
$$M_1 := \{f \in M; (\rho f, Z_1) = 0\}.$$

In other words, M_1 consists of functions that are orthogonal to the first eigenfunction $Z_1(z)$. Repeating the above arguments, one can easily show that
$$\inf_{f \in M_1} \frac{(Lf, f)}{\|f\|^2} = \lambda_2 > \lambda_1,$$

and that the above minimum is attained at the second-smallest eigenfunction $f(z) = Z_2(z) \in M_1$ of the Sturm–Liouville problem. Continuing this line of reasoning, one obtains the following extremal property of the eigenfunctions and eigenvalues of the Sturm–Liouville problem:

Property 5. *The eigenvalues of the Sturm–Liouville problem form an increasing infinite sequence*
$$\lambda_1 < \lambda_2 < \cdots < \lambda_n < \cdots \tag{48}$$

of the minimal values of the functionals

$$\lambda_n = \inf_{f \in M_{n-1}} \frac{(Lf, f)}{\|f\|^2}, \tag{49}$$

9.8. Sturm–Liouville Problem

where

$$M_k := \{f \in M : (\rho f, Z_j) = 0, j = 1, 2, \ldots, k\}.$$

Moreover, the minima are attained at the corresponding eigenfunctions

$$Z_1(z), Z_2(z), \ldots, Z_n(z), \ldots.$$

9.8.5 Comparison Theorems for Eigenvalues

The goal of this subsection is to establish connections between the behavior of the coefficients $p(z), q(z)$ of the Sturm–Liouville operator (4) and the weight function $\rho(z)$ appearing in the Sturm–Liouville equation (1), and the related problem of the magnitude of the corresponding eigenvalues.

For example, assume that $p_1(z) \geq p_2(z)$, for all $z \in [0, h]$. It turns out that with the same functions $q(z)$ and $\rho(z)$, the corresponding eigenvalues satisfy the inequality $\lambda_n^{(1)} \geq \lambda_n^{(2)}$. Here, the upper indices indicate that the eigenvalues correspond to the Sturm–Liouville operators $L_n^{(1)}$ and $L_n^{(2)}$ with coefficients $p = p_1(z)$ and $p = p_2(z)$, respectively. For the sake of simplicity, we shall prove this property only for the first eigenvalue λ_1.

The proof depends on a simple observation to the effect that for every function $f \in M$, we have the inequality

$$\frac{(L^{(1)}f, f)}{\|f\|^2} \geq \frac{(L^{(2)}f, f)}{\|f\|^2}.$$

Thus the same inequality is preserved for the minimal values of the left and right sides, i.e., for the first eigenvalues of the operators $L^{(1)}, L^{(2)}$. In other words, $\lambda_1^{(1)} \geq \lambda_1^{(2)}$.

Now let us keep the functions $p(z)$ and $q(z)$ unchanged and consider different weight coefficients $\rho_1 \geq \rho_2$ in (1). Then for every $f \in M$,

$$\|f\|_{\rho_1}^2 = \int_0^h f^2(z)\rho_1(z)\,dz \geq \int_0^h f^2(z)\rho_2(z)\,dz = \|f\|_{\rho_2}^2.$$

Consequently,

$$\frac{(L^{(1)}f, f)}{\|f\|_{\rho_1}^2} \leq \frac{(L^{(2)}f, f)}{\|f\|_{\rho_2}^2},$$

so that

$$\lambda_1^{(1)} = \inf_{f \in M} \frac{(L^{(1)}f, f)}{\|f\|_{\rho_1}^2} \leq \inf_{f \in M} \frac{(L^{(2)}f, f)}{\|f\|_{\rho_2}^2} = \lambda_1^{(2)}.$$

In other words, if the weight function ρ is replaced by a larger one, then the first eigenvalue gets smaller. The same observation remains valid for other eigenvalues. ∎

The above two results imply the following fundamental property of the Sturm–Liouville problem:

Property 6. The eigenvalues λ_n of the Sturm–Liouville problem increase asymptotically like n^2 as $n \to \infty$.

To prove this asymptotic result, let us consider, together with the Strum–Liouville equation

$$\frac{d}{dz}(p(z)Z') - q(z)Z + \lambda\rho(z)Z = 0, \tag{50}$$

the equations

$$\begin{aligned} p_1 Z'' + (\lambda\rho_2 - q_1)Z &= 0, \\ p_2 Z'' + (\lambda\rho_1 - q_2)Z &= 0, \end{aligned} \tag{51}$$

where p_1, q_1, ρ_1 are the maximal values of the functions $p(z), q(z), \rho(z)$ on the interval $z \in [0, h]$, and p_2, q_2, ρ_2 are their minimal values. Denote the eigenvalues of the Sturm–Liouville problems corresponding to (50)–(51) by $\lambda_n, \lambda_n', \lambda_n''$, respectively. The above comparison theorems imply that $\lambda_n' \leq \lambda_n \leq \lambda_n''$, and the extreme terms in these inequalities can be easily determined because of the simplicity of the corresponding equations (51). For the sake of concreteness we will consider the above equations with boundary conditions of the first kind (30). In this case, in view of the solutions found in 9.8.3, we already know the eigenvalues of the Sturm–Liouville problem (51), (30), so the above inequalities give the following inequalities for the eigenvalues of the problem (50), (30):

$$\frac{p_1}{\rho_2}\left(\frac{\pi n}{h}\right)^2 + \frac{q_1}{\rho_2} \leq \lambda_n \leq \frac{p_2}{\rho_1}\left(\frac{\pi n}{h}\right)^2 + \frac{q_2}{\rho_1}. \tag{52}$$

These inequalities imply the claimed asymptotics $\lambda_n \sim n^2$ $(n \to \infty)$. ∎

9.8.6 Expansion of Functions with Respect to Eigenfunctions of the Sturm–Liouville Problem

It is a remarkable property of the eigenfunctions of the Sturm–Liouville problem that every function $f \in M$ can be represented as a series with respect

9.8. Sturm–Liouville Problem

to those eigenfunctions. We shall prove this fact assuming that the eigenfunctions $Z_1(z), Z_2(z), \ldots$ have already been normalized with respect to the weight function $\rho(z)$, i.e.,

$$(\rho Z_n, Z_m) = \int_0^h Z_n Z_m \rho(z)\, dz = \delta_{nm}, \tag{53}$$

where the Kronecker symbol δ_{nm} is equal to 1 if $n = m$, and to 0 if $n \neq m$. Recall that in view of the linearity of the Sturm–Liouville problem, the orthogonal system of eigenfunctions can always be normalized by replacing $Z_n(z)$ by $Z_n(z)/\|Z_n(z)\|_\rho$.

A few definitions regarding general orthogonal expansions are now in order. The series

$$\sum_{k=1}^{\infty} C_k Z_k(z), \tag{54}$$

with coefficients

$$C_k = (f, \rho Z_k) = \int_0^h f(z) Z_k(z) \rho(z)\, dz, \tag{55}$$

is called the *Fourier series* of the function $f(z)$ with respect to the orthonormal system $Z_1(z), Z_2(z), \ldots$. The sums

$$S_n(z) = \sum_{k=1}^{n} C_k Z_k(z) \tag{56}$$

are called the partial sums of the Fourier series, and

$$R_n(z) = f(z) - S_n(z)$$

is called the remainder term of the series. We shall say that the Fourier series *converges in the mean square* to the function $f(z)$ if

$$\|R_n\|_\rho^2 = \left\| f(z) - \sum_{k=1}^{n} C_k Z_k(z) \right\|_\rho^2 \longrightarrow 0, \quad (n \to \infty). \tag{57}$$

We observe a few properties of the remainder $R_n(z)$:

Property 7. The remainder $R_n(z)$ is orthogonal, with weight $\rho(z)$, to the eigenfunctions $Z_1(z), \ldots, Z_n(z)$.

Indeed, for all $i = 1, 2, \ldots, n$,

$$(R_n, \rho Z_i) = \left(f - \sum_{k=1}^{n} C_k Z_k, \rho Z_i\right) = (f, \rho Z_i) - \sum_{k=1}^{n} C_k \delta_{ik} = C_i - C_i = 0,$$

which gives the desired orthogonality property. ∎

Property 8. For all $n = 1, 2, \ldots,$

$$\|f\|_\rho^2 = (f, \rho f) = \sum_{k=1}^{n} C_k^2 + \|R_n\|_\rho^2. \tag{58}$$

To prove this property, observe that in view of the definition of the remainder of the Fourier series,

$$f(z) = S_n(z) + R_n(z), \tag{59}$$

and since by Property 7, R_n and S_n are orthogonal $((R_n, \rho S_n) = 0)$,

$$\|f\|_\rho^2 = (S_n + R_n, \rho(S_n + R_n)) = \|S_n\|_\rho^2 + \|R_n\|_\rho^2 = \sum_{k=1}^{n} C_k^2 + \|R_n\|_\rho^2, \tag{60}$$

which concludes the proof of the Property 8. ∎

Equality (58), together with the fact that $\|R_n\|_\rho^2 > 0$, also immediately implies the *Bessel inequality*

$$\sum_{k=1}^{n} C_k^2 \leq \|f\|_\rho^2. \tag{61}$$

In fact, the validity of *Parseval's equality*,

$$\sum_{k=1}^{\infty} C_k^2 = \|f\|_\rho^2, \tag{62}$$

is a necessary and sufficient condition for the Fourier series of f to converge to f in the mean square. These facts lead to the following property:

Property 9 (Steklov's theorem). For every function $f \in M$, its Fourier series expansion with respect to the eigenfunctions of a Sturm–Liouville problem converges to f in the mean square.

9.8. Sturm–Liouville Problem

To verify Steklov's theorem, let us represent f in the form of the sum (59) and consider the quadratic functional

$$(Lf, f) = (LS_n + LR_n, S_n + R_n) = (LS_n, S_n) + (LR_n, R_n) + 2(LS_n, R_n).$$

The last term on the right vanishes in view of Property 7 of the remainder term, and evaluation of the first term yields

$$(LS_n, S_n) = \left(L\sum_{k=1}^n C_k Z_k, \sum_{k=1}^n C_k Z_k\right) = \left(\sum_{k=1}^n \lambda_k C_k Z_k, \sum_{k=1}^n C_k Z_k\right) =$$

$$\sum_{k=1}^n \sum_{i=1}^n \lambda_k C_k C_i (\rho Z_k, Z_i) = \sum_{k=1}^n \lambda_k C_k^2.$$

The above two equalities yield

$$(LR_n, R_n) = (Lf, f) - \sum_{k=1}^m \lambda_k C_k^2 \leq (Lf, f). \tag{63}$$

We shall use this inequality to obtain an estimate for the last term in the equality (58), keeping in mind that the remainder R_n belongs to the set of functions

$$M_n = \{f \in M : (\rho f, Z_j) = 0, \ j = 1, 2, \ldots, n\},$$

and as such, satisfies the inequality

$$\lambda_{n+1} = \inf_{f \in M_n} \frac{(Lf, f)}{\|f\|_\rho^2} \leq \frac{(Lf, f)}{\|R_n\|_\rho^2}.$$

This inequality and (63) imply

$$\|R_n\|_\rho^2 \leq \frac{1}{\lambda_{n+1}}(Lf, f).$$

Since we showed earlier that $\lambda_{n+1} \to \infty$ as $n \to \infty$, we get

$$\lim_{n \to \infty} \|R_n\|_\rho = 0. \qquad \blacksquare$$

Remark 1. Utilizing more powerful mathematical tools, one can prove that if $f \in M$, then we have the pointwise convergence of the Fourier series, i.e., for all $z \in [0, h]$,

$$f(z) = \sum_{k=1}^\infty C_k Z_k(z), \quad \text{for every } z \in [0, h], \tag{64}$$

and moreover, the Fourier series on the right-hand side converges absolutely and uniformly.

Remark 2. Fourier series can be used to represent not only functions from the set M, but also distributions, such as continuous linear functionals on the set of test functions $\phi \in M$. Let T_f be such a distribution. Let us start with describing the algorithm of its action on an arbitrary test function ϕ. In other words, let us evaluate the functions $T_f[\phi]$. In view of the above Steklov's theorem, the test function can be represented by an absolutely and uniformly convergent series,

$$\phi(z) = \sum_{k=1}^{\infty} D_k Z_k(z).$$

Substituting this representation into the functional, we get

$$T_f[\phi] = \sum_{k=1}^{\infty} D_k C_k, \tag{65}$$

where $C_k = T_f[Z_k]$, $k = 1, 2, \ldots$, are coefficients in the expansion of the distribution T_f into its Fourier series. These coefficients are well defined because $Z \in M$ is a test function for the distribution under consideration. We shall show that this expansion is

$$T_f = \sum_{k=1}^{\infty} C_k Z_k. \tag{66}$$

Indeed, if we take the scalar product of both sides with the test function ϕ, we arrive at (64), which shows the validity of the equality (65) in the weak distributional sense.

Example. The generalized Fourier series for the Dirac delta

$$\delta(z - z_0), \qquad z_0 \in (0, h),$$

is of the form

$$\delta(z - z_0) = \sum_{k=1}^{\infty} Z_k(z_0) Z_k(z), \tag{67}$$

and the functional (65) represents the convergent series

$$(\delta(z - z_0), \phi) = \sum_{k=1}^{\infty} D_k Z_k(z). \qquad \blacksquare$$

9.9 Waves in Waveguides Revisited

We can now return to the calculation of the Green's function of a monochromatic wave in a flat cylindrical waveguide that we began in Sect. 9.7; in Sect. 9.8, we developed the necessary tools to finish the job. Let us briefly recall that the problem in Sect. 9.7 was to find the Green's function $G(\boldsymbol{y}, z, z_0)$ satisfying the equation

$$\Delta_\perp G + \frac{\partial^2 G}{\partial z^2} + k^2(z)G = \delta(\boldsymbol{y})\delta(z - z_0) \tag{1}$$

and the homogeneous boundary conditions

$$\left.\frac{\partial}{\partial z} G(\boldsymbol{y}, z, z_0)\right|_{z=0} = 0, \quad \left.\frac{\partial}{\partial z} G(\boldsymbol{y}, z, z_0)\right|_{z=h} = 0. \tag{2}$$

For this purpose, we constructed a solution

$$u(\boldsymbol{y}, z) = \sum_\mu A(\mu) Y_\mu(\boldsymbol{y}) Z_\mu(z) \tag{3}$$

of the (slightly different from (1)) equation

$$\Delta_\perp u + \frac{\partial^2 u}{\partial z^2} + k^2(z)u = \delta(\boldsymbol{y}) \sum_\mu A(\mu) Z_\mu(z). \tag{4}$$

The functions $Y_\mu(\boldsymbol{y})$ and $Z_\mu(z)$ enter into (3) and (4). The former are given by integrals (9.7.9) and (9.7.11), while the latter solve the boundary value problem

$$Z'' + [k^2(z) - \mu^2]Z = 0, \quad Z'(0) = 0, \quad Z(h) = 0. \tag{5}$$

From now on, to simplify our notation we will drop the index μ. In view of Sect. 9.8, problem (5) reduces to the Sturm–Liouville problem (9.8.1)–(9.8.4). Indeed, it suffices to take there

$$q(z) = k^2 - k^2(z), \quad p(z) = 1, \quad \rho(z) = 1, \quad \lambda = k^2 - \mu^2. \tag{6}$$

Here k^2 is the maximal value of the coefficient $k^2(z)$ inside the ocean layer $z \in [0, h]$.

It follows from the results of the preceding section that the Sturm–Liouville problem (5)–(6) has an infinite number of solutions (eigenfunctions) $Z_1(z), Z_2(z), \ldots$ corresponding to an increasing sequence of eigenvalues

$\lambda_1, \lambda_2, \ldots$. Those eigenvalues form an orthonormal system, i.e., in our present case,

$$(Z_n, Z_m) = \int_0^h Z_n(z) Z_m(z) \, dz = \delta_{nm}, \tag{7}$$

and in particular,

$$\|Z_n\|^2 = \int_0^h Z_n^2(z) \, dz = 1. \tag{8}$$

The system of eigenfunctions of the Sturm–Liouville problem was useful to us because it could serve as a basis for Fourier expansions of functions and distributions. In particular, we obtained the remarkable relation

$$\delta(z - z_0) = \sum_n Z_n(z_0) Z_n(z). \tag{9}$$

In this context, it is clear that an abstract index μ in solution (3) can be replaced by an index labeling the eigenfunctions, and (3) itself can be rewritten in the form

$$u(\boldsymbol{y}, z) = \sum_n A_n Y_{\mu(n)}(\boldsymbol{y}) Z_n(z), \tag{10}$$

where

$$\mu^2(n) = k^2 - \lambda_n^2 \tag{11}$$

expresses the old parameter μ via the new index of summation n corresponding to the numbering of the eigenfunction. In the new notation, (4) takes the form

$$\Delta_\perp u + \frac{\partial^2 u}{\partial z^2} + k^2(z) u = \delta(\boldsymbol{y}) \sum_n A_n Z_n(z). \tag{12}$$

Comparing its right-hand side with the distributional equality (9), we conclude that (12) is identical with (1), and that if we select $A_n = Z_n(z_0)$, then equality (10) provides the solution of the boundary value problem (1)–(2). Consequently, the Green's function of the acoustic waves inside the ocean layer is of the form

$$G(\boldsymbol{y}, z, z_0) = \sum_{\mu(n)} Y_{\mu(n)}(\boldsymbol{y}) Z_n(z_0) Z_n(z). \tag{13}$$

The explicit form of the Green's function will become available once we find all the eigenfunctions $Z_n(z)$ and substitute them into (13). In this calculation let us restrict our attention to the case of the homogeneous ocean, where

$$k^2(z) = k^2 = const.$$

9.9. Waves in Waveguides Revisited

As often happens, an analysis of the simplest case provides insight into characteristic features of the general Green's function for the stratified waveguide.

In the homogeneous ocean, (5) takes the form

$$Z'' + \lambda Z = 0, \qquad \lambda = k^2 - \mu^2. \tag{14}$$

Its general solution, for $\lambda > 0$, is well known:

$$Z(z) = C_1 \cos(\sqrt{\lambda} z) + C_2 \sin(\sqrt{\lambda} z). \tag{15}$$

To satisfy the first boundary condition in (5), we find the derivative

$$Z'(z) = -C_1 \sqrt{\lambda} \sin(\sqrt{\lambda} z) + C_2 \sqrt{\lambda} \cos(\sqrt{\lambda} z). \tag{16}$$

Substituting $z = 0$, we get $C_2 = 0$, so that

$$Z(z) = C_1 \cos(\sqrt{\lambda} z). \tag{17}$$

This function satisfies the second boundary condition of (5) at $z = h$ only at discrete values of

$$\lambda = \lambda_n = \frac{\pi^2}{4h^2}(2n+1)^2, \qquad n = 0, 1, 2, \ldots, \tag{18}$$

which are the eigenvalues of the Sturm–Liouville problem (5). The reader should check that the corresponding orthonormalized eigenfunctions are

$$Z_n(z) = \sqrt{\frac{2}{h}} \cos\left(\frac{\pi}{2h} z (2n+1)\right). \tag{19}$$

Substituting them into (13), we arrive at the following expression for the Green's function:

$$G(\boldsymbol{y}, z, z_0) = k_0 \frac{4}{\pi} \sum_{n=0}^{\infty} Y_{\mu(n)}(\boldsymbol{y}) \cos\Big(z k_0 (2n+1)\Big) \cos\Big(z_0 k_0 (2n+1)\Big), \tag{20}$$

where

$$\mu^2(n) = k^2 - k_0^2 (2n+1)^2, \qquad k_0 = \frac{\pi}{2h}. \tag{21}$$

Now it is clear that $\mu^2(n)$ is a decreasing function of the integer-valued variable n, whose the maximum value is

$$\mu_0^2 = \mu^2(0) = k^2 - k_0^2. \tag{22}$$

The analysis of the function $Y_\mu(\boldsymbol{y})$ carried out in Sect. 9.7 shows that if μ_0^2 is negative, then all $Y_\mu(\boldsymbol{y})$ decay exponentially as $|\boldsymbol{y}|$ increases. In physics jargon, one often speaks of such Green's functions (and thus the hydroacoustic wave described by it) as being *localized*. In particular, at large distances (as $|\boldsymbol{y}| \to \infty$), one can with good accuracy approximate (20) by its first term,

$$G(\boldsymbol{y}, z, z_0) \approx \frac{k_0}{\pi}\sqrt{\frac{2}{\pi|\mu_0|r}} \exp\!\left(-|\mu_0|r\right) \cos(k_0 z)\cos(k_0 z_0). \qquad (23)$$

Here we used the asymptotic formula (9.7.12) and introduced the radial coordinate

$$r = \sqrt{y_1^2 + y_2^2}. \qquad (24)$$

If $\mu_0^2 > 0$ ($k > k_0$), the Green's function behaves in a completely different fashion. In that case, there exists

$$N = \left\lfloor \frac{1}{2}(\epsilon^2 - 1) \right\rfloor \geq 0 \qquad (25)$$

such that for all $n \leq N$, we have $\mu^2 > 0$. In (25) we use the *floor* function notation $\lfloor x \rfloor$ (which means the greatest integer not exceeding x), and also introduce the dimensionless parameter

$$\epsilon = \frac{k}{k_0} = \frac{2}{\pi} hk. \qquad (26)$$

It follows from (9.7.10) that for $\mu > 0$, the function $Y_\mu(\boldsymbol{y})$ decays relatively slowly as r increases. Thus, the terms in the sum (20) can be split into two types: the slowly decreasing terms with indices $n \leq N$, and the rapidly exponentially decreasing terms with $n > N$. For large values of r, the exponentially decreasing terms can be dropped, giving the following approximation for (20):

$$G(\boldsymbol{y}, z, z_0) \approx k_0 \frac{4}{\pi} \sum_{n=0}^{N} Y_{\mu(n)}(\boldsymbol{y}) \cos\!\left(z k_0(2n+1)\right)\cos\!\left(z_0 k_0(2n+1)\right). \qquad (27)$$

Physicists call the summands in the above formula *modes*, and say that $N+1$ modes are propagating in the waveguide.

The final intrinsically consistent asymptotic (as $r \to \infty$) expression for the Green's function can be obtained by substituting asymptotics (9.7.10) into (27):

$$G(\boldsymbol{y}, z, z_0) = \frac{k_0}{\pi} \sum_{n=0}^{N} f(r, n) \cos\!\left(z k_0(2n+1)\right)\cos\!\left(z_0 k_0(2n+1)\right), \qquad (28)$$

where
$$f(r,x) = \sqrt{\frac{2}{\pi i \mu(x) r}} \exp\left(-i\mu(x)r\right). \tag{29}$$

9.10 Exercises

1. Find the Green's function satisfying the equation $G'' = \delta(x-y)$ for $0 < x, y < l$ and the homogeneous boundary conditions $G(x=0, y) = G(x=l, y) = 0$.

2. Find the solution of the equation $G'' + k^2 G = \delta(x-y)$ for $x, y > 0$ that satisfies the boundary condition $G(x = 0, y) = 0$ and the radiation condition.

3. Solve the following boundary problem for the 2-D Laplace equation: $\Delta u = 0$, $x_1 \in \mathbf{R}$, $x_2 > 0$; $u(x_1, x_2 = 0) = f(x_1)$.

4. What boundary value problem in the half-space $x_2 < 0$ has the solution found in Exercise 3?

5. Using the method of reflection, solve the 2-D Laplace equation $\Delta u = 0$ inside the unit disk (for $\rho < 1$) with the boundary condition $u(\rho = 1, \varphi) = f(\varphi)$.

6. Find the electrostatic potential of a dipole satisfying the equation
$$\Delta U = -p(\mathbf{n} \cdot \nabla)\delta(\mathbf{x} - \mathbf{y}), \quad \mathbf{x} \in \mathbf{R}^3,$$
where $\mathbf{p}$ is the dipole vector, $\mathbf{n} = \mathbf{p}/p$, and $\mathbf{y}$ is the coordinate indicating the position of the dipole.

7. Find a solution of the 2-D Helmholtz equation
$$\frac{\partial^2 u}{\partial x^2} + \frac{\partial^2 u}{\partial y^2} + k^2 u = 0$$
for $x > 0$ with the boundary condition $u(x = 0, y) = f(y)$, assuming that it satisfies the radiation condition and converges to 0 as $x \to \infty$. (*Hint.* Use the Fourier transform in y to transform the original partial differential equation into an ordinary differential equation.)

Chapter 10

Diffusions and Parabolic Evolution Equations

We begin with a study of the classic 1-D diffusion equation (also called heat equation) and its self-similar solutions. This is the simplest example of a linear parabolic partial differential equations. Well-posedness of an initial value problem with periodic data is then discussed. Subsequently, the exposition switches to the complex domain, and we introduce a simple version of the general Schrödinger equation. This makes it possible to study the diffraction problem and the so-called Fresnel zones. Multidimensional parabolic equation follow, and the general reflection method is explained. The chapter concludes with a study of the moving boundary problem and the standard physical problem of particle motion in a potential well.

10.1 Diffusion Equation and Its Green's Function

The standard homogeneous *diffusion (heat)* equation

$$\frac{\partial u}{\partial t} = \frac{D}{2}\frac{\partial^2 u}{\partial x^2}, \tag{1}$$

where $u = u(x,t)$, is the simplest example of a parabolic equation; it contains the first derivative in the time variable $t \in \mathbf{R}$, and the second derivative in the space variable $x \in \mathbf{R}$. The parameter $D > 0$ is called the *diffusion coefficient*.

Consider the Green's function (fundamental solution) for this simplest 1-D parabolic problem, that is, the function $G = G(x,t)$ that satisfies the

equation
$$\frac{\partial G}{\partial t} = \frac{D}{2}\frac{\partial^2 G}{\partial x^2} + \delta(x)\delta(t). \tag{2}$$

We will ask for only a forward-in-time, $t \geq 0$, solution of (2), since in this case, the problem is *well posed*. Recall that a problem is said to be well posed if it has a unique solution that changes little under small perturbations of the initial conditions. The fact that the initial value problem for (2) is well posed only in the forward-in-time direction has profound physical significance. Equations of this type describe *irreversible* physical processes such as heat flow and molecular diffusion. For the Green's function problem for the diffusion equation to be well posed, the Green's function itself has to satisfy the causality condition

$$G(x, t < 0) = 0.$$

Hence, proceeding as in the case of the fundamental solution (equation (2.2.9) in Volume 1) to the initial value problem for an ordinary differential equation, we shall be seeking the Green's function of the diffusion equation (2) in the product form

$$G(x,t) = u(x,t)\chi(t), \tag{3}$$

where $\chi(t)$ is the familiar Heaviside function, and $u(x,t)$ is an unknown function. Substitute (3) into (2) and recall that $\chi'(t) = \delta(t)$. As a result, we obtain that

$$\chi(t)\left[\frac{\partial u}{\partial t} - \frac{D}{2}\frac{\partial^2 u}{\partial x^2}\right] + u(x,0)\delta(t) = \delta(x)\delta(t),$$

from which it is evident that if we take

$$u(x, 0) = \delta(x), \tag{4}$$

then $u(x,t)$ satisfies the homogeneous diffusion equation (1) with the initial condition (4).

10.2 Self-Similar Solutions

Consider an auxiliary function

$$v(x,t) = \int_{-\infty}^{x} u(y,t)\,dy,$$

where $u(x,t)$ is a solution of the diffusion equation given by (10.1.1)–(4). If u is interpreted as the *density* of diffusing particles, then v is usually called

10.2. Self-Similar Solutions

the *cumulative distribution* of the particles. The function $v(x,t)$ solves the initial value problem

$$\frac{\partial v}{\partial t} = \frac{D}{2}\frac{\partial^2 v}{\partial x^2}, \qquad v(x,0) = \chi(x). \tag{1}$$

Remarkably, problem (1) has a *self-similar* solution, by which we mean here a solution determined by a function f of only one argument

$$\rho = t^n x,$$

which is the product of power functions of the original variables t and x. The self-similarity property of v can then be stated as follows: for every x and $t > 0$,

$$v(x,t) = f(\rho) = f(t^n x) = v(xt^n, 1).$$

Remark 1. The analysis of possible self-similar solutions is explored in this chapter in the case of linear equations, but as we shall see in Part IV, it is also among the most efficient tools in the study of nonlinear partial differential equations. If such solutions exist, then they have to satisfy ordinary differential equations that are, as a rule, simpler to handle than the original nonlinear partial differential equation.

To find a self-similar solution to problem (1), observe that by the chain rule,

$$\frac{\partial v}{\partial t} = n t^{n-1} x f'(\rho), \qquad \frac{\partial^2 v}{\partial x^2} = t^{2n} f''(\rho),$$

where primes denote derivatives with respect to the variable ρ. Substituting these expressions into (1) and dividing both sides by t^{2n}, we obtain

$$\frac{nx}{t^{n+1}} f' = \frac{D}{2} f''.$$

This equation is consistent only if the fraction on the left-hand side is a power of the self-similar variable ρ, i.e., if

$$\frac{x}{t^{n+1}} = \rho^\alpha = t^{n\alpha} x^\alpha,$$

for some exponent α. A comparison of the powers of the variables x and t on the left- and right-hand sides, respectively, yields the coupled equations

$1 = \alpha$ and $-n - 1 = n\alpha$. Hence, $\alpha = 1$, $n = -1/2$, $\rho = x/\sqrt{t}$, and the equation for $f(\rho)$ takes the form

$$f'' = -\frac{\rho}{D} f'.$$

This equation is easy to solve, since clearly, the function $g(\rho) = f'(\rho)$ satisfies a first-order ordinary differential equation with the general solution

$$g(\rho) = f'(\rho) = C \exp\left(-\frac{\rho^2}{2D}\right),$$

where C is an arbitrary constant. Integrating this expression with respect to ρ and taking into account the obvious requirements

$$f(-\infty) = 0, \qquad f(+\infty) = 1,$$

which fit the initial condition in (1), we get that

$$f(\rho) = \frac{1}{\sqrt{2\pi D}} \int_{-\infty}^{\rho} \exp\left(-\frac{\tau^2}{2D}\right) d\tau = \frac{1}{2}\left[1 + \operatorname{erf}\left(\frac{\rho}{\sqrt{2D}}\right)\right]. \qquad (2)$$

The special function

$$\operatorname{erf}(z) = \frac{2}{\sqrt{\pi}} \int_0^z \exp(-\tau^2) d\tau, \qquad (3)$$

which appears in the solution (2), is called the *error function*, and its graph is shown in Fig. 10.2.1.

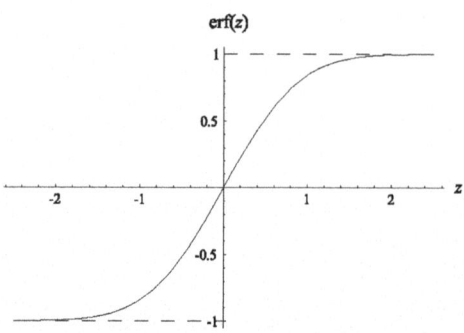

FIGURE 10.2.1
The plot of the error function erf z.

10.2. Self-Similar Solutions

Substituting in (2) the explicit expression for ρ in terms of x and t, we arrive at the following solution to (1):

$$v(x,t) = f(\rho) = \frac{1}{2}\left[1 + \mathrm{erf}\left(\frac{x}{\sqrt{2Dt}}\right)\right]. \tag{4}$$

As $t \to 0$, $v(x,t)$ converges pointwise to $\chi(x)$ and thus satisfies the initial condition in (1). Differentiating (4) with respect to x yields the function

$$u(x,t) = \frac{1}{\sqrt{2\pi Dt}} \exp\left(-\frac{x^2}{2Dt}\right), \tag{5}$$

which is a solution of the diffusion equation (10.1.1). Of course, it weakly converges to $\delta(x)$ as $t \to 0$. The solution (5) is also self-similar in the sense that for every x and e every $a, t > 0$,

$$u(a^{1/2}x, at) = a^{-1/2}u(x,t),$$

or equivalently,

$$u(x,t) = t^{-1/2}u(xt^{-1/2}, 1).$$

The above discussion, in view of (10.1.3), implies that the fundamental solution of the diffusion equation has the form

$$G(x,t) = \chi(t)\frac{1}{\sqrt{2\pi Dt}} \exp\left(-\frac{x^2}{2Dt}\right).$$

Once the fundamental solution is known, one can easily write the solution of an arbitrary *inhomogeneous diffusion equation*

$$\frac{\partial u}{\partial t} = \frac{D}{2}\frac{\partial^2 u}{\partial x^2} + \varphi(x,t), \qquad u(x, t=0) = u_0(x).$$

Indeed, it is of the form

$$u(x,t) = \int u_0(y)G(x-y,t)\,dy + \int_0^t \int \varphi(y,\tau)G(x-y, t-\tau)\,dy\,d\tau. \tag{6}$$

The first convolution integral takes into account the initial condition, while the presence of a source $\varphi(x,t)$ is reflected in the second convolution integral.

Remark 2. Mathematicians are accustomed to working with *dimensionless variables*, but for physicists, the dimensions of the variables play an essential role, and *dimensional analysis* is an effective tool in the search for self-similar solutions. This can be illustrated with the example of the diffusion equation (10.1.1), which governs functions of two variables: t measured in units of time, and the spatial coordinate x measured in units of length—information usually encapsulated in the dimensional formulas $[\![t]\!] = T$ and $[\![x]\!] = L$. Additionally, the diffusion equation (10.1.1) contains a diffusion coefficient D with dimension $[\![D]\!] = L^2/T$, so that the density function u can have the proper dimensionality of, say, number of particles per unit length. Also, it is clear that any solution of the diffusion equation, with the exception of those that have power scaling, has to be a function of a dimensionless argument; otherwise, there would be no way to match the dimensions on both sides of the equation. The simplest dimensionless combination of variables x, t, and the coefficient D is given by the expression

$$x^2/Dt.$$

As a result, self-similar solutions have to be of the form

$$u = A x^n t^m f(x^2/Dt),$$

where $f(z)$ is a dimensionless function of a dimensionless argument. The Green's function appearing in the above diffusion problem is self-similar, since the Dirac delta source introduces no additional temporal and spatial scales.

10.3 Well-Posedness of Initial Value Problems with Periodic Data

Consider a simple parabolic initial value problem

$$\frac{\partial u}{\partial t} = \frac{D}{2}\frac{\partial^2 u}{\partial x^2}, \qquad u(x, t=0) = ae^{ikx}, \tag{1}$$

with periodic initial data. Substituting the above initial condition into the first integral in (10.2.6) and recalling that

$$\int \exp(-bx^2 + ikx)dx = \sqrt{\frac{\pi}{b}} \exp\left(-\frac{k^2}{4b}\right), \qquad \operatorname{Re} b > 0, \tag{2}$$

10.4. Complex Parabolic Equations

we obtain a solution to (1) of the form

$$u(x,t) = a \exp\left(ikx - \frac{D}{2}k^2 t\right). \tag{3}$$

This solution decays as t increases, which reflects the physical phenomenon of equalization of spatial temperature inhomogeneities due to heat diffusion. Moreover, as the time variable t increases, any perturbation of the initial field is smoothed out in the same manner, thus ensuring a continuous dependence of the solution on the initial conditions.

However, for negative t, the picture is quite different. The solution field grows exponentially with the growth of $|t|$, and that growth is stronger for small-scale initial fields, i.e., for large values of the wavenumber parameter k. As a result, arbitrarily small variations of the initial condition $u_0(x)$ can lead, as time $t < 0$ recedes, to extremely fast—and arbitrarily large—variations of the field $u(x,t)$; for time t running backward, the parabolic initial value problem (1) is not well posed.

10.4 Complex Parabolic Equations

Introduction of an "imaginary" diffusion coefficient in the parabolic equation (10.3.1) significantly changes the situation. In this section, we will consider only the simple equation,

$$i\hbar \frac{\partial \psi}{\partial t} = -\frac{\hbar^2}{2m} \frac{\partial^2 \psi}{\partial x^2}. \tag{1}$$

Its more general version

$$i\hbar \frac{\partial \psi}{\partial t} = -\frac{\hbar^2}{2m} \Delta \psi + U\psi$$

is called the *Schrödinger equation*. It describes the quantum-mechanical laws of particle motion in a force field with potential $U = U(\boldsymbol{x})$. The causality principle is assumed to be satisfied. The constant $\hbar$ is called *Planck's constant*, and m denotes the particle's mass. We shall discuss this more general equation in greater detail in Sect. 10.9.

Here, we will seek a solution of (1) satisfying the initial condition

$$\psi(x, t = 0) = \psi_0(x).$$

By definition, the Green's function $G(x,t)$ of (1) solves the initial value problem

$$i\hbar \frac{\partial G}{\partial t} = -\frac{\hbar^2}{2m}\frac{\partial^2 G}{\partial x^2}, \qquad G(x,t=0) = \delta(x),$$

and it can be obtained from the Green's function (10.2.5) of the diffusion equation by a formal substitution of the quantity $i\hbar/m$ in place of the diffusion coefficient D. Thus

$$G(x,t) = \sqrt{\frac{m}{2\pi i\hbar t}} \exp\left(\frac{imx^2}{2\hbar t}\right). \tag{2}$$

As $t \to 0$, the Green's function (2) weakly converges to $\delta(x)$, regardless of the fact that the modulus of the function is independent of x and diverges to infinity as $t \to 0$.

In contrast to the diffusion equation, we deliberately omitted the Heaviside function on the right-hand side of (2) because the Schrödinger equation describes phenomena that are reversible in time. Thus, the initial value problem for the Schrödinger equation is correctly posed both forward and backward in time, and the Green's function (2), for $t < 0$, reflects the past history of the particle. Mathematically, the reversibility of the quantum-mechanical particle motion is expressed by the equality

$$\psi(x,t) = \int \psi(y,\tau) G(x-y, t-\tau) dy, \tag{3}$$

which, for $\tau > t$, means that the wave function $\psi(x,t)$ at time t can be reconstructed from its value $\psi(x,\tau)$ at a later time. This can be accomplished by solving (1) backward in time. Substituting $t = 0$ in (3) and expressing $\psi(y,\tau)$ through the initial state $\psi(x,0)$ by the same integral, we discover that the Green's function has to satisfy the equation

$$\int G^*(x-y,\tau) G(y-p,\tau) dy = \delta(x-p), \tag{4}$$

where the asterisk denotes complex conjugation, and where we have taken into account that

$$G(x,-\tau) = G^*(x,\tau).$$

Let us prove the validity of (4) by considering an auxiliary "regularized" integral

$$J_\varepsilon(x,p) = \int G^*(x-y,\tau) G(y-p,\tau) \exp(-\varepsilon^2 y^2)\, dy. \tag{5}$$

10.4. Complex Parabolic Equations

Substituting the explicit expression (2) for the Green's function of (1), we get that

$$J_\varepsilon(x, p) = \frac{m}{2\pi\hbar\tau} \exp\left(\frac{im}{2\hbar\tau}(p^2 - x^2)\right) \int \exp\left(\frac{im}{\hbar\tau}(x - p)y - \varepsilon^2 y^2\right) dy.$$

Evaluating this integral with the help of formula (10.3.2), we arrive at the expression

$$J_\varepsilon(x, p) = \frac{1}{\sqrt{\pi\alpha}} \exp\left(\frac{im}{2\hbar\tau}(p^2 - x^2) - \frac{(x - p)^2}{\alpha^2}\right),$$

where $\alpha = 2\hbar\tau\varepsilon/m$. As $\varepsilon \to 0$, α converges to 0, and $J_\varepsilon(x, p)$ weakly converges to $\delta(x - p)$.

Example. A diffraction problem. Complex parabolic equations are frequently encountered in the theory of wave propagation. The following diffraction problem is an example of such an application. Recall that a complex monochromatic wave field propagating along the z-axis is expressed as a functional of the initial field $u_0(\boldsymbol{y})$, given in the plane $z = 0$, via the integral (9.5.2):

$$u(\boldsymbol{y}, z) = \iint u_0(\boldsymbol{p}) g(\boldsymbol{y} - \boldsymbol{p}, z) d^2 p, \tag{6}$$

where

$$g(\boldsymbol{y}, z) = \frac{z}{2\pi r^3}(1 + ikr)e^{-ikr}, \qquad r = |\boldsymbol{x}| = \sqrt{|\boldsymbol{y}|^2 + z^2}, \tag{7}$$

represents the *transfer function*, or *propagator*, of the vacuum.

Let us check the behavior of this field far away from the radiation plane $z = 0$. Assume that $u_0(\boldsymbol{y})$ is equal to zero outside the disk of radius a with center at $\boldsymbol{y} = 0$, and that the observation point $(\boldsymbol{y}, z)$ is close to the z-axis ($|\boldsymbol{y}|^2 \ll z^2$) and is located in the *wave zone* ($kz \gg 1$). In such a case, we are justified in replacing the factor in front of the exponential function in (7) by a term with the same principal asymptotics, and we can rewrite (7) in the form

$$g(\boldsymbol{y}, z) = \frac{ik}{2\pi z} \exp\left(-ikz\sqrt{1 + \frac{|\boldsymbol{y}|^2}{z^2}}\right). \tag{8}$$

If, additionally, the condition

$$3k|\boldsymbol{y}|^4/8z^3 \ll 1$$

is satisfied, then we are justified (see Sect. 9.5.1) in retaining just the first two terms in the Taylor expansion of the radical in the exponent of (7). As a result, after dropping the nonessential factor $\exp(-ikz)$, we get that

$$g(\boldsymbol{y}, z) = \frac{ik}{2\pi z} \exp\left(-\frac{ik|\boldsymbol{y}|^2}{2z}\right). \tag{9}$$

As $z \to 0$, this expression weakly converges to the 2-D Dirac delta. Consequently, the field $u(\boldsymbol{y}, z)$ defined by (6) satisfies the so-called *parabolic equation of quasioptics*

$$2ik\frac{\partial u}{\partial z} = \Delta_\perp u, \qquad u(\boldsymbol{y}, z = 0) = u_0(\boldsymbol{y}), \tag{10}$$

which describes wave diffraction in the *Fresnel approximation*. Here, $\Delta_\perp$ denotes the Laplacian in transverse coordinates $\boldsymbol{y}$.

10.5 Fresnel Zones

In this section we discuss further the solution (10.4.9) of the diffraction problem in the context of a concrete example: the divergent integral that appeared first in Sect. 8.11 of Volume 1. This analysis provides a physical argument in support of the separation of scales condition of Sect. 8.5, and provides an example of a situation in which it is violated.

Assume that a point source of a monochromatic wave is located at the origin of the coordinate system $(\boldsymbol{y}, z)$, where z is the longitudinal coordinate and $\boldsymbol{y}$ represents the transverse coordinates. In the Fresnel approximation, the complex amplitude of the wave is given by the Green's function (10.4.9),

$$g(\boldsymbol{y}, z) = \frac{ik}{2\pi z} \exp\left(-\frac{ik|\boldsymbol{y}|^2}{2z}\right). \tag{1}$$

In particular, at the point of observation with the coordinates $z = l$ and $\boldsymbol{y} = 0$, the complex amplitude of the wave is equal to

$$g(0, l) = \frac{ik}{2\pi l}. \tag{2}$$

We will try to obtain this equality by other means, utilizing the *Huygens–Fresnel principle*, which asserts that a wave in the plane $z = l$ can be represented as a superposition of fields of secondary sources, which are determined

10.5. Fresnel Zones

by the wave incident on a certain auxiliary surface, say the plane $z = l/2$. Hence,
$$g(0,l) = \iint g^2(\mathbf{y}, l/2) d^2y.$$
Note that the above equality is a direct consequence of the parabolic equation of quasioptics (10.4.10). Substituting expression (1) for the Green's function, we get
$$g(0,l) = -\left(\frac{k}{\pi l}\right)^2 \iint \exp\left(-\frac{i2k|\mathbf{y}|^2}{l}\right) d^2y. \tag{3}$$
Changing to polar coordinates ρ and φ and using the radial symmetry of the integrand, we can reduce the above double integral to the single integral
$$g(0,l) = -\frac{k}{2\pi l} \int_0^\infty e^{-ix} dx, \tag{4}$$
where we have introduced the dimensionless variable of integration $x = 2k\rho^2/l$. Observe that although each of the double integrals in (3) converges, the integral (4) is a typical divergent integral. The situation is analogous to a well-known phenomenon in the theory of infinite series: the product of two convergent series may turn out to be divergent.[1]

In the case of integral (4), physicists save the day by employing a procedure that essentially amounts to an application of the generalized summation method satisfying the separation of scales condition. The Abel summation method of Volume 1, in the continuous case, would require us to replace (4) by
$$\lim_{\alpha \to 0} \frac{-k}{2\pi l} \int_0^\infty e^{-ix} e^{-\alpha x} dx.$$
This attenuation is implemented in the general case by introduction of a smoothly decreasing damping factor $f(x)$ such that
$$f(0) = 1, \quad \lim_{x \to \infty} f(x) = 0.$$
After inclusion of the attenuation factor, the integral (4) can be rewritten in the form
$$g(0,l) = -\frac{k}{2\pi l} \int_0^\infty f(x) e^{-ix} dx. \tag{5}$$

---

[1]Of course, (3) contains just an ordinary Gaussian integral, which can be computed by means of analytic continuation from the case of a real-valued exponent. However, our reasoning emphasizes the usefulness of Abel summation in applied physics problems.

As we have already observed in discussing the integral (8.11.2), for sufficiently smooth $f(x)$, the integral in (5) actually does not depend on the shape of $f(x)$ and is equal to

$$\int_0^\infty f(x)e^{-ix}dx = \frac{1}{i}f(0) = -i. \tag{6}$$

Substituting this value into (5), we arrive at the correct answer, (2).

The result obtained in (6) can be elucidated with the help of the following vector diagrams. Consider the auxiliary interference integral

$$I(t) = \int_0^t e^{-ix}dx = \text{``}\sum_0^t e^{-ix}dx\text{,''}$$

which we have heuristically replaced by the sum of infinitesimally small summands in the complex plane (see Fig. 10.5.1a).

As t increases, the point representing the integral in the complex plane moves—in the direction indicated by the arrow—along the unit circle shown in Fig. 10.5.1a. The situation is slightly reminiscent of the oscillatory motion of Achilles in Fig. 8.3.1 of Volume 1. Values $t_n - n\pi$, representing points where the value $I(t)$ of the integral is most distant from the origin, divide—in the terminology of a physicist—the plane $z = l/2$ of secondary sources into *Fresnel zones*. In that plane, the Fresnel zones are concentric annuli with circular boundaries with radii $\rho_n = \sqrt{\pi n l/2k}$.

Now we are able to construct a vector diagram of the integral

$$I_f(t) = \int_0^t f(x)e^{-ix}dx$$

that includes the attenuation factor $f(x)$. Its smooth decay to zero causes the representing point $I_f(t)$ to move along a spiral instead of a circle. As $t \to \infty$, the spiral converges to the center of the original circle (see Fig. 10.5.1b). Hence, the vector diagram accounting for the smoothly decaying attenuation factor again leads to the answer $I_f(\infty) = -i$, which coincides with (6).

The attenuating factor is not necessary if we place a screen in the plane $z = l/2$ that is transparent at the point of intersection with the z-axis but becomes more and more opaque as we move away from the center. Then, instead of the attenuating factor $f(x)$, it suffices to insert into the integral (6) a function $P(x)$ that describes the screen's transparency. This leads to a convergent (if $\lim_{x\to\infty} P(x) = 0$ at a fast enough rate) integral

$$I = \int_0^\infty P(x)e^{-ix}dx. \tag{7}$$

10.5. Fresnel Zones

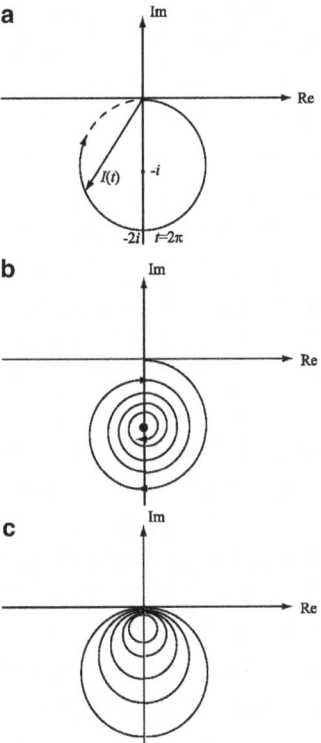

FIGURE 10.5.1
Vector diagrams for the interference integral $I(t)$.

Now suppose that the screen's transparency varies in a discrete fashion according to the function

$$P(x) = e^{-\alpha m}, \quad 2\pi m < x < 2\pi(m+1). \tag{8}$$

Then the integral (7) is representable in the form of a convergent series

$$I = \sum_{m=0}^{\infty} e^{-\alpha m} \int_{2\pi m}^{2\pi(m+1)} e^{-ix} dx = 0, \tag{9}$$

each term of which is equal to 0. As $\alpha \to 0$, the screen (8) becomes completely transparent. However, the amplitude of waves $g(0, l)$ turns out to be zero. This phenomenon is related to the fact that the function (8) regularizing the integral (4) does not satisfy the separation of scales condition. Nevertheless, (9) describes a real physical effect: the existence of shadows at the center of the screen placed in the plane $z = l/2$ when the light of a point source passes through a practically transparent (small α) screen with the transparency expressed by the formula (8). In the language of vector diagrams, this can be explained by the fact that the auxiliary integral

$$J(t) = \int_0^t P(x)e^{-ix}dx$$

travels, as t increases, along a family of circles with smaller and smaller radii, with the common point of the circles located at the origin of the coordinate system (see Fig. 10.5.1c).

10.6 Multidimensional Parabolic Equations

The parabolic equation of quasioptics (10.4.10) is analogous to the 2-D version of the complex parabolic equation (10.4.1) in y-space, with the role of time played by the z-coordinate. Since many physically motivated applied problems require multidimensional considerations, we shall devote this section to a study of their properties using the example of the N-dimensional diffusion equation

$$\frac{\partial G}{\partial t} = \frac{1}{2}\sum_{n=1}^{N} D_n \frac{\partial^2 G}{\partial x_n^2}, \qquad G(\boldsymbol{x}, t=0) = \delta(\boldsymbol{x}). \tag{1}$$

In contrast to the elliptic equations, whose fundamental solutions in spaces of different dimensions are drastically different, the fundamental solutions of parabolic equations in spaces of several dimensions are simply products of 1-D fundamental solutions:

$$G(\boldsymbol{x}, t) = \prod_{n=1}^{N} G_n(x_n, t), \quad G_n(x_n, t) = \frac{1}{\sqrt{2\pi D_n t}} \exp\left(-\frac{x_n^2}{2D_n t}\right).$$

The validity of such a separation of variables can be directly verified by substituting the above product formula into (1). If the diffusion coefficients D_n are the same for all the axes, then the solution is radially symmetric, as was the case for the Green's function (10.4.9) of the parabolic equation of quasioptics.

In particular, for a 2-D space $\boldsymbol{x} = (x_1, x_2)$, if $D_1 = D_2 = D$, the Green's function of the parabolic equation is of the form

$$G(\boldsymbol{x}, t) = \frac{1}{2\pi Dt} \exp\left(-\frac{\boldsymbol{x}^2}{2Dt}\right).$$

10.6. Multidimensional Parabolic Equations

Recall that the usefulness of the Green's function arises from the fact that one can recover with its help any solution of the equation by a straightforward integration process. If at $t = 0$, the initial profile of the field is $u_0(\boldsymbol{x})$, then for $t > 0$, the corresponding solution of the 2-D parabolic equation (1) is given by the convolution integral

$$u(\boldsymbol{x},t) = \iint u_0(\boldsymbol{p}) G(\boldsymbol{x}-\boldsymbol{p},t) d^2p = \frac{1}{2\pi Dt} \iint u_0(\boldsymbol{p}) \exp\left(-\frac{(\boldsymbol{x}-\boldsymbol{p})^2}{2Dt}\right) d^2p.$$

Let us evaluate this convolution for the Gaussian initial profile

$$u_0(\boldsymbol{x}) = u_0 \exp\left(-\frac{\mu x^2}{2}\right).$$

In this case, the convolution integral takes the form

$$u(\boldsymbol{x},t) = u_0 \frac{1}{2\pi Dt} \iint \exp\left(-\frac{\mu p^2}{2} - \frac{(\boldsymbol{x}-\boldsymbol{p})^2}{2Dt}\right) d^2p,$$

which can be evaluated by completing the square in the exponent, so that

$$\frac{\mu p^2}{2} + \frac{(\boldsymbol{x}-\boldsymbol{p})^2}{2Dt} = \frac{\mu x^2}{2(1+\gamma)} + \frac{1+\gamma}{2Dt}\left(\boldsymbol{p} - \frac{\boldsymbol{x}}{1+\gamma}\right)^2,$$

where

$$\gamma = \mu Dt$$

is a dimensionless parameter. Substituting this identity into the integral and passing to new variables of integration

$$\boldsymbol{q} = \boldsymbol{p} - \frac{\boldsymbol{x}}{1+\gamma},$$

we get that

$$u(\boldsymbol{x},t) = u_0 \frac{1}{2\pi Dt} \exp\left(-\frac{\mu x^2}{2(1+\gamma)}\right) \iint \exp\left(-\frac{1+\gamma}{2Dt} q^2\right) d^2q.$$

The remaining integral splits into two *Poisson integrals*

$$\int \exp(-bx^2) dx = \sqrt{\frac{\pi}{b}}.$$

The more general integral (10.3.2) reduces to the same Poisson integral for $k = 0$. Note that in our case, $b = (1+\gamma)/2Dt$, so that

$$u(\boldsymbol{x},t) = \frac{u_0}{1+\gamma} \exp\left(-\frac{\mu x^2}{2(1+\gamma)}\right). \tag{2}$$

The above solution of the 2-D diffusion equation indicates that if the profile $u_0(\boldsymbol{x})$ of the initial temperature field is Gaussian at $t = 0$, then it remains Gaussian for all $t > 0$. Thus we have discovered a kind of "Gaussian invariance principle" for solutions of parabolic equations in unbounded homogeneous space. Also, note that the relation (2) contains a much richer lode of information for the complex equation than was the case for (10.4.1), or the parabolic equation of quasioptics. We shall exploit this fact to analyze diffraction of a *Gaussian beam*.

Example 1. Parabolic equation of quasioptics. Consider a solution of the parabolic equation of quasioptics: substitute $D = 1/ik$ in (2), and replace t by the longitudinal coordinate z and $\boldsymbol{x}$ by the transverse vector $\boldsymbol{y}$. The amplitude of the wave beam, which in the radiation plane $z = 0$ has the Gaussian profile

$$u_0(\boldsymbol{y}) = u_0 \exp\left(-\frac{\mu \boldsymbol{y}^2}{2}\right),$$

is given, according to formula (2), by the expression

$$u(\boldsymbol{y}, z) = \frac{u_0}{1+\gamma} \exp\left(-\frac{\mu \boldsymbol{y}^2}{2(1+\gamma)}\right),$$

where now $\gamma = \mu z/ik$.

In the physical context considered here, the parameter μ can also be complex. In a physically suggestive notation it is often written in the form

$$\mu = \frac{1}{a^2} - i\frac{k}{F},$$

where a is the beam's initial effective radius, and $F > 0$ is the distance from the radiation plane to the focal plane where the beam is to be focused.

From the complex wave amplitude u one can obtain a description of the power characteristics of the wave in terms of the *wave intensity*

$$I(\boldsymbol{y}, z) = |u(\boldsymbol{y}, z)|^2.$$

In our case,

$$I(\boldsymbol{y}, z) = I_0 \frac{a^2}{a^2(z)} \exp\left(-\frac{\boldsymbol{y}^2}{a^2(z)}\right),$$

where $I_0 = |u_0|^2$, and

$$a(z) = a|1+\gamma| = a\sqrt{\left(1-\frac{z}{F}\right)^2 + \left(\frac{\delta z}{F}\right)^2}, \qquad \delta = \frac{F}{ka^2},$$

10.7. The Reflection Method

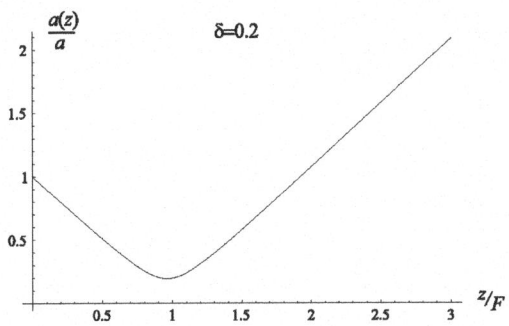

FIGURE 10.6.1
Diffracting beam's effective radius $a(z)/a$ as a function of the distance from the radiation plane $z = 0$. For $\delta \ll 1$, its minimum actually corresponds to the beam's focusing plane $z = F$.

is the effective radius of the Gaussian beam at distance z from the radiation plane. For $z = 0$, it is equal to the initial radius a, which reflects the weak convergence, as $z \to 0$, of the Green's function of the parabolic equation of quasioptics (10.4.9) to the Dirac delta. A typical graph of $a(z)$ as a function of z is shown in Fig. 10.6.1.

10.7 The Reflection Method

Thus far, we have mostly studied equations of mathematical physics in the whole space, i.e., in domains without boundaries. In practice, it is often necessary to take into account the influence of boundaries that may have different physical properties, e.g., they may be reflecting, absorbing, etc. For a few special cases, the problem in domains with boundaries can be reduced to the previously considered problems in the whole space by an application of a seemingly superficial but quite effective *reflection method*, which already made its appearance in the elliptic problems of Chap. 9.

10.7.1 Diffusion in the Presence of an Absorbing Barrier

Let us illustrate the reflection method by way of the example of the 1-D diffusion equation

$$\frac{\partial f}{\partial t} = \frac{D}{2}\frac{\partial^2 f}{\partial x^2}. \tag{1}$$

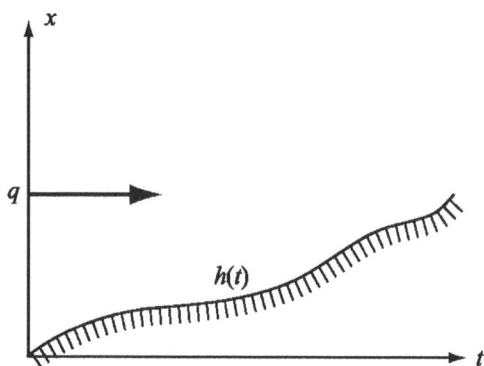

FIGURE 10.7.1
The domain of diffusion with a moving absorbing barrier.

To be more specific, we will interpret the function $f(x,t)$ as the 1-D density of a cloud of diffusing (Brownian) particles that move randomly along the x-axis.[2] If the initial position of the particles is known to be q, then the solution should additionally satisfy the initial condition

$$f(x, t = 0) = \delta(x - q). \qquad (2)$$

Now assume that an *absorbing barrier* is located at the point $x = h$, annihilating particles that reach it. Mathematically, this fact can be expressed as the *Dirichlet boundary condition*

$$f(x = h, t) = 0. \qquad (3)$$

The question of finding a solution $f(x,t)$ to (1) in the presence of an absorbing barrier constitutes a *mixed boundary problem* consisting of (1), initial condition (2), and boundary condition (3). If the position $h = h(t)$ of the barrier changes in time, then the boundary condition (3) remains in force with the constant h replaced by the function $h(t)$.

We shall find a solution of the problem for $x > h(t)$ (and $q > h(0)$). The domain of the diffusion restricted by an absorbing barrier is shown in Fig. 10.7.1. For a general function $h(t)$ describing the motion of an absorbing barrier, no analytic solution is known. However, if that motion is uniform,

$$h = \alpha t, \qquad \alpha = \text{const}, \qquad (4)$$

---

[2]For more information about Brownian motion, see Volume 3.

10.7. Reflection Method

a situation shown in Fig. 10.7.2, then one can construct a solution to the mixed problem (1)–(3) using the already known Green's function

$$G(x,t) = \frac{1}{\sqrt{2\pi Dt}} \exp\left(-\frac{x^2}{2Dt}\right) \qquad (5)$$

for the diffusion equation on the whole x-axis.

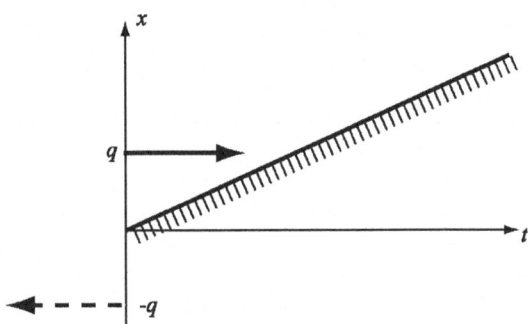

FIGURE 10.7.2
A uniformly moving absorbing barrier.

Indeed, consider a linear combination of Green's functions

$$f(x,t) = G(x-q,t) - AG(x+q,t) =$$
$$\frac{1}{\sqrt{2\pi Dt}} \exp\left(-\frac{x^2+q^2}{2Dt}\right) \left[\exp\left(\frac{xq}{Dt}\right) - A\exp\left(-\frac{xq}{Dt}\right)\right]. \qquad (6)$$

For $x = \alpha t$, the expression in the square brackets is independent of time, and there exists a constant

$$A = \exp(2q\alpha/D) \qquad (7)$$

for which the expression (6) becomes zero at the barrier $x = \alpha t$. Consequently, (6) gives a solution of the initial value problem

$$\frac{\partial f}{\partial t} = \frac{D}{2}\frac{\partial^2 f}{\partial x^2}, \qquad f(x,t=0) = \delta(x-q) - \exp\left(\frac{2\alpha q}{D}\right)\delta(x+q), \qquad (8)$$

and for $x \geq \alpha t$, it simultaneously solves the boundary problem (1)–(3) with a uniformly moving barrier. The last Dirac delta in (8), which is positioned below the barrier $x = \alpha t$, serves as a sort of "mirror image" of the "real" Dirac delta in the original initial condition (2); hence the name "reflection method" for the above procedure.

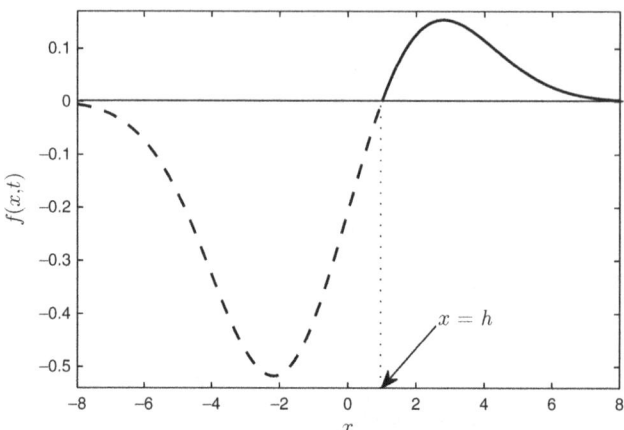

FIGURE 10.7.3
A typical solution of the diffusion equation with an absorbing barrier. The parameter values in this figure are $D = 2$, $\alpha = 0.5$, $q = 2$, and $t = 2$ ($h = 1$).

Figure 10.7.3 shows a typical graph of the solution (6) of the diffusion equation for $t > 0$. The solid line shows the part of the solution that is physically meaningful; the dashed line shows its Alice-in-Wonderland-like "behind-the-mirror" part.

10.7.2 The Reflection Method and the Invariance Property

What is the rationale behind the success of the reflection method? To answer this question, it will be useful to pass to the coordinate system $y = x - h(t)$, moving together with the absorbing barrier. The adjusted function $g(y, t) = f(y + h(t), t)$ satisfies the equation

$$\frac{\partial g}{\partial t} = \alpha(t) \frac{\partial g}{\partial y} + \frac{D}{2} \frac{\partial^2 g}{\partial y^2}, \qquad y > 0, t > 0, \tag{9}$$

where $\alpha(t) = h'(t)$, with the initial condition

$$g(y, 0) = \delta(y - q) \tag{10}$$

10.7. Reflection Method

and the boundary condition
$$g(0, t) = 0 \tag{11}$$
on the fixed boundary $y = 0$.

Now consider the case $\alpha(t) = \alpha = const$. According to (6), the solution of the auxiliary initial value problem
$$\frac{\partial g}{\partial t} = \alpha \frac{\partial g}{\partial y} + \frac{D}{2}\frac{\partial^2 g}{\partial y^2}, \qquad y \in \mathbf{R}, t > 0, \tag{12}$$
$$g(y, t = 0) = \delta(y - q) - \exp\left(\frac{2q\alpha}{D}\right)\delta(y + q),$$

with the initial condition obtained by the reflecting condition (10) into the "behind-the-mirror" region of the negative y's, also solves the boundary problem (9)–(10). The solution can be written in the form
$$g(y,t) = \frac{1}{\sqrt{2\pi Dt}}\left[\exp\left(-\frac{(y-q+\alpha t)^2}{2Dt}\right) - \exp\left(\frac{2q\alpha}{D} - \frac{(y+q+\alpha t)^2}{2Dt}\right)\right]. \tag{13}$$

Note its special feature that for every $t > 0$, it enjoys the spatial *symmetry property*
$$g(y, t) = -g(-y, t) \exp\left(-\frac{2\alpha y}{D}\right), \tag{14}$$
which is a result of the *invariance property* of (12). Here, invariance means that the equation for the function
$$w(y, t) = g(-y, t) \exp(-2\alpha y/D) \tag{15}$$
is identical to the equation for g. Hence, if the initial condition $g_0(y)$ of (12) enjoys the symmetry property
$$g_0(-y) = -g_0(y) \exp(2\alpha y/D),$$
then the solution $g(y, t)$ will enjoy the same property (14) at every time $t > 0$. It is clear from (14) that if the function g is continuous in y, then $g(0, t) \equiv 0$, and the boundary condition (11) is automatically satisfied.

The reflection method applied to the boundary problem
$$\frac{\partial g}{\partial t} = \alpha \frac{\partial g}{\partial y} + \frac{D}{2}\frac{\partial^2 g}{\partial y^2}, \qquad y > 0, t > 0, \tag{16}$$
$$g(y, t = 0) = g_0(y), \qquad y > 0, \tag{17}$$
$$g(0, t) = 0, \qquad t > 0, \tag{18}$$

with an arbitrary function $g_0(y)$, $y > 0$, can be summarized as follows: in view of the equation's invariance with respect to the transformation (15), the boundary problem automatically corresponds to an auxiliary initial value problem consisting of (12) and the initial condition

$$g_0(y) = \begin{cases} g_0(y), & \text{for } y > 0, \\ -g_0(-y) \exp(-2\alpha y/D), & \text{for } y < 0, \end{cases} \quad (19)$$

symmetrically augmented to the semiaxis $y < 0$.

Now we are ready to return to the original boundary value problem (9)–(11). Thus far, we have augmented the definition of the initial condition for the semiaxis $y < 0$. However, it will be beneficial to extend the equation itself to the same negative semiaxis, so that it satisfies the necessary invariance condition. Let us carry out this operation on (9) in the case of an arbitrary function

$$\alpha(t) = h'(t).$$

Initially, consider a piecewise linear function

$$h(t) = \begin{cases} \alpha t, & 0 < t < \tau, \\ \alpha\tau + \alpha_1(t - \tau), & t > \tau, \end{cases} \quad (20)$$

with

$$\alpha(t) = \begin{cases} \alpha, & 0 < t < \tau, \\ \alpha_1, & t > \tau. \end{cases} \quad (20')$$

In this case, for $0 < t < \tau$, the solution of the boundary value problem (9)–(11) is identical the the solution of the auxiliary initial value problem (12) enjoying the symmetry property (14). At time $t = \tau$, the coefficient of (12) jumps from the value α to α_1, and the equation for $g(y,t)$ assumes the form

$$\frac{\partial g}{\partial t} = \alpha_1 \frac{\partial g}{\partial y} + \frac{D}{2} \frac{\partial^2 g}{\partial y^2}, \quad y > 0, \ t > \tau.$$

In order to apply the reflection method for $t > \tau$, one has to augment this equation by the initial condition

$$g(y, t = \tau) = g(y, \tau) \exp[(\alpha_1 - \alpha)(|y| - y)/D],$$

which, for $y > 0$, is equal to the solution $g(y, \tau)$ of the initial value problem (12), and for $y < 0$, is constructed in such a way that the new symmetry

10.8. Moving Boundary: The Detonating Fuse Problem

condition (14) is satisfied with α replaced by α_1. The indicated correction is carried out automatically if the equation for the auxiliary initial condition is written in the form

$$\frac{\partial g}{\partial t} = \alpha(t)\frac{\partial g}{\partial y} + \frac{D}{2}\frac{\partial^2 g}{\partial y^2} + \frac{\alpha_1 - \alpha}{D}\delta(t-\tau)(|y|-y)g,$$

where $\alpha(t)$ is given by the formula 20'. For $y > 0$, the above equation coincides with the equation of the boundary value problem, and for $y < 0$, it contains a term that at time $t = \tau$, repairs the symmetry of the solution when the time t crosses the discontinuity of the function $\alpha(t)$.

Similarly, for an arbitrary continuous function $h(t)$, the desired correction of symmetry of the solution at any time is accomplished by the equation

$$\frac{\partial g}{\partial t} = \alpha(t)\frac{\partial g}{\partial y} + \frac{D}{2}\frac{\partial^2 g}{\partial y^2} + \frac{\beta(t)}{D}(|y|-y)g, \tag{21}$$

where

$$\beta(t) = \alpha'(t) = h''(t).$$

It is easy to show that (21) is invariant under the transformation

$$w(y,t) = -g(-y,t)\exp(-2a(t)y/D),$$

similar to (15). So, the solution of (21), with the initial condition (19), where α is replaced by $\alpha(0)$, satisfies the symmetry condition

$$g(y,t) = -g(-y,t)\exp\left(-\frac{2\alpha(t)y}{D}\right), \tag{22}$$

for all $t > 0$. One can also show that $g(y, t > 0)$ is continuous in y and that as a result, it satisfies the boundary condition (11) of the original boundary value problem. Thus, by correcting the equation, we have managed to reduce the boundary value problem to an auxiliary initial value problem. However, in the process, we lost the main advantage of the standard reflection method, since the equation of the resulting initial value problem became much more complicated than the original equation (9).

10.8 Moving Boundary: The Detonating Fuse Problem

The reflection method also makes it possible to find solutions of more realistic problems than those discussed above. The detonating fuse problem is one such example.

Suppose that a fuse of length l is stretched along the x-axis over the interval $[0, l]$. Initially, its temperature is equal to zero: $u(x, t = 0) = 0$. At time $t = 0$, the left endpoint of the fuse is ignited, and thereafter, the temperature of the moving left endpoint is equal to the "burning" temperature v. The temperature at the right endpoint is kept at zero at all times. The side surface of the fuse is assumed to be insulated.

Under these circumstances, the temperature distribution $u(x, t)$ in the fuse is a solution of the following mixed moving boundary problem:

$$\frac{\partial u}{\partial t} = \frac{D}{2}\frac{\partial^2 u}{\partial^2 x}, \qquad t > 0, \quad h(t) < x < l, \tag{1}$$

$$u(x, 0) = 0, \qquad 0 < x < l,$$

$$u(h(t), t) = v(t), \qquad u(l, t) = 0, \qquad t > 0.$$

Here $h(t)$ represents the time-dependent coordinate of the fuse's left endpoint, which moves to the right as the fuse burns. In our concrete calculations we shall put $h = \alpha t$, where α is the constant speed of burning. However, for now, let us keep $h(t)$ to be a general smooth function equal to zero at $t = 0$. Also, for the sake of generality, we shall allow the burning temperature $v(t)$ of the fuse's left endpoint to vary in time.

Just as was the case in other linear problems considered thus far, the solution of (1) is given by an integral of the corresponding Green's function $G(x, q, t)$ solving the boundary value problem with homogeneous boundary conditions

$$\frac{\partial G}{\partial t} = \frac{D}{2}\frac{\partial^2 G}{\partial^2 x}, \qquad t > 0, \quad h(t) < x < l, \tag{2}$$

$$G(x, q, 0) = \delta(x - q), \qquad 0 < q < l,$$

$$G(h(t), q, t) = G(l, q, t) = 0, \qquad t > 0.$$

We will derive a formula that expresses $u(x, t)$ through $G(x, q, t)$ by first considering the expression

$$\frac{\partial}{\partial \tau}\Big(u(q, \tau)G(q, x, t - \tau)\Big).$$

Calculating the derivative with respect to τ, and taking into account equations satisfied by functions $u(q, \tau)$ and $G(q, x, t - \tau)$, we get that

$$\frac{\partial}{\partial \tau}(uG) = \frac{D}{2}\left[G(q, x, t - \tau)\frac{\partial^2}{\partial q^2}u(q, \tau) - u(q, \tau)\frac{\partial^2}{\partial q^2}G(q, x, t - \tau)\right].$$

10.8. Moving Boundary: The Detonating Fuse Problem

Now integrate both sides of the above equality over the entire length $(h(\tau), l)$ of the fuse. The integration by parts formula applied to the right-hand side gives

$$\int_{h(\tau)}^{l} \frac{\partial}{\partial \tau}(uG)\,dq = \frac{D}{2}\left[G\frac{\partial u}{\partial q} - u\frac{\partial G}{\partial q}\right]\bigg|_{q=h(\tau)}^{l}.$$

Taking into account the boundary conditions in problems (1) and (2), the equality reduces to the form

$$\int_{h(\tau)}^{l} \frac{\partial}{\partial \tau}(uG)\,dq = \frac{D}{2}v(\tau)\frac{\partial}{\partial q}G(q,x,t-\tau)\bigg|_{q=h(\tau)}.$$

The left-hand side can be transformed to yield

$$\int_{h(\tau)}^{l} \frac{\partial}{\partial \tau}(uG)\,dq = \frac{\partial}{\partial \tau}\int_{h(\tau)}^{l} uG\,dq + \frac{dh}{d\tau}u(h(\tau),\tau)G(h(\tau),x,t-\tau),$$

where the last term on the right-hand side vanishes in view of the boundary condition in (2). Hence, we get the equality

$$\frac{\partial}{\partial \tau}\int_{h(\tau)}^{l} uG\,dq = \frac{D}{2}v(\tau)\frac{\partial}{\partial q}G(q,x,t-\tau)\bigg|_{q=h(\tau)}.$$

Integrating with respect to τ over the interval $[0, t]$, utilizing the initial condition in problem (1) and the probing property of the Dirac delta appearing in the initial condition in (2), we finally obtain

$$u(x,t) = \frac{D}{2}\int_0^t v(\tau)\frac{\partial}{\partial q}G(q,x,t-\tau)\bigg|_{q=h(\tau)}\,d\tau. \tag{3}$$

Thus we have demonstrated that the integral of the Green's function gives the sought solution of the boundary value problem under consideration. Our next job is to find the Green's function itself using the reflection method. We shall do so only in the case of constant burning speed $h(t) = \alpha t$. As a matter of fact, the constant α can be either positive or negative. In the latter case, our moving boundary problem can be interpreted, e.g., as a description of the temperature field of a crystallizing rod, its the length increasing as the crystallization process progresses.

It turns out that here it is easier to first guess a solution of a more general problem and then to obtain the desired solution as a special case. Consider a function $f(x,t)$ that solves the boundary value problem

$$\frac{\partial f}{\partial t} = \frac{D}{2}\frac{\partial^2 f}{\partial x^2}, \qquad t > 0, \quad \alpha t < x < l, \tag{4}$$

$$f(x, t = 0) = f_0(x), \qquad 0 < x < l,$$
$$f(\alpha t, t) = f(l, t) = 0.$$

An expression for the Green's function will be obtained by taking $f_0(x) = \delta(x - q)$.

Following the reflection method, the original boundary value problem (4) will now be replaced by an auxiliary initial value problem

$$\frac{\partial f}{\partial t} = \frac{D}{2} \frac{\partial^2 f}{\partial x^2}, \qquad t > 0, \quad x \in \mathbf{R}, \tag{5}$$

$$f(x, t = 0) = f(x),$$

where inside the interval $(0, l)$, the function $f(x)$ coincides with the initial condition

$$f(x) = f_0(x), \qquad 0 < x < l, \tag{6}$$

of the original boundary value problem (4), and outside is extended so that $f(x, t)$ automatically satisfies the boundary conditions (4) at $x = \alpha t$ and $x = l$. It follows from the discussion of the previous section that it is sufficient for $f(x)$ to satisfy the symmetry condition (10.6.14) relative to the point $x = 0$, i.e.,

$$f(x) = -f(-x) \exp(-2\alpha x/D), \tag{7}$$

and also the "antisymmetry" condition

$$f(x) = -f(2l - x) \tag{8}$$

relative to the point $x = l$. Combining (7) and (8), we arrive at the following useful equality:

$$f(x) = f(x + 2l) \exp(-2\alpha x/D), \tag{9}$$

which has to be satisfied by the function $f(x)$. It allows us to extend the definition of $f(x)$ onto the whole x-axis once its values are known inside any interval of length $2l$. The function $f(x)$ can be defined on an interval of length $2l$ by either of the formulas (7) and (8). Before doing so, let us extend the definition of the original initial condition function $f_0(x)$ from (4) to the whole x-axis by making it equal to zero outside the interval $(0, l)$. Then it follows from (8) that the values of $f(x)$ on the interval $(0, 2l)$ are given by the formula

$$f^0(x) = f_0(x) - f_0(2l - x). \tag{10}$$

For $x \in (0, 2l)$, the new function $f^0(x)$ coincides with $f(x)$, and it is equal to zero outside that interval. Denote by $f^m(x)$ a similarly defined function on

10.8. Moving Boundary: The Detonating Fuse Problem

the mth interval $(-2lm, 2l - 2lm)$. It is obtained by applying the operation (9) m times to $f^0(x)$, i.e., by shifting m times the function $f^0(x)$ by $-2l$. In other words,

$$f^m(x) = f^0(x + 2ml) \prod_{r=0}^{m-1} \exp\left(-\frac{2\alpha}{D}(x + 2rl)\right). \tag{11}$$

Taking into account the arithmetic progression formula

$$\sum_{r=1}^{m-1} r = \frac{m(m-1)}{2},$$

we get that

$$f^m(x) = f^0(x + 2ml) \exp\left(-\frac{2\alpha}{D} m(x + ml - l)\right). \tag{12}$$

Hence, the final form of the initial condition for the auxiliary initial value problem (5) is

$$f(x) = \sum_{m=-\infty}^{\infty} f^m(x) = \sum_{m=-\infty}^{\infty} f_0(x + 2ml) \exp\left(-\frac{2\alpha}{D} m(x + ml - l)\right) \tag{13}$$
$$- \sum_{n=-\infty}^{\infty} f_0(2l - x + 2nl) \exp\left(\frac{2\alpha}{D} n(x - nl - l)\right).$$

In deriving this formula, we used equalities (10), (11), and (12), and replaced the last sum's index of summation m by $-n$.

The solution of the auxiliary initial value problem with the above initial condition is of the form

$$f(x, t) = \int f(r) G_0(x - r, t) dr, \tag{14}$$

where

$$G_0(x - r, t) = \frac{1}{\sqrt{2\pi Dt}} \exp\left(-\frac{(x - r)^2}{2Dt}\right). \tag{15}$$

Substituting in (14) the series (13) for $f(r)$, and changing the variables of integration to $q = r + 2ml$ and $q = 2l - r + 2nl$, we obtain

$$f(x, t) = \int_0^l f_0(q) \left[G_1(x, q, t) - G_2(x, q, t)\right] dq, \tag{16}$$

where

$$G_1(x,q,t) = \sum_{m=-\infty}^{\infty} G_0(x-q+2ml,t)\exp\left(-\frac{2\alpha}{D}m(q-ml-l)\right) \quad (17)$$

and

$$G_2(x,q,t) = \sum_{n=-\infty}^{\infty} G_0(x+q-2l-2nl,t)\exp\left(-\frac{2\alpha}{D}n(q-nl-l)\right).$$

Therefore, in view of (16), the sought Green's function is

$$G(x,q,t) = G_1(x,q,t) - G_2(x,q,t). \quad (18)$$

The above tortuous mathematical manipulations might have diverted our attention from the fundamental problem. So let us return to basics and check that the Green's function (18) obtained above indeed solves the boundary value problem (2).

For $t \to 0$, the terms of the series (17) weakly converge to Dirac deltas. For $0 < q < l$, the support of only one of them, namely the one corresponding to $m = 0$, is located inside the interval $0 < x < l$. This means that the function (18) satisfies the initial condition of the boundary value problem (2). Substituting $x = l$ in the series (17) and taking into account the evenness of the fundamental solution $G_0(x,t)$, we discover that the terms of the series (17) with indices $m = n$ coincide. Thus the boundary condition $G(l,q,t) = 0$ is satisfied.

It remains to prove that for every $t > 0$, the boundary condition $G(\alpha t, q, t) = 0$ is satisfied as well. To accomplish this goal, it is convenient to insert in (17) and (18) the explicit expression (15) for the function $G(x,t)$. Also, we write the Green's function in a form more convenient for analysis by passing to the dimensionless spatial and time coordinates

$$s = x/l, \quad p = q/l, \quad \tau = Dt/l^2. \quad (19)$$

In the process, we encounter the dimensionless parameter

$$\gamma = \alpha l/D, \quad (20)$$

which characterizes the competition between two processes: the fuse burning and the heat diffusion. We shall also pass to the dimensionless Green's function

$$g(s,p,\tau) = lG(sl,pl,\tau l^2/D), \quad (21)$$

which, in view of (17) and (18), can be expressed by the formula

$$g(s,p,\tau) = g_1(s,p,\tau) - g_2(s,p,\tau), \tag{22}$$

where

$$g_1(s,p,\tau) = \frac{1}{\sqrt{2\pi\tau}} \sum_{m=-\infty}^{\infty} \exp\left[-\frac{(s-p+2m)^2}{2\tau} - 2\gamma m(p-m-1)\right] \tag{23}$$

and

$$g_2(s,p,\tau) = \frac{1}{\sqrt{2\pi\tau}} \sum_{n=-\infty}^{\infty} \exp\left[-\frac{(s+p-2-2n)^2}{2\tau} - 2\gamma n(p-n-1)\right].$$

In the new dimensionless coordinates, the left endpoint of the fuse moves according to the law

$$s = \gamma\tau. \tag{24}$$

Substituting (24) in (23), it is not difficult to check that the terms of the series with indices m and $n = m - 1$ coincide. So the function g from (22) satisfies the required boundary condition $g(\gamma\tau, p, \tau) = 0$.

As a final step, let us investigate the convergence of the series (23). For this purpose we shall group together, in the first series' exponential, the terms containing m^2 to get

$$-\frac{2}{\tau}m^2 + 2\gamma m^2 = 2\left(\gamma - \frac{1}{\tau}\right)m^2.$$

Clearly, the series converges if and only if the coefficient in front of m^2 is negative. This indeed is the case as long as $\tau < 1/\gamma$, or equivalently, as long as the fuse has not entirely burned out. As τ gets closer to $1/\gamma$, in order to guarantee the required accuracy in computation, one has to take into account more and more terms of the series (23). For $\tau = 1/\gamma$, both series diverge, and the moment of the complete burnout of the fuse corresponds, so to speak, to a mathematical catastrophe: divergence of the series (23).

10.9 Particle Motion in a Potential Well

Physicists often complain, with some justification, that mathematicians tend to ignore the physical essence of the problem under consideration. For a pure mathematician, the physical phenomenon described by a mathematical model is usually irrelevant. Basically, he studies the formal properties of the model and does not worry whether the model adequately reflects physical reality. Such a disregard of the physical side of the problem sometimes exacts a heavy price. Generalizing his model with abandon, a mathematician usually does not realize that he may have gone beyond a psychological threshold that is very difficult for the physicist to cross. For the latter, even a slight modification of the model often means a departure beyond the realm of physically realizable phenomena.

On the other hand, the history of science reminds us that similar violations of physical taboos led to the positron's discovery and the special theory of relativity, and opened other avenues of fruitful application of already developed mathematical apparatus to analysis of new physical phenomena. We have already run into similar situations ourselves, and in this section, we will encounter another one. The Green's function of the heat equation, which was found by the method of reflection, will be used to describe a qualitatively different phenomenon: particle motion in a potential well bounded by ideal reflecting barriers.

Recall that if the real diffusion coefficient in (10.1.1) is replaced by the imaginary quantity $D = i\hbar/2m$, then the diffusion equation is transformed into the Schrödinger-like equation

$$i\hbar \frac{\partial G}{\partial t} = -\frac{\hbar^2}{2m}\frac{\partial^2 G}{\partial x^2}, \qquad \alpha t < x < l,$$

$$G(x, q, t = 0) = \delta(x - q), \qquad 0 < x < l, \qquad (1)$$

$$G(\alpha t, q, t) = 0, G(l, q, t) = 0, \qquad t > 0.$$

The above boundary value problem describes the wave function of a quantum-mechanical particle located, at time $t = 0$, at the point $x = q$ and positioned between the ideal reflecting barriers at $x = \alpha t$ and $x = l$.

Having already found the Green's function of the heat equation, we obtain—without any extra effort—the solution of problem (1):

$$g(s, p, \tau) = g_1(s, p, \tau) - g_2(s, p, \tau),$$

$$g_1(s, p, \tau) = \sqrt{\frac{i}{2\pi\tau}} \sum_{k=-\infty}^{\infty} \exp\left[i\frac{(s - p + 2k)^2}{2\tau} + 2i\gamma k(p - k - 1)\right], \qquad (2)$$

$$g_2(s, p, \tau) = \sqrt{\frac{i}{2\pi\tau}} \sum_{n=-\infty}^{\infty} \exp\left[i\frac{(s + p - 2 - 2n)^2}{2\tau} + 2i\gamma n(p - n - 1)\right].$$

Here, in analogy with (10.7.19)–(10.7.23), we introduce a dimensionless function $g = lG$, with parameters and coordinates

$$s = x/l, \quad p = q/l, \quad \tau = \hbar t/ml^2, \quad \gamma = aml/\hbar. \qquad (3)$$

The series (2) themselves do not converge absolutely. Physically, this is related to the fact that in view of the uncertainty principle, the total localization of the particle implies the total indeterminacy of the particle's momentum. However, if the initial wave function $\psi_0(s)$ of the particle is sufficiently smooth, then the integrated series

$$\Psi(s,t) = \int_0^1 \Psi_o(p) g(s,p,\tau) ds \qquad (4)$$

becomes absolutely convergent, and its terms describe multiple reflections of particles from the barriers.

10.10 Exercises

1. Solve the initial value problem for the parabolic equation

$$\frac{\partial f}{\partial t} + \alpha(t)\frac{\partial f}{\partial x} = \frac{\beta(t)}{2}\frac{\partial^2 f}{\partial x^2}, \quad t \geq 0, \, x \in \mathbf{R}, \qquad (1)$$

$$f(x, t = 0) = \delta(x),$$

where $\alpha(t)$ and $\beta(t)$ are arbitrary integrable functions on the semiaxis $t \geq 0$. *Hint:* Rewrite the initial value problem in the form of an inhomogeneous equation and with the help of the Fourier transform with respect to the coordinate x, find its solution satisfying the causality condition.

2. Suppose that the function $f(x,t)$ satisfies the following initial value problem:

$$\frac{\partial f}{\partial t} = \frac{\partial}{\partial x}[a(x)f] + \frac{1}{2}\frac{\partial}{\partial x}\left[\beta(x)\frac{\partial f}{\partial x}\right], \quad t > 0, \, x \in \mathbf{R},$$

$$f(x, t = 0) = \delta(x), \qquad (2)$$

where $a(x)$ and $\beta(x)$ are everywhere differentiable functions. Additionally, to ensure that the initial value problem under consideration is correctly posed, assume that $\beta(x) > 0$. Check that

$$\mathcal{N}(t) = \int f(x,t)\, dx \equiv 1, \quad t > 0. \qquad (3)$$

In other words, the integral of the solution to (10) is a conserved quantity.

3. Using the Fourier transform with respect to x, solve the parabolic equation
$$\frac{\partial f}{\partial t} = \alpha(t)\frac{\partial}{\partial x}(xf) + \frac{\beta(t)}{2}\frac{\partial^2 f}{\partial x^2}, \qquad t \geq 0, x \in \mathbf{R}, \tag{4}$$
with the initial condition
$$f(x, t=0) = \delta(x-y). \tag{5}$$

4. Find an explicit expression for the solution of the initial value problem (4)–(5) with constant coefficients $\alpha(t) = \alpha > 0$ and $\beta(t) = \beta > 0$. Explore the asymptotic behavior of these solutions as $t \to \infty$.

5. Find the stationary solution
$$f_{st}(x) = \sqrt{\frac{\alpha}{\pi\beta}} \exp\left(-\frac{\alpha}{\beta}x^2\right) \tag{6}$$
by directly solving the original parabolic equation (4) in the case of constant α and β.

6. Find the normalized stationary solution of the parabolic equation
$$\frac{\partial f}{\partial t} = \frac{\partial}{\partial x}[\alpha x f] + \frac{1}{2}\frac{\partial}{\partial x}(\beta + \gamma x^2)\frac{\partial f}{\partial x}, \tag{7}$$
with $\alpha \geq 0$, $\beta > 0$, and $\gamma \geq 0$. State conditions for the existence of a stationary solution.

7. In probability theory, the nth moment of a time-dependent probability density function $f(x,t)$ is defined by the formula
$$m_n(t) = \int x^n f(x,t)\, dx. \tag{8}$$
Find the first two moments of the probability density function $f(x,t)$ that obeys (7) and the initial condition (5). Discuss their behavior in time.

8. Find a stationary solution of the parabolic equation
$$\frac{\partial f}{\partial t} = \alpha\frac{\partial f}{\partial x} + \frac{\beta}{2}\frac{\partial^2 f}{\partial x^2}, \qquad t > 0, 0 < x < \ell, \tag{9}$$

10.10. Exercises

with the initial condition (5) ($0 < y < \ell$) and the periodic boundary condition
$$f(x = 0, t) = f(x = \ell, t), \qquad t > 0. \tag{10}$$

9. The Kolmogorov–Feller equation

$$\frac{\partial f}{\partial t} + \nu f = \nu \int w(y) f(x-y, t)\, dy, \qquad t > 0, x \in \mathbf{R}, \tag{11}$$

$$f(x, t = 0) = \delta(x),$$

plays an important role in probability theory. Using the Fourier transform, solve the Kolmogorov–Feller equation and explore the asymptotic properties (as $t \to \infty$) of its solutions in the case that the kernel $w(x)$ is Gaussian, i.e.,

$$w(x) = \frac{1}{\sqrt{2\pi}} \exp\left(-\frac{x^2}{2}\right). \tag{12}$$

10. Determine the asymptotic behavior (for $\nu t \gg 1$) of the solution of the Kolmogorov–Feller equation (12) with Cauchy kernel

$$w(x) = \frac{1}{\pi} \frac{1}{1 + x^2}. \tag{67}$$

11. Verify numerically the correctness of the asymptotics of the exact continuous part of the solution of the Kolmogorov–Feller equation with Cauchy kernel. Utilize the solution to Problem 10 provided at the end of this volume.

12. Using results of Chap. 4, Volume 1, examine the main asymptotics (for an arbitrary $t > 0$ and $x \to \infty$) of the continuous part of the solution of the Kolmogorov–Feller equation with Cauchy kernel.

Chapter 11
Waves and Hyperbolic Equations

Waves are everywhere, literally, and our senses are acutely attuned to waves of various types: mechanical, water, acoustic, electromagnetic, optical, etc. Moreover, quantum mechanics tells us that matter itself is, in a sense, indistinguishable from waves. In this chapter we concentrate on properties of *linear waves* in dispersive media; a discussion of nonlinear waves will be postponed until Chaps. 12–14. Here the methods of choice are integral transforms and asymptotic relations, especially the Fourier transform and the stationary phase method. They will occupy a central role in what follows. Their main advantage is that they reduce relevant partial differential, or even integral, equations to algebraic, or in some cases transcendental, equations. For example, the 1-D wave equation

$$\frac{\partial^2 u}{\partial t^2} = c^2 \frac{\partial^2 u}{\partial x^2},$$

already familiar to the readers of this book, is reduced via the Fourier transform to the simple algebraic equation

$$\omega = \pm ck.$$

As a result, in this chapter, partial differential and integral equations describing wave propagation will be mostly hidden behind their corresponding, and much more convenient and elementary, *dispersion relations*.

Recall that the above wave equation has the general solution

$$u(x,t) = f(x - ct) + g(x + ct),$$

where $f(x)$ and $g(x)$ are arbitrary functions. This formula hints at two remarkable properties of such 1-D waves:

- Waves propagating in opposite directions do not interact with each other;

- Waves propagating in a given direction do not change their shape.

The first property is shared by all waves propagating in homogeneous linear dispersive media and allows us to concentrate our attention on waves propagating in one direction only. The second property fails in general dispersive media, because the velocities of propagation of different harmonic components of the wave are different. However, the stationary phase and related asymptotic methods will permit us to study distortion of the wave's shape at large times and at long distances.

The dispersive properties are a consequence of the medium's inner structure, which responds differently to different wave scales. But similar effects can also occur in 2-D and 3-D nondispersive media (such as a vacuum in the case of electromagnetic waves); such is the case of beams for which the velocities of the plane-wave component have different projections on the main direction of beam propagation. As a result, such multidimensional beams undergo distortions similar to those experienced by waves in 1-D dispersive media. The case of such *geometric dispersion* will also be discussed in the present chapter.

11.1 Dispersive Media

This section studies wave propagation in dispersive media. Initially, the study will be conducted at a physical level of rigorousness, and we shall restrict our attention to waves depending only on the x-coordinate and the time t.

11.1.1 Media Properties and Corresponding Dispersion Relations

Each action on a medium causes a reaction. In a linear homogeneous medium, the response $u(x,t)$ to forcing $g(x,t)$ is a linear convolution-type functional

$$u(x,t) = \iint h(s,\tau)g(x-s,t-\tau)ds d\tau. \qquad (1)$$

11.1. Dispersive Media

Its Fourier image

$$\tilde{u}(\omega, \kappa) = \left(\frac{1}{2\pi}\right)^2 \iint u(x,t) \exp(i\omega t - i\kappa x)\, dt\, dx,$$

taken either in the classical or in the distributional sense, satisfies the equation

$$\tilde{u}(\omega, \kappa) = (2\pi)^2 \tilde{h}(\omega, \kappa) \tilde{g}(\omega, \kappa), \qquad (2)$$

where $\tilde{h}$ and $\tilde{g}$ are the Fourier images of the functions appearing on the right-hand side of (1). Let us rewrite (2) in the form

$$L(\omega, \kappa)\tilde{u} = \tilde{g},$$

where $L(\omega, \kappa) = 1/(4\pi^2 \tilde{h}(\omega, \kappa))$.

The function $\tilde{h}$ (or equivalently, $L(\omega, \kappa)$) reflects properties of the medium. For real physical media, the function $h(x,t)$ is real-valued. Therefore, $\tilde{h}$ and L satisfy equalities

$$\tilde{h}(-\omega, -\kappa) = \tilde{h}^*(\omega, \kappa), \qquad L(-\omega, -\kappa) = L^*(\omega, \kappa), \qquad (3)$$

similar to (3.1.4) of Volume 1. Recall that the asterisk denotes the complex conjugate. If the medium is isotropic, then a spatially symmetric action will generate a spatially symmetric reaction. In the 1-D case, this means that

$$L(\omega, \kappa) = L(\omega, -\kappa). \qquad (4)$$

If the medium is *time-reversible*, then

$$L(\omega, \kappa) = L(-\omega, \kappa). \qquad (5)$$

It follows from (3)–(5) that in an isotropic, time-reversible medium, the function $L(\omega, \kappa)$ is real-valued and symmetric (even) in the sense that

$$L(\omega, \kappa) = L(-\omega, -\kappa). \qquad (6)$$

Now assume that the above medium was perturbed in the very distant past. The wave u thus generated will keep propagating forever only if its Fourier transform satisfies the homogeneous equation

$$L(\omega, \kappa)\tilde{u} = 0. \qquad (7)$$

Equation (7) has nontrivial solutions if the equation

$$L(\omega, \kappa) = 0,$$

which traditionally is called the *dispersion relation*, has real roots.

11.1.2 Propagation of Waves in Dispersive Media

For the sake of simplicity, let us assume that for every κ, the function L has two roots,
$$\omega = \pm W(\kappa).$$
Their symmetry (signs $\pm$) is a reflection of the medium's isotropy. Consider a wave moving to the right, which corresponds to the selection of the root
$$\omega = W(\kappa). \tag{8}$$
In this case, (7) has a singular solution
$$\tilde{u}(\omega, \kappa) = \tilde{f}(\kappa)\,\delta(\omega - W(\kappa)), \tag{9}$$
similar to the solution (1.5.5) of (1.5.6) of Volume 1, where $\tilde{f}(\kappa)$ is an arbitrary function of its argument. In concrete problems this function is determined by the initial or boundary conditions. In view of the probing property of the Dirac delta, substitution of (9) into the inverse Fourier integral
$$u(x, t) = \iint \tilde{u}(\omega, \kappa) \exp(-i\omega t + i\kappa x)\, d\omega\, d\kappa$$
gives us the expression
$$u(x, t) = \int \tilde{f}(\kappa) \exp\Big(i(\kappa x - W(\kappa)t)\Big) d\kappa \tag{10}$$
for a wave freely propagating in a medium. Formula (10) describes the so-called *wave packet*, which contains a continuum of harmonic waves whose contributions to the packet are measured by the complex *amplitude* $\tilde{f}(\kappa)$.

Recall that if $u(x, 0) = f(x)$ is a real function, then $\tilde{f}(-\kappa) \equiv \tilde{f}^*(\kappa)$. Accordingly, to make the field $u(x,t)$ in (10) real-valued for every $t \neq 0$, we shall assume below that $W(\kappa)$ is an odd functions of κ, i.e., $W(-\kappa) = -W(\kappa)$.

The sum
$$S = \kappa x - W(\kappa)t + \arg[\tilde{f}(\kappa)]$$
appearing in the exponent in (10) is called the *phase* of the harmonic wave. Note that the phase is constant on straight lines $x = ct + d$. The constant
$$c = \frac{W(\kappa)}{\kappa} \tag{11}$$
is called the *phase velocity* of the harmonic wave.

Observe that the function W has the dimension of ω but depends on an argument κ of a different dimension. So $W(\kappa)$ has to include dimensional physical parameters (spatial and temporal scales) reflecting the inner *dispersion properties* of the medium. In the simplest case of a scale-invariant or *nondispersive* medium, the phase velocities of different harmonic waves are the same, and
$$W(\kappa) = c\kappa. \tag{12}$$
Substituting (12) into (10), we discover that in such a medium,
$$u(x,t) = f(x - ct), \tag{13}$$
and the wave propagates preserving its shape. If W depends not only on c but also on the spatial scale l, then the medium is said to exhibit *spatial dispersion*. Similarly, one can define media with temporal or spatiotemporal dispersions. Even in the case of a single spatial scale, $W(\kappa)$ can have a structure much more complex than (12). A good example is provided by the function
$$W(\kappa) = c\kappa\,\varphi(l\kappa), \tag{14}$$
where the function $\varphi(z)$ is an arbitrary function of a dimensionless argument defined by concrete properties of the medium.

11.2 Examples of Dispersive Media

In this section we shall provide several examples of dispersive media. The first illustrates spatial dispersion due to the geometry of the domain wherein the waves propagate. Consider waves in a 2-D dispersionless medium depending on Cartesian coordinates x and y. The general 2-D harmonic wave is of the form
$$\tilde{f}(\boldsymbol{k},\omega)\exp[i(\boldsymbol{k}\cdot\boldsymbol{r}) - i\omega t], \tag{1}$$
where $\boldsymbol{r}$ is the position vector with Cartesian coordinates x and y, and $\boldsymbol{k}$ is the *wavevector*, whose components will be denoted by κ and μ. In such a medium, the wavevector $\boldsymbol{k}$ and the wave frequency ω are related via the equality
$$\omega^2 = c^2 \boldsymbol{k}^2 = c^2(\kappa^2 + \mu^2). \tag{2}$$

Example 1. Wave in a waveguide. Suppose that the wave $u(\boldsymbol{r},t)$ propagates in a *waveguide* bounded by walls at $y = 0$ and $y = l$. The wave is assumed to vanish on the boundary. In other words,
$$u(x, y = 0, t) = u(x, y = l, t) = 0. \tag{3}$$

Waves (1) cannot satisfy these boundary conditions, but their superpositions

$$u(x,y,t) = \tilde{f}(\kappa,\mu,\omega)\sin(\mu y)\exp(i\kappa x - i\omega t) \qquad (4)$$

$$= \frac{1}{2i}\tilde{f}(\kappa,\mu,\omega)\left[\exp\Big(i(\kappa x + \mu y) - i\omega t\Big) - \exp\Big(i(\kappa x - \mu y) - i\omega t\Big)\right]$$

can. Indeed, it suffices to guarantee that $\sin(\mu l) = 0$, i.e., to select the waves corresponding to the wavenumbers

$$\mu_n = \frac{\pi n}{l}, \qquad n = 1, 2, \ldots.$$

For a given n, such a wave is called the nth *waveguide mode*, and it is of the form

$$u_n(x,y,t) = \tilde{f}_n(\kappa)\sin(\mu_n y)\exp\Big[i\kappa x - iW_n(\kappa)t\Big],$$

where in view of (2), the wave frequency

$$\omega = W_n(\kappa) = c\sqrt{\kappa^2 + \mu_n^2} \qquad (5)$$

depends also on the spatial parameter l and the mode number n.

Thus, each mode has its own dispersion relation (5), and different monochromatic components of the same mode have different phase velocities. The *dispersion curves* (5) of the first four modes are shown in Fig. 11.2.1.

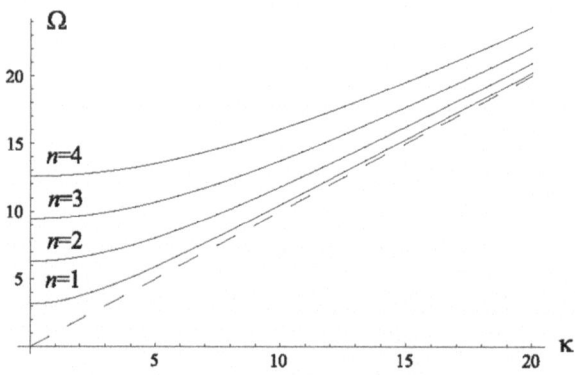

FIGURE 11.2.1
Dispersion curves for the first four modes of a waveguide. The vertical axis represents the dimensionless frequency $\Omega = \omega l/c$, and the horizontal axis, the wavenumber $\kappa = kl$. The **dashed line** represents the nondispersive asymptotics $\omega = ck$.

11.2. Examples of Dispersive Media

Example 2. Vibrations of a rod. Small transverse vibrations of a rod are another typical example of dispersive waves. They are described by an equation of the form
$$\frac{\partial^2 u}{\partial t^2} + \gamma^2 \frac{\partial^4 u}{\partial x^4} = 0.$$
The Fourier transform of the above equation yields the condition
$$(\omega^2 - \gamma^2 \kappa^4)\tilde{u} = 0.$$
Solving the resulting equation for ω, we arrive at the dispersion relation
$$\omega = \gamma \kappa^2. \tag{6}$$

Example 3. Gravitational surface water waves. It is known that the gravitational surface water waves in a basin of depth h satisfy the dispersion relation
$$\omega^2 = g\kappa \tanh(\kappa h),$$
where g is the gravitation constant. In the case of a deep basin (or small-scale waves, i.e., when the inequality $\kappa h \gg 1$ holds), one often uses the approximate dispersion law
$$\omega = \sqrt{g\kappa}. \tag{7}$$
For large-scale waves, whose Fourier image is concentrated in the domain $\kappa h \ll 1$, a different approximation for the dispersion relation is useful. It is obtained by expanding the exact relation $\omega = \sqrt{g\kappa \tanh(\kappa h)}$ in a Taylor series and keeping only the first two nonzero terms. Thus, we get
$$\omega = c\kappa - \nu\kappa^3, \tag{8}$$
where
$$c = \sqrt{gh}, \qquad \nu = ch^2/6. \tag{9}$$
These dispersion relations correspond to what is commonly called the *shallow water approximation*; it is equivalent to the condition that the Fourier image of the wave satisfies the equation
$$(\omega - c\kappa + \nu\kappa^3)\tilde{u}(\omega, \kappa) = 0.$$
Taking its inverse Fourier transform (see (3.2.1) of Volume 1) results in the so-called *linearized Korteweg–de Vries equation*
$$\frac{\partial u}{\partial t} + c\frac{\partial u}{\partial x} + \nu\frac{\partial^3 u}{\partial x^3} = 0, \tag{10}$$
which describes, for shallow-water waves, the small surface deviations $u(x, t)$ from the equilibrium level. See Fig. 11.2.2 (For analysis of the nonlinear Korteweg–de Vries equation, see Chap. 14.)

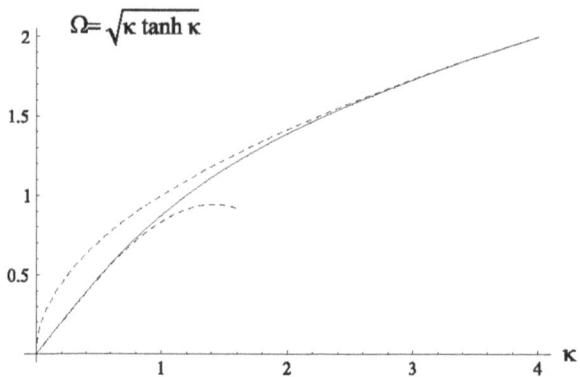

FIGURE 11.2.2
Plot of the dispersion relation for gravitational water waves. The axes represent, respectively, the dimensionless frequency $\Omega = \omega\sqrt{h/g}$ and the wavenumber $\kappa = kh$. The *dashed lines* represent the curves corresponding to the deep- (*top*) and shallow-water (*bottom*) approximations.

11.3 Integral Laws of Motion for Wave Packets

Wave packets (11.1.10) possess infinitely many invariants, i.e., quantities that are conserved in time. One of them is energy. A wave packet's energy is a quadratic functional of $u(x,t)$, but its structure depends on the physical context. In this section we shall restrict our analysis to functionals that are linear functionals of the simplest quadratic form,

$$I(x,t) = u^2(x,t) = |u(x,t)|^2. \tag{1}$$

We shall call $I(x,t)$ the *wave intensity*, leaving aside a discussion of the actual physical energetic properties of waves. Consequently, in our context, a wave's *energy* is defined by the simplest quadratic integral functional,

$$E = \int |u(x,t)|^2 \, dx. \tag{2}$$

Substituting the right-hand side of (11.1.10) into (2), we arrive at the triple integral

$$E = \int \left[\iint \tilde{f}(\kappa)\tilde{f}^*(\mu) \exp\left[i(\kappa-\mu)x - it\left(W(\kappa) - W(\mu)\right)\right] d\kappa \, d\mu \right] dx.$$

11.3. Integral Laws of Motion for Wave Packets

Changing the order of integration, noticing that

$$\int \exp[i(\kappa - \mu)x]dx = 2\pi\delta(\mu - \kappa), \tag{3}$$

and using the probing property of the Dirac delta, we arrive at *Parseval's equality*

$$E = 2\pi \int |\tilde{f}(\kappa)|^2 d\kappa = const, \tag{4}$$

which in our context expresses the energy conservation law.

If we consider a more general quadratic functional,

$$\langle g(x) \rangle = \frac{1}{E} \int g(x) I(x,t) dx,$$

which "averages" the function $g(x)$ over the whole wave packet with the weight proportional to the wave intensity, then arguments similar to those used above to obtain the energy invariance principle lead to the equality

$$\langle g(x) \rangle = \frac{2\pi}{E} \iint \tilde{g}(\mu - \kappa) \tilde{f}(\kappa) \tilde{f}^*(\mu) \exp\left(it[W(\mu) - W(\kappa)]\right) d\kappa d\mu. \tag{6}$$

Then, if one takes $g(x) = x$, one can think of $\langle x \rangle$ as the coordinate of the wave packet's "center of mass." In view of (3.3.5) of Volume 1, the singular Fourier image of this function is

$$\tilde{g}(\mu - \kappa) = i\frac{\partial}{\partial \mu}\delta(\mu - \kappa).$$

A substitution of this formula into (6) permits evaluation of the integral with respect to μ, which gives

$$\langle x \rangle = -\frac{2\pi i}{E} \int \tilde{f}(\kappa) \exp\left(-itW(\kappa)\right) \frac{d}{d\kappa}\left[\tilde{f}^*(\kappa) \exp\left(itW(\kappa)\right)\right] d\kappa. \tag{7}$$

Calculating the derivative inside the integral, we finally get

$$\langle x \rangle = a + \bar{v}t, \tag{8}$$

where

$$a = -\frac{2\pi i}{E} \int \tilde{f}(\kappa) \frac{d}{d\kappa} \tilde{f}^*(\kappa) d\kappa$$

is the coordinate of the wave packet's "center of mass" at $t = 0$, and

$$\bar{v} = \frac{2\pi}{E} \int |\tilde{f}(\kappa)|^2 W'(\kappa)\, d\kappa \tag{9}$$

is the velocity of propagation of the wave packet as a whole.

Observe that $\bar{v}$ is the average of the function $W'(\kappa)$, with respect to the wave's normalized spectral density $2\pi|\tilde{f}(\kappa)|^2/E$. The function $W'(\kappa)$ is called the *group velocity* of the wave and is naturally viewed as the velocity of propagation of the harmonic component of the wave corresponding to value κ. Note that in a dispersive medium, it does not coincide with the phase velocity (11.1.11) of the same harmonic component. The group velocity plays a fundamental role in the theory of waves in dispersive media, and in numerous realistic situations it is the only velocity of the wave's propagation that is easily measurable.

11.4 Asymptotics of Waves in Dispersive Media

Let us return to the integral

$$u(x,t) = \int \tilde{f}(\kappa) \exp\Big(i(\kappa x - W(\kappa)t)\Big) d\kappa,$$

first introduced in (11.1.10). For $t = 0$, $u(x, t = 0) = f(x)$, where the function $f(x)$ describes the initial shape of the wave. In a dispersive medium, in contrast to a nondispersive one, the wave's shape evolves over time, and the problem of quantitative description of this evolution arises. The practical importance of this problem became clear more than a hundred years ago, after the first transatlantic telephone cable had been laid. Dispersion effects accumulating over long distances distorted transmitted signals to the point of being unintelligible.

In this section we shall evaluate the integral (11.1.10) via the stationary phase method of Chap. 5, Volume 1, which was developed by Lord Kelvin specifically to evaluate these types of integrals. The calculation will be carried out under the assumption that $\tilde{f}(\kappa)$ and $W(\kappa)$ have all the necessary smoothness properties. In contrast to the canonical integral (5.2.1), the integral (11.1.10) contains two independent parameters x and t. Let us eliminate one of them, letting the experimenter follow the wave with velocity v. This corresponds to a substitution $x = vt$. At the point of observation, we have

$$U(t) = u(x = vt, t) = \int \tilde{f}(\kappa) \exp[-it\, p(\kappa)]\, d\kappa,$$

11.4. Asymptotics of Waves in Dispersive Media

where
$$p(\kappa) = W(\kappa) - v\kappa.$$

It is the asymptotic behavior ($t \to \infty$) of this integral that we shall study via the stationary phase method. As usual, we begin by looking for the stationary points, which are roots of the equation

$$p'(\kappa) = W'(\kappa) - v = 0. \tag{1}$$

Suppose that the equation has only one root, equal to k. Then the sum corresponding to (5.2.3) reduces to a single term, and

$$U(t) \sim \sqrt{\frac{2\pi}{itW''(k)}} \tilde{f}(k) \exp\left[it(vk - W(k))\right], \qquad (t \to \infty), \tag{2}$$

assuming that
$$p''(k) = W''(k) \neq 0.$$

In view of (2) and (1), the astonished experimenter "sees" a unique component of the wave's Fourier image $\tilde{f}(k)$ running with the *group velocity*

$$v(k) = W'(k). \tag{3}$$

Example. Transverse vibration of a rod. For the transverse vibration of a rod (Example 11.2.2), the dispersion relation is (11.2.6), and the group velocity,

$$v(k) = 2c(k) = 2\gamma k, \tag{4}$$

is twice as large as the phase velocity. For deep-water waves, the dispersion law (11.2.7) holds, and the group velocity

$$v(k) = \frac{1}{2}c(k) = \frac{1}{2}\sqrt{\frac{g}{k}} \tag{5}$$

is only half as large as the phase velocity.

Substituting $v = x/t$ in (2), we get the following asymptotic expression for the wave at large x and t:

$$U(x,t) \sim \sqrt{\frac{2\pi}{itW''(k)}} \tilde{f}(k) \exp\left[i(kx - W(k)t)\right]. \tag{6}$$

The dependence of k on x and t can be found by solving the equation

$$x - W'(k)t = 0 \tag{7}$$

for the unknown k.

11.5 Energy Conservation Law in the Stationary Phase Method

Approximate methods carry the potential danger of "throwing the baby out with the bathwater"—that is, losing some fundamental physical properties of a phenomenon under consideration as a byproduct of simplifying the analytic problem at hand. Thus a "reality check" is always in order: do the approximate solutions preserve the crucial known properties of the exact solutions? Various invariants are often used for this purpose. What is at stake here is not just a pragmatic test of the accuracy of the proposed approximate method, but also a deeper level of understanding as to whether the approximation preserves the essence of the physical phenomenon being investigated.

Let us apply the above philosophy to the method of the stationary phase. We shall test whether it preserves the energy conservation law by checking whether the approximate expression (11.4.6) agrees with the "energy" integral (11.3.3),

$$E = \int u^2(x,t)dx = 2\pi \int |\tilde{f}(\kappa)|^2 d\kappa = const. \tag{1}$$

First of all, note that in the derivation of (11.4.6), we did not take into account the fact that by (11.1.4), in addition to the stationary point $\kappa = k > 0$, there always exists a conjugate point $\kappa = -k$. Both correspond to the same group velocity. Hence, in the analysis of the energetic properties of the wave, one has to replace (11.4.6) by

$$u(x,t) = U(x,t) + U^*(x,t) = 2\operatorname{Re} U(x,t).$$

The corresponding wave intensity is then equal to

$$I(x,t) = u^2(x,t) = 2|U(x,t)|^2 + U^2(x,t) + U^{*2}(x,t).$$

We shall split it into two components,

$$I(x,t) = I_0(x,t) + I_1(x,t), \tag{2}$$

where

$$I_0(x,t) = 2|U(x,t)|^2, \quad \text{and} \quad I_1(x,t) = U^2(x,t) + U^{*2}(x,t).$$

By (11.4.6), the first component satisfies

$$I_0(x,t) = \frac{4\pi}{t|W''(k)|}|\tilde{f}(k)|^2 \quad (k > 0). \tag{3}$$

11.5. Energy Conservation Law in the Stationary Phase Method

The probing property (1.7.2), Volume 1, of the Dirac delta of a composite argument permits us to replace (3) by the equivalent expression

$$I_0(x,t) = 4\pi \int_0^\infty |\tilde{f}(\kappa)|^2 \delta(x - W'(\kappa)t) d\kappa. \qquad (4)$$

This indicates that at large times, the Fourier components of the wave behave as if they had separated spatially. Integrating (4) over the entire x-axis, we obtain the conservation law

$$\int I_0(x,t) dx = 4\pi \int_0^\infty |\tilde{f}(\kappa)|^2 d\kappa = const.$$

Since the function $f(x)$ is real-valued, the modulus squared of its Fourier transform $|\tilde{f}(\kappa)|^2$ is even. Hence, the last expression is equivalent to the energy invariant (11.3.3), and we have discovered that the entire energy of the wave is contained in the first component of the intensity, $I_0(x,t)$.

So, what happened to the energy contained in the second component, $I_1(x,t)$? The answer to this question will be provided by investigating the integral

$$\int I_1(x,t) dx = 4\pi \mathrm{Im} \left[\frac{1}{t} \int_0^\infty \tilde{f}^2(k) \exp\left[i(kx - W(k)t)\right] \frac{dx}{v'(k)} \right].$$

Introducing the new variable of integration k via the substitution $x = v(k)t$, we get

$$\int I_1(x,t) dx = 4\pi \mathrm{Im} \int_0^\infty \tilde{f}^2(k) \exp\left[2ik(v(k) - c(k))t\right] dk.$$

This is a typical integral of a rapidly oscillating function of the form (5.1.1), whose main asymptotics, for $t \to \infty$, are described by formulas (5.1.5)–(5.1.6). In the particular case of the transverse vibration of a rod, (11.2.6), when the difference of the group and phase velocities appearing in the exponent is $v(k) - c(k) = \gamma k$, the integral acquires the form

$$\int_0^\infty \tilde{f}^2(k) \exp(i2\gamma t k^2) \, dk.$$

The stationary phase method predicts the vibration's decay to zero with the asymptotics $1/\sqrt{t}$ as $t \to \infty$. It is also possible to obtain analogous results about the asymptotic decay of similar integrals for other dispersion laws.

Hence, the conclusion is that the approximate intensity (2) of the wave does not satisfy the energy conservation law (11.3.4). However, that law is satisfied by the main component $I_0(x,t)$ of the intensity, while the contribution of the term $I_1(x,t)$ decays to zero as $t \to \infty$. One could eliminate the contribution from $I_1(x,t)$ and preserve a more accurate energy conservation law by going beyond the framework of the stationary phase method and keeping further terms of the asymptotic expansion of the original integral (11.1.10).

11.6 Wave as Quasiparticle

In this section we shall construct an instructive analogy between the wave intensity (11.5.4) and the particle density in a gas of particles.

11.6.1 The Density of the Particle Flow

Consider particles traveling along the x-axis, and denote respectively by $X(t;y,r)$ and $V(t;y,r)$ the position and velocity at time t of a particle that at $t=0$ was located at y and moved with constant velocity r. The singular density of a single particle of unit mass in the *phase space* (x,v) is equal to (see Chap. 2, Volume 1)

$$f_1(x,v,t) = \delta(X(t;y,r) - x)\delta(V(t;y,r) - v). \tag{1}$$

Multiplying (1) by the initial density $f_0(y,r)$ of particles in the phase space and integrating the result over all y's and r's gives the current density of the gas of particles:

$$f(x,v,t) = \iint f_0(y,r)\delta(X(t;y,r) - x)\delta(V(t;y,r) - v)\,dy\,dr. \tag{2}$$

Example 1. Uniformly moving particles. Consider the case of particles moving uniformly with constant velocity, i.e., $V = r$. Then $X = y + rt$, and the density is expressed by the formula

$$f(x,v,t) = \int f_0(y,v)\delta(x - y - vt)\,dy. \tag{3}$$

Example 2. Particles in hydrodynamic flow. Particles in *hydrodynamic flow* move with velocities depending on their locations only: particles found at a given point in space have identical velocities. We assume here that the initial density of the flow,

$$f_0(x,v) = \rho_0(x)\delta(v - v(x)), \tag{4}$$

11.6. Wave as Quasiparticle

is singular. Here, $\rho_0(x)$ is the initial density of particles, and $v(x)$ is the velocity of particles located originally at the point x. Substituting (4) into (3), we obtain

$$f(x, v, t) = \int \rho_0(y)\delta(v - v(y))\delta(x - y - vt)dy. \tag{5}$$

Integrating the particle density $f(x, v, t)$ over all v's, we obtain the time evolution of the spatial density of the particles:

$$\rho(x, t) = \int f(x, v, t)dv. \tag{6}$$

In particular, in the hydrodynamic flow,

$$\rho(x, t) = \int \rho_0(y)\delta(x - y - v(y)t)dy. \tag{7}$$

Example 3. Particle density in an explosion. The evolution of matter density after an explosion of a (one-dimensional) bomb is described by (7), with $\rho_0(x)$ the original matter density in the explosive device. Standing far away from the explosion site, one can assume that all the explosive material was originally concentrated at a single point $y = 0$ and ignore the component y in the argument of the Dirac delta. In this "central explosion" approximation, the expression (7) takes the form

$$\rho(x, t) = \int \rho_0(y)\delta(x - v(y)t)dy, \tag{8}$$

similar to the expression for the wave intensity in the framework of the stationary phase approximation (11.5.4).

11.6.2 Waves as Flows of Quasiparticles

The previous example establishes the promised analogy between the wave intensity and the hydrodynamic flow's density. Since the expressions (8) and (11.5.4) are identical, the following analysis of the flow density automatically applies to the wave intensity (11.5.4).

Let us begin with the observation that the density (8) is a superposition of singular densities

$$\delta(x - v(y)t)$$

of microparticles following each other. By analogy, the Dirac-delta-like components
$$\delta(x - W'(\kappa)t)$$
of the intensity (11.5.4) will be called *quasiparticles*.

Suppose that the velocity $v(y)$ is a monotone function of y (an analogous assumption about $W'(\kappa)$ is satisfied in most dispersive media). Then the mass situated between two particles located initially at points y_1 and y_2, $y_1 < y_2$, does not change, since

$$\Delta m = \int_{v_1 t}^{v_2 t} \rho(x,t) dx = \int_{y_1}^{y_2} \rho_0(y) dy.$$

Here, v_1 and v_2 stand for velocities of the extreme particles ($v_1 = v(y_1)$, $v_2 = v(y_2)$ if $v(y)$ increases monotonically, and $v_1 = v(y_2)$, $v_2 = v(y_1)$ if it decreases). In addition, in view of the mass conservation law, the density of a gas of diverging particles should decrease. Indeed, in view of the probing property of the Dirac deltas of a composite argument, (8) implies that

$$\rho(x,t) = \frac{\rho_0(y)}{t|v'(y)|}, \qquad (9)$$

where $y = y(x,t)$ is the initial coordinate of the particle, which at time t, is located at the point x. Formula (9) repeats, up to a change of notation, the expression for intensity (11.5.3), and implies that in a neighborhood of the particle arriving at the point x, the density is proportional to the initial density $\rho_0(y)$. The factor $1/t$ guarantees mass conservation for the diverging particles. The factor $1/|v'(y)|$ describes variations of the particle density caused by the relative rarefaction of the flow. Let us illustrate these effects with the following examples.

Example 4. Rod vibration and deep-water waves. Consider two velocity laws
$$v^1(y) = y \quad \text{and} \quad v^2(y) = 1/\sqrt{y}.$$

The first corresponds to the model of the group velocity (see (11.4.4)) of the transverse vibration of a rod, while the second (see (11.4.5)) applies to the group velocity of deep-water waves. The variables y, x, and t are assumed to be dimensionless. In the first case, the explosion disperses the matter uniformly (see Example 3), and its density at time t,

$$\rho^1(x,t) = \frac{1}{t}\rho_0\left(\frac{x}{t}\right),$$

11.6. Wave as Quasiparticle

retains the shape of the initial matter density. In the second case, the nonuniformity of the flow of matter and a change in the order in which the particles are arranged leads to a qualitatively different law,

$$\rho^2(x,t) = \frac{t^2}{x^3}\rho_0\left(\frac{t}{x}\right).$$

11.6.3 Continuity Equation for Quasiparticle Flows

Like the density of the hydrodynamic flow, the intensity (11.5.4) satisfies the *continuity equation*. We shall demonstrate this fact using the example of a single quasiparticle,

$$I_s(x,t) = \delta(x - v(\kappa)t).$$

The "gas" of such quasiparticles represents the intensity of the wave packet (11.5.4). Differentiating the last equality with respect to t and applying the chain rule, we get

$$\frac{\partial}{\partial t}I_s = -\frac{\partial}{\partial x}\Big(v(\kappa)\delta(x - v(\kappa)t)\Big).$$

Now, in view of the probing property of the Dirac delta, we can replace the multiplier $v(\kappa)$ by

$$v(x,t) = v[k(x,t)],$$

the group velocity of the quasiparticle, which at time t, is located at the point x. As a result, we arrive at the continuity equation

$$\frac{\partial}{\partial t}I_s + \frac{\partial}{\partial x}\Big(v(x,t)I_s\Big) = 0.$$

Its linearity permits an application of the superposition law, so that the full intensity of the wave packet (11.5.4) satisfies the equation

$$\frac{\partial}{\partial t}I + \frac{\partial}{\partial x}(v(x,t)I) = 0. \qquad (10)$$

Equations for $k(x,t)$ and $v(x,t)$ can be found by differentiating the equation

$$x = v(k)t$$

with respect to t and x. Thus

$$v(k) + tv'(k)\frac{\partial k}{\partial t} = 0, \quad \text{and} \quad 1 = tv'(k)\frac{\partial k}{\partial x}.$$

Elimination of $v'(k)$ leads to the following equation for $k(x,t)$:

$$\frac{\partial k}{\partial t} + v(k)\frac{\partial k}{\partial x} = 0. \tag{11}$$

Multiplying this equation by $v'(k)$, we arrive at an equation for $v(x,t)$:

$$\frac{\partial v}{\partial t} + v\frac{\partial v}{\partial x} = 0. \tag{12}$$

The latter can also be derived, in a physically more natural fashion, from the group velocity conservation law for a quasiparticle. Solutions of nonlinear partial differential equations like (11) and (12) will be discussed in Chap. 12.

11.7 Wave Packets with Narrow-Band Spectrum

The asymptotic stationary phase method gave us an opportunity to introduce the fundamental concept of group velocity. However, the group velocity of many physical waves meaningfully determines their behavior long before the asymptotic method itself is applicable. This is the case for the *narrow-band wave packets* often encountered in physical applications.

A wave packet is said to be narrow-band if its Fourier image $\tilde{f}(\kappa)$ differs from zero only in a small vicinity of some wavenumber k. A narrow-band wave packet may be conveniently described by the formula

$$u(x,t) = \operatorname{Re} U(x,t), \tag{1}$$

where

$$U(x,t) = \int \tilde{f}(\kappa - k) \exp\Big(i(\kappa x - W(\kappa)t)\Big) d\kappa,$$

and where the support of $\tilde{f}(\kappa)$ contains the origin $\kappa = 0$ and has width $\Delta \ll k$. As a rule, $W(\kappa)$ is a smooth function of κ, slowly varying on ab interval of length Δ.

In such a case, $W(\kappa)$ in (1) can be replaced, with a negligible error, by the first three terms of its Taylor expansion,

$$W \approx W(k) + v(k)(\kappa - k) + \frac{1}{2}r(\kappa - k)^2, \qquad r = W''(k).$$

This gives

$$U(x,t) = \exp[i(kx - \omega t)] \int \tilde{f}(\mu) \exp\Big(i\mu(x - vt) - irt\frac{\mu^2}{2}\Big) d\mu, \tag{2}$$

where $\mu = \kappa - k$ is a new variable of integration, $\omega = W(k)$, and $v = v(k)$. According to (2), the wave packet is a harmonic wave $\exp[i(kx - \omega t)]$ propagating with the phase velocity and modulated by the "shape" of the wave (moving with the group velocity)

$$f(z,t) = \int \tilde{f}(\mu) \exp\left(i\mu z - irt\frac{\mu^2}{2}\right) d\mu, \tag{3}$$

where $z = x - vt$. As long as $\Delta^2 rt \ll 1$, the function $f(z,t) = f(x - vt)$ retains the initial shape of the wave packet. At times t, with $\Delta^2 rt \sim 1$, the function $f(z,t)$ begins to "flatten out," and for large times, when $\Delta^2 rt \gg 1$, it is described by the asymptotic formulas of the stationary phase method:

$$f(z,t) = \sqrt{\frac{2\pi}{irt}} \tilde{f}\left(\frac{z}{rt}\right).$$

11.8 Optical Wave Behind a Phase Screen

Optics is one of the most voracious customers of asymptotic methods, including the stationary phase method. In this section we shall use this method to study the behavior of optical waves behind a phase screen.

Consider a monochromatic optical wave propagating in the direction of the z-axis. The transverse coordinate vector will be denoted by $\boldsymbol{x}$. In the Fresnel approximation (see Sect. 5.3, Volume 1), the amplitude $u(\boldsymbol{x}, z)$ of a complex wave satisfies (see (10.4.10)) the parabolic equation of quasioptics,

$$2ik\frac{\partial u}{\partial z} = \Delta_\perp u, \quad u(\boldsymbol{x}, z = 0) = u_0(\boldsymbol{x}). \tag{1}$$

Suppose that a phase screen is placed in the plane $z = 0$ and that it changes the incident wave's phase by $k\psi(\boldsymbol{x})$. If the incident wave is planar, propagating along the z-axis, then its complex amplitude just in front of the screen is

$$u(\boldsymbol{x}, z = 0_+) = \exp[-ik\psi(\boldsymbol{x})], \tag{2}$$

and the solution of the boundary value problem (1) is of the form

$$u(\boldsymbol{x}, z) = \frac{ik}{2\pi z} \iint \exp\left(-ik\psi(\boldsymbol{y}) - \frac{ik(\boldsymbol{y} - \boldsymbol{x})^2}{2z}\right) d^2 y. \tag{3}$$

In optics, the role of very large parameter is played by the wavenumber k. In comparison with 1-D waves in media with dispersion, the novel element here is that the integral (3) is two-dimensional; its analysis requires a 2-D version of the stationary phase method.

Let us begin by rewriting (3) in a form more convenient for analysis,

$$u(\boldsymbol{x}, z) = \frac{ik}{2\pi z} \iint \exp\left(-ik\, G(\boldsymbol{x}, \boldsymbol{y}, z)\right) d^2 y, \qquad (4)$$

where

$$G(\boldsymbol{x}, \boldsymbol{y}, z) = \psi(\boldsymbol{y}) + \frac{(\boldsymbol{x} - \boldsymbol{y})^2}{2z} \qquad (5)$$

is a function independent of k. By analogy with the 1-D case, we need to find stationary points of G in the $\boldsymbol{y}$-plane, i.e., to solve the equation

$$\nabla_y G = 0,$$

which in our case is of the form

$$\boldsymbol{x} = \boldsymbol{y} + \boldsymbol{v}(\boldsymbol{y}) z, \qquad (6)$$

where

$$\boldsymbol{v}(\boldsymbol{y}) = \nabla \psi(\boldsymbol{y}) \qquad (7)$$

is a vector function.

For simplicity's sake, assume that for a given $\boldsymbol{x}$, there exists only one stationary point $\boldsymbol{y}^*(\boldsymbol{x}, z)$ in the $\boldsymbol{y}$-plane. In a neighborhood of $\boldsymbol{y}^*(\boldsymbol{x}, z)$, a twice continuously differentiable function G has the asymptotic form

$$G \sim G(\boldsymbol{x}, \boldsymbol{y}^*, t) + \frac{1}{2z}(y_i - y_i^*)(y_j - y_j^*) r_{ij}, \qquad (8)$$

where

$$r_{ij} = \delta_{ij} + z \tau_{ij}. \qquad (9)$$

In the above formula, $\tau_{ij} = \tau_{ji}$,

$$\tau_{ij}(\boldsymbol{x}, z) = \left. \frac{\partial^2}{\partial y_i \partial y_j} \psi(\boldsymbol{y}) \right|_{\boldsymbol{y}=\boldsymbol{y}^*},$$

and δ_{ij} is the Kronecker symbol.

Let us express the variables of integration $\boldsymbol{y}$ in terms of the main axes (eigenvectors) of the symmetric matrix r_{ij} (or equivalently, of τ_{ij}). In the new coordinate system, the quadratic form (8) has a diagonal representation,

$$G \sim G(\boldsymbol{x}, \boldsymbol{y}^*, t) + \frac{1}{2z}(y_i - y_i^*)^2 [1 + z\tau_i], \qquad (10)$$

11.8. Optical Wave Behind a Phase Screen

where τ_1, τ_2 are the eigenvalues of r_{ij}. In a small vicinity of the stationary point, the 2-D integral (4) splits into a product of one-dimensional integrals. The asymptotics of each of them is given by formula (5.3.1) of Volume 1, so that the final stationary phase approximation for (4) is

$$u(\boldsymbol{x}, z) = \exp[-ikG(\boldsymbol{x}, \boldsymbol{y}^*, t)] \frac{1}{\sqrt{|(1 + z\tau_1)(1 + z\tau_2)|}}.$$

The above formula for the wave's complex amplitude u immediately gives the following expression for the wave's intensity:

$$I(\boldsymbol{x}, z) = |u(\boldsymbol{x}, z)|^2 = \frac{1}{(1 + z\tau_1)(1 + z\tau_2)}. \tag{11}$$

The latter has a transparent geometric interpretation, similar to the quasiparticle concept. Notice that the equality (6) defines a mapping of the $\boldsymbol{x}$-plane into the $\boldsymbol{y}$-plane. The Jacobian of the $\boldsymbol{y}$-to-$\boldsymbol{x}$ transformation is given by the denominator in (11), i.e.,

$$J(\boldsymbol{x}, z) = (1 + z\tau_1)(1 + z\tau_2).$$

Thanks to formula (1.9.1), Volume 1, for the Dirac delta of a composite multidimensional argument, the wave intensity (11) can be written in the integral form

$$I(\boldsymbol{x}, z) = \iint \delta\Big(\boldsymbol{x} - \boldsymbol{y} - \boldsymbol{v}(\boldsymbol{y})z\Big) d^2\boldsymbol{y},$$

similar to the integral (11.5.4).

So, the optical wave intensity also satisfies the continuity equation

$$\frac{\partial I}{\partial z} + \operatorname{div}(\boldsymbol{v}(\boldsymbol{x}, z)I) = 0, \tag{12}$$

where the field $\boldsymbol{v}(\boldsymbol{x}, z)$ satisfies the vector equation

$$\frac{\partial \boldsymbol{v}}{\partial z} + (\boldsymbol{v} \cdot \nabla)\boldsymbol{v} = 0. \tag{13}$$

The operators div and ∇ in (12) and (13) are respectively the divergence and gradient in the transverse plane $\boldsymbol{x}$. These equations are similar to the equations satisfied by the wave intensity and the group velocity in dispersive media in the stationary phase approximation, and they constitute the well-known *equations of geometric optics*. They are identical in form to the equations of the 2-D hydrodynamic flow of uniformly moving particles, where the role of the longitudinal coordinate z is played by the time t.

11.9 One-Dimensional Phase Screen

This section provides further analysis of the wave distortion behind the phase screen in the case that the wave depends only on one transverse coordinate, which we shall denote by x. In this case, formula (11.8.4) is reduced to a 1-D integral,

$$u(x,z) = \sqrt{\frac{ik}{2\pi z}} \int \exp\bigl(-ikG(x,y,z)\bigr) dy, \qquad (1)$$

where

$$G(x,y,z) = \psi(y) + \frac{(x-y)^2}{2z}. \qquad (2)$$

The geometric optics approximation for the wave intensity is then also reduced to the familiar single integral

$$I(x,z) = \int \delta(x - y - v(y)z) dy. \qquad (3)$$

The above Dirac delta probes values $y = y(x,t)$ that are roots of the equation

$$x = y + v(y)z. \qquad (4)$$

This fact has a clearcut geometric significance, with x, z the coordinates of the ray radiating from the screen point y. In the small angle approximation considered here, in which the parabolic equation of quasioptics is valid, the function $v(y) = \psi'(y)$ can be viewed as the angle between the ray and the z-axis, and $v'(y)$ as the curvature of the wavefront in the ray's vicinity. If the latter is positive, then the neighboring rays diverge with the growth of the z-coordinate. If $v'(y) < 0$, then the rays converge.

If the minimal curvature is equal to $-1/R$ (i.e., $v'(y) \geq -1/R$), then at small distances from the screen ($z < R$), (4) has a unique root $y(x,z)$, for any given x. This means that at an arbitrary point (x, z) close enough to the screen, only one ray can be detected, and in view of (3), the wave intensity is equal to

$$I(x,z) = \frac{1}{1 + v'(y)z}\bigg|_{y=y(x,z)}. \qquad (5)$$

Example 1. Sinusoidal screen. Let $v(x) = \sin(x)$. Then (4) takes the form

$$x = y + z \sin y, \qquad (6)$$

with curvature radius $R = 1$. The plot of the intensity (5) in this case, at the distance $z = 1/2$, is shown in Fig. 11.9.1 (right).

11.10. Caustics

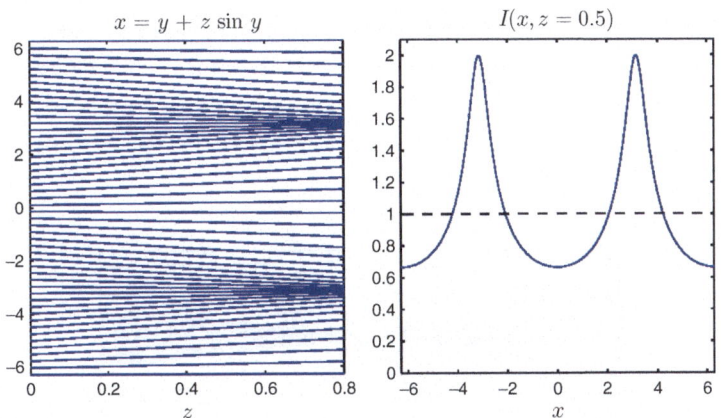

FIGURE 11.9.1
Left: **Rays behind the phase screen.** *Right:* **Intensity behind the phase screen.**

The plot clearly demonstrates that focusing of the rays by the screen creates areas of increased intensity. The reader may have seen such patterns created on a wall by sun rays passing through a window pane. A transparent, seemingly smooth sheet of glass always has some slight thickness fluctuations and acts as a 1-D phase screen, which explains the creation of luminous zones on the wall facing the window.

11.10 Caustics

At larger distances from the screen $(z > R)$, the *multiray regime* sets in, whereby several rays radiating from different screen points can meet at the same point (x, z). In this case, the intensity (11.9.3) becomes the sum

$$I(x,z) = \sum_i \frac{1}{1 + v'(y)z}\bigg|_{y=y_i(x,z)}, \qquad (1)$$

where the summation extends over all the roots of (11.9.4). This sum describes the intensity in the *incoherent-superposition-of-waves approximation*; all rays arriving at the point (x, z) are taken into account. The areas of the multiray regimes are surrounded by *caustic surfaces* (caustic curves in the (x, z)-plane in the case of a 1-D phase screen), or simply by *caustics*. The equation

$$1 + v'(y)z = 0 \qquad (2)$$

is satisfied on caustics, which means that the corresponding stationary points are no longer simple: the function G from the formula (11.9.2) is $G \sim o[(y - y^*)^2]$ in a neighborhood of the stationary point y^*. As a result, the stationary phase method is not applicable to caustics, where it predicts infinite intensity values.

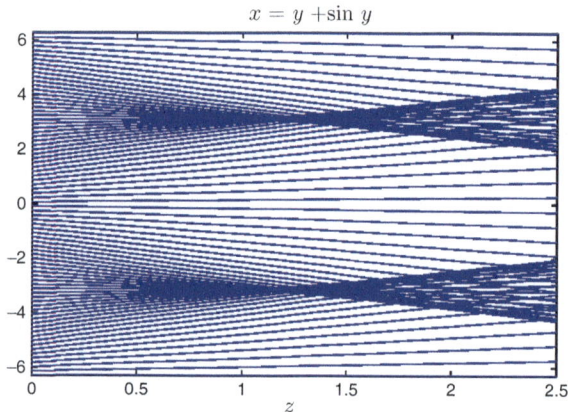

FIGURE 11.10.1
A set of rays, caustics, and multiray regions.

The typical behavior of the intensity (1) in the vicinity of caustics is illustrated in Fig. 11.10.1, for the special case described by formula (11.9.6), with $z = 2.5$. Graphs of all the components of the sum (1) are shown in the multiray zones between the caustics.

Let us find an analytic approximation for the intensity values in the caustics' vicinity via an approach more subtle than the method of the stationary phase (Fig. 11.10.2). Instead of (11.9.1), we shall study the wave intensity in the direct form

$$I(x, z) = |u(x, z)|^2 = \frac{k}{2\pi z} \times$$

$$\iint \exp\left(-ik[\psi(y_1) - \psi(y_2)] - \frac{ik}{2z}\left[(x - y_1)^2 - (x - y_2)^2\right]\right) dy_1 dy_2.$$

If we introduce new variables of integration

$$\rho = (y_1 + y_2)/2, \quad s = y_1 - y_2,$$

and express the old coordinates through the new ones by the inverse relations

$$y_1 = \rho + s/2, \quad y_2 = \rho - s/2,$$

11.10. Caustics

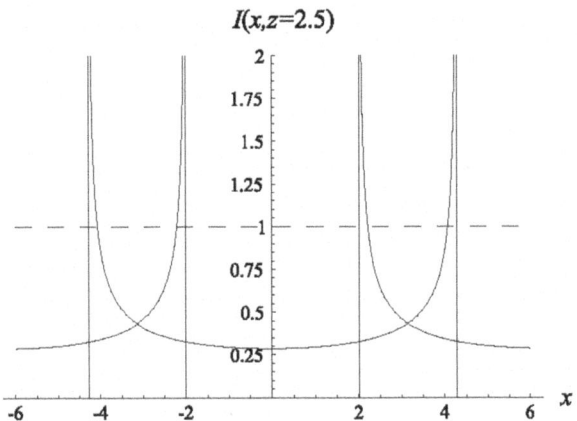

FIGURE 11.10.2
Intensity in the multiray area.

then the integral takes the form

$$I(x,z) = \frac{k}{2\pi z} \iint \exp\left(-ik\left[\psi\left(\rho+\frac{s}{2}\right) - \psi\left(\rho-\frac{s}{2}\right)\right] - i\frac{k}{z}(\rho-x)s\right) d\rho\, ds.$$

Changing the variable of integration s to the normalized variable $l = ks/z$, we obtain

$$I = \frac{1}{2\pi} \iint \exp\left(-ik\left[\psi\left(\rho+\frac{lz}{2k}\right) - \psi\left(\rho-\frac{lz}{2k}\right)\right] - i\rho l\right) e^{ixl} dl\, d\rho. \quad (3)$$

Let us take a closer look at the inner integral

$$L(x,\rho,z) = \frac{1}{2\pi} \int \exp\left(-ik\left[\psi\left(\rho+\frac{lz}{2k}\right) - \psi\left(\rho-\frac{lz}{2k}\right)\right] - i\rho l\right) e^{ixl} dl. \quad (4)$$

Searching for its asymptotics ($k \to \infty$) the way a physicist would, let us expand the difference of functions ψ into a series in powers of lz/k, and—without further ado—keep only as many terms as are needed to obtain a reasonable result.

Retaining only the component of the expansion linear in l gives

$$L = \frac{1}{2\pi} \int \exp\left(-i[v(\rho)z + \rho - x]l\right) dl.$$

In view of (3.3.3) in Volume 1, this integral is equal to

$$L = \delta(x - \rho - v(\rho)z). \quad (5)$$

Substituting this expression into the integral (3) yields

$$I(x,z) = \int L(x,\rho,z)d\rho = \int \delta(x - \rho - v(\rho)z)\, d\rho, \tag{6}$$

and we recover the familiar geometric optics approximation for the wave intensity (11.9.3). Clearly, it is not good enough here, since it predicts infinite values of the intensity on the caustics. Hence, the conclusion is that the expansion should include the next nonvanishing term, which is of the third degree in l:

$$k\left[\psi\left(\rho + \frac{lz}{2k}\right) - \psi\left(\rho - \frac{lz}{2k}\right)\right] \cong zv(\rho)l + \frac{z^3}{24k^2}v''(\rho)l^3.$$

Inserting this approximation into (4), we obtain

$$L = \frac{1}{2\pi}\int \exp\left(-i(a-x)l - i\frac{1}{3}b^3 l^3\right) dl, \tag{7}$$

where

$$a(\rho, z) = v(\rho)z + \rho, \qquad b(\rho, z) = \frac{z}{2k}\sqrt[3]{kv''(\rho)}. \tag{8}$$

The integral (7) can be expressed via the special function

$$\text{Ai}(x) = \frac{1}{\pi}\int_0^\infty \cos\left(xt + \frac{1}{3}t^3\right) dt, \tag{9}$$

called *Airy's function*. Its graph is shown in Fig. 11.10.3.

Although as $t \to \infty$, the integrand does not converge to zero, for large t, the crests and troughs compensate each other sufficiently to guarantee the integral's convergence. As a result, the integral of Airy's function over the entire x-axis is well defined and equal to 1.

Clearly,

$$L(x,\rho,z) = \frac{1}{b}\text{Ai}\left(\frac{a-x}{b}\right), \tag{10}$$

which weakly converges to (5) as $b \to 0$. Thus in the present approximation, the final expression for the wave intensity is

$$I(x,z) = \int \frac{1}{b(\rho,z)}\text{Ai}\left(\frac{\rho + v(\rho)z - x}{b(\rho,z)}\right) d\rho. \tag{11}$$

A detailed analysis shows that for sufficiently large k, the formula (11) actually coincides with the geometric optics approximations (11.9.5) and (11.10.1) in the zones between caustics. Close to the caustics, Airy's function eliminates the singularities by limiting the maximal value of the intensity.

11.11. Telegrapher's Equation

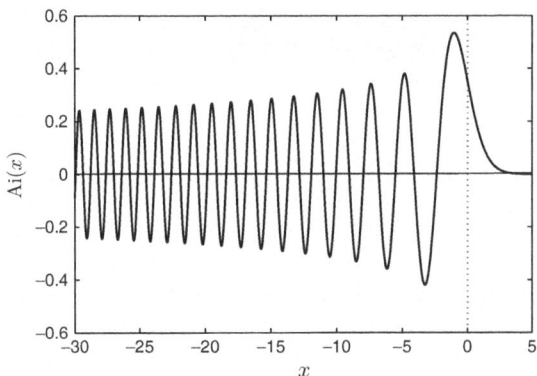

FIGURE 11.10.3
Graph of Airy's function.

11.11 Telegrapher's Equation

The telegrapher's equation was first studied by William Thompson (Lord Kelvin) in connection with the difficulties encountered in the exploitation of the first transatlantic cable laid in 1855. It describes how electrical signals behave as they propagate along long transmission lines. The current version of the equation is due to a later (1885) study by Heaviside.

11.11.1 Derivation of the Telegrapher's Equation

Current intensity $i(x,t)$ and voltage $v(x,t)$ in long transmission lines depend on the time t and spatial coordinate x, and they satisfy a hyperbolic equation traditionally called the *telegrapher's equation*. We shall derive it assuming that the line has capacitance C, inductance L, and resistance R, and that the insulation has conductivity G, each per unit length.

So let us select a segment dx of the line (see Fig. 11.11.1) and calculate the voltage drop over the length of this segment:

$$-dv = v(x,t) - v(x+dx,t) = -\frac{\partial v}{\partial x}\,dx\,.$$

The drop has a resistive component $Ri\,dx$ and an inductive component $Li_t\,dx$, so that

$$-\frac{\partial v}{\partial x}\,dx = R\,i\,dx + L\frac{\partial i}{\partial t}\,dx\,.$$

The quantities $R\,dx$ and $L\,dx$ are respectively the resistance and the inductance of the selected line segment. Dividing both sides of the above

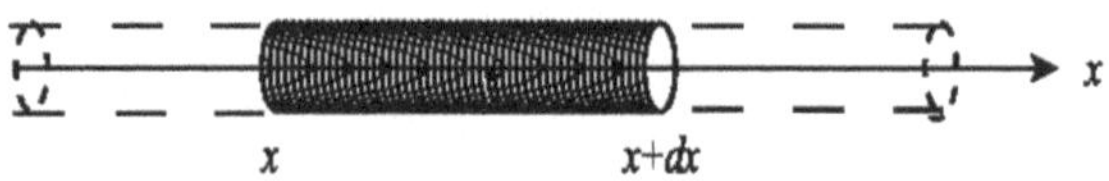

FIGURE 11.11.1
Schematic illustration of an infinitesimal section of a long transmission line.

equality by dx, we obtain the first equation linking voltage and current in the transmission line:
$$\frac{\partial v}{\partial x} + L\frac{\partial i}{\partial t} + Ri = 0. \tag{1}$$
The second equation for the two unknown functions v and i will be obtained by writing the balance of currents entering and leaving the line segment $[x, x+dx]$:
$$-di = i(x,t) - i(x+dx, t) = -\frac{\partial i}{\partial x}\,dx,$$
which is determined by the capacitative charging $C\,dx\,v_t$ of this line segment and the current drain $G\,dx\,v$ due to the conductivity of the insulation. Consequently,
$$-\frac{\partial i}{\partial x}\,dx = C\frac{\partial v}{\partial t}\,dx + G\,v\,dx,$$
or
$$\frac{\partial i}{\partial x} + C\frac{\partial v}{\partial t} + Gv = 0. \tag{2}$$

The system of two first-order equations (1) and (2) can now be reduced to a single second-order equation for voltage $v(x,t)$ or current $i(x,t)$ as follows. Differentiate (1) with respect to x and (2) with respect to t, and multiply the latter by L to obtain
$$\frac{\partial^2 v}{\partial x^2} + L\frac{\partial^2 i}{\partial t\partial x} + R\frac{\partial i}{\partial x} = 0,$$
$$L\frac{\partial^2 i}{\partial x\partial t} + LC\frac{\partial^2 v}{\partial t^2} + LG\frac{\partial v}{\partial t} = 0.$$
Subtracting the second equation from the first one, we get
$$\frac{\partial^2 v}{\partial x^2} - LC\frac{\partial^2 v}{\partial t^2} - LG\frac{\partial v}{\partial t} + R\frac{\partial i}{\partial x} = 0.$$
Now we can eliminate the derivative i_x using (2), arriving at the final form of the telegrapher's equation:
$$\frac{\partial^2 v}{\partial x^2} = LC\frac{\partial^2 v}{\partial t^2} + (CR + LG)\frac{\partial v}{\partial t} + RGv. \tag{3a}$$

11.11. Telegrapher's Equation

Similarly, one can obtain the following equation for current $i(x,t)$:

$$\frac{\partial^2 i}{\partial x^2} = LC \frac{\partial^2 i}{\partial t^2} + (CR + LG) \frac{\partial i}{\partial t} + RG\, i \,. \tag{3b}$$

Thus, voltage and current satisfy the same hyperbolic telegrapher's equation.

The telegrapher's equation may be simplified by the substitution

$$v(x,t) = e^{-\mu t} u(x,t) \,. \tag{4}$$

Indeed, substituting (4) into (3) and taking into account that

$$\frac{\partial^2 v}{\partial x^2} = e^{-\mu t} \frac{\partial^2 u}{\partial x^2}, \qquad \frac{\partial v}{\partial t} = e^{-\mu t} \left(\frac{\partial u}{\partial t} - \mu u \right)$$

and

$$\frac{\partial^2 v}{\partial t^2} = e^{-\mu t} \left(\frac{\partial^2 u}{\partial t^2} - 2\mu \frac{\partial u}{\partial t} + \mu^2 u \right),$$

we get an equation for the function $u(x,t)$:

$$\frac{\partial^2 u}{\partial x^2} = LC \frac{\partial^2 u}{\partial t^2} + (CR + LG - 2\mu LC) \frac{\partial u}{\partial t}$$

$$+ [LC\mu^2 - (CR + LG)\mu + RG]\, u \,.$$

So far, μ was arbitrary, but now we set

$$\mu = \frac{CR + LG}{2LC}, \tag{5}$$

so that the term containing u_t in the above equation disappears.

The equation for $u(x,t)$ then takes the simplified form

$$\frac{\partial^2 u}{\partial t^2} = a^2 \frac{\partial^2 u}{\partial x^2} + b^2 u, \tag{6}$$

where

$$a^2 = \frac{1}{LC}, \qquad b^2 = \left(\frac{CR - LG}{2LC} \right)^2 \,. \tag{7}$$

11.11.2 Distortionless Line

The propagation of electrical waves in long transmission lines in the case $b \neq 0$ is qualitatively different from that in the case $b = 0$, the latter corresponding to the parameters satisfying the relation

$$CR = LG. \tag{8}$$

Lines satisfying condition (8) were called by Heaviside *distortionless lines*. For such lines, $\mu = R/L = G/C$, and (6) becomes the standard hyperbolic equation

$$\frac{\partial^2 u}{\partial t^2} = a^2 \frac{\partial^2 u}{\partial x^2}, \qquad -\infty < x < +\infty, \quad 0 < t < +\infty. \tag{9}$$

Its general solution is

$$u(x,t) = \varphi(x - at) + \psi(x + at),$$

where $\varphi(x)$ and $\psi(x)$ are arbitrary functions. Consequently, the voltage in a long transmission line without distortion is of the form

$$v(x,t) = e^{-\frac{R}{L}t}\left[\varphi(x - at) + \psi(x + at)\right], \tag{11}$$

and the wave propagates along the line at speed a, decaying to zero in the course of time but without changing its shape, whence the name "distortionless line,"

Given the formula (11) for voltage in a long distortionless line, we can now calculate the behavior of the current by substituting the expression (11) into (2):

$$\frac{\partial i}{\partial x} = -C\frac{\partial v}{\partial t} - Gv$$

$$= e^{-\frac{R}{L}t}\left(Ca\frac{\partial \varphi}{\partial x} - Ca\frac{\partial \psi}{\partial x} + C\frac{R}{L}\varphi + C\frac{R}{L}\psi - G\varphi - G\psi\right). \tag{12}$$

Since the nondistortion condition (8) is satisfied, the last four terms in (12) simplify, and we obtain

$$\frac{\partial i(x,t)}{\partial x} = e^{-\frac{R}{L}t} Ca \left[\frac{\partial \varphi(x - at)}{\partial x} - \frac{\partial \psi(x + at)}{\partial x}\right].$$

Integrating this equality with respect to x and bearing in mind that $Ca = \sqrt{C/L}$, we get

$$i(x,t) = e^{-\frac{R}{L}t}\sqrt{\frac{C}{L}}[\varphi(x - at) - \psi(x + at) + \alpha(t)]. \tag{13}$$

11.11. Telegrapher's Equation

To explain the nature of the dependency of the integration constant $\alpha(t)$ on time, notice that $i(x,t)$ in (13) and $v(x,t)$ in (11) must satisfy (1), whence we have
$$\alpha'(t) = 0.$$
Consequently, $\alpha(t) = \alpha = const$. Hence the voltage and current in distortionless lines are given by
$$v(x,t) = e^{-\frac{R}{L}t}[\varphi(x-at) + \psi(x+at)]$$
and
$$i(x,t) = \sqrt{\frac{C}{L}} e^{-\frac{R}{L}t} [\varphi(x-at) - \psi(x+at) + \alpha].$$

Finally, it is not difficult to see that the introduction of an arbitrary integration constant α is not necessary, since a suitable redefinition of the arbitrary functions φ and ψ produces the same effect. Indeed, one can take
$$\Phi(x) = \varphi(x) + \frac{\alpha}{2}, \qquad \Psi(x) = \psi(x) - \frac{\alpha}{2},$$
which gives the solutions in the final form
$$\begin{aligned} v(x,t) &= e^{-\frac{R}{L}t}\left[\Phi(x-at) + \Psi(x+at)\right], \\ i(x,t) &= \sqrt{\frac{C}{L}} e^{-\frac{R}{L}t}\left[\Phi(x-at) - \Psi(x+at)\right]. \end{aligned} \tag{14}$$

11.11.3 Initial and Boundary Conditions for Telegrapher's Equations

If the endpoints of a long transmission line are located far enough from the point where the initial perturbations occur, then the influence of the boundaries on the evolution of the processes in the line during the "small" time interval under consideration can be neglected. In such a case, wave propagation in the line is uniquely determined by the initial distribution of the voltage and current:
$$v(x,0) = v_0(x), \qquad i(x,0) = i_0(x). \tag{15}$$

If we try to solve the telegrapher's equation (3) for voltage only, then instead of the condition (15), it is more convenient to impose initial conditions specifying the voltage and its first derivative in time. It follows from (15) and (2) that
$$v_t(x,0) = -\frac{1}{C}\left[i_{0x}(x) + Gv_0(x)\right]. \tag{16}$$

If, using equality (4), we transform (3) into (6), then it is necessary to formulate initial conditions for (6) that are equivalent to the initial conditions (15).

To accomplish this, note that

$$u_t = e^{\mu t}\left(v_t + \mu v\right).$$

Combining the above equality with equalities (5), (15), and (16), we obtain the following initial conditions for (6):

$$u(x,0) = v_0(x), \qquad \frac{\partial u(x,0)}{\partial t} = -\frac{1}{C}\frac{\partial i_0(x)}{\partial x} + \frac{RC - GL}{2LC}v_0(x). \qquad (17)$$

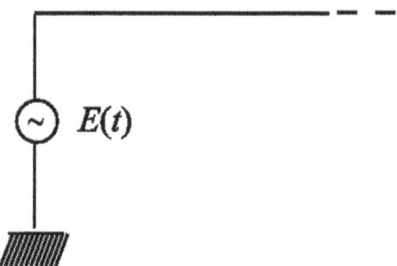

FIGURE 11.11.2
Voltage $v(x,0) = E(t)$ applied at the point $x = 0$ of a long transmission line.

In particular, for a distortionless line, these conditions are simplified, and we have

$$u(x,0) = v_0(x), \qquad \frac{\partial u(x,0)}{\partial t} = -\frac{1}{C}\frac{\partial i_0(x)}{\partial x}.$$

At this point, it will be useful to consider a few examples of typical boundary conditions for the telegrapher's equation. Assume that at the left endpoint of the line, i.e., for $x = 0$, we switch on a voltage source $E(t)$ (see Fig. 11.11.2), i.e.,

$$v(0,t) = E(t). \qquad (18)$$

Before we proceed with the presentation of examples of practical importance that may reflect situations when the line is grounded, say, via a resistor, a capacitor, or an inductance, let us explain how to handle mathematically the presence of a concentrated grounding resistance using the example of a damaged line.

11.11. Telegrapher's Equation

Assume that a damaged segment $[x_0, x_0 + d]$ of a line has length d and resistivity R_0. The corresponding resistance per unit length is $R = R_0/d$. For the sake of simplicity, suppose that in the damaged area of the line, inductance, capacitance, and the insulation's conductivity are all zero. Then the current in every section is the same, and $i(x_0, t) = i(x_0 + d, t)$, but the voltage drop over this segment, which is caused only by resistivity, is

$$v(x_0 + d, t) - v(x_0) = -i(x_0, t)\, Rd = -i(x_0, t) R_0\,.$$

If we let $d \to 0$, then the distributed resistance of the damaged segment of the line shrinks to a resistance concentrated at the point x_0, and the voltage at this point has a jump

$$v(x_0 + 0, t) - v(x_0 - 0, t) = -i(x_0, t)\, R_0\,. \tag{19}$$

The voltage jump on a concentrated resistivity is also present in the case of nonzero capacitance C, inductance L, and insulation conductivity G of the segment $[x_0, x_0 + d]$, as long as the total capacitance Cd, inductance Ld, and insulation conductivity Gd tend to zero as $d \to 0$.

FIGURE 11.11.3
A long transmission line with two endpoints grounded through concentrated resistivities.

Now let us assume that one endpoint (at $x = 0$) of the line is grounded through a concentrated resistance R_0 (see Fig. 11.11.3). Then $v(-0, t) = 0$, but the jump condition gives the boundary value

$$v(+0, t) = -R_0\, i(0, t)\,. \tag{20}$$

Similarly, if the other end of the line (at $x = l$) is grounded through a concentrated resistance R_l, then the jump condition (19) yields the boundary value

$$v(l - 0, t) = R_l\, i(l, t)\,. \tag{21}$$

In a similar fashion one can introduce concentrated capacitance, inductance, and insulation conductivity and the corresponding jump conditions for current and voltage. In particular, if one endpoint of the line is grounded through a concentrated inductance and the other through a grounded capacitance (see Fig. 11.11.4), then the boundary conditions are as follows:

$$v(0,t) = -L_0\, i_t(0,t), \qquad i(l,t) = C_l\, v_t(l,t). \tag{22}$$

FIGURE 11.11.4
A long transmission line grounded at the left endpoint by a concentrated inductance and at the right endpoint by a concentrated capacitance.

11.11.4 Practical Examples

We shall conclude our discussion of the telegrapher's equation by providing a number of concrete examples. Simpler situations are described in the exercises in Sect. 11.12.

Example 1. Infinite transmission line grounded through a concentrated resistivity. Let us find the time-dependent distribution of voltage and current in an infinite transmission line grounded at $x = 0$ through a concentrated resistance R_1, assuming that for $t < 0$, a wave was present propagating according to the relations

$$v(x,t) = e^{-\frac{R}{L}t} f(x-at), \quad i(x,t) = \sqrt{\frac{C}{L}}\, e^{-\frac{R}{L}t} f(x-at),$$

for $x < 0$, but where $f(x) = 0$ for $x > 0$ (see Fig. 11.11.5).

A mathematical reformulation of the problem requires that we state the jump condition at the grounding point $x = 0$. Obviously, at that point,

11.11. Telegrapher's Equation

current has a jump due to drainage through the grounding resistivity,

$$i(-0, t) - i(+0, t) = i_1 , \qquad (23)$$

where i_1 is the current through resistance R_1. The value i_1 is connected with the voltage at the point of grounding by Ohm's law:

$$v(0, t) = i_1 R_1 .$$

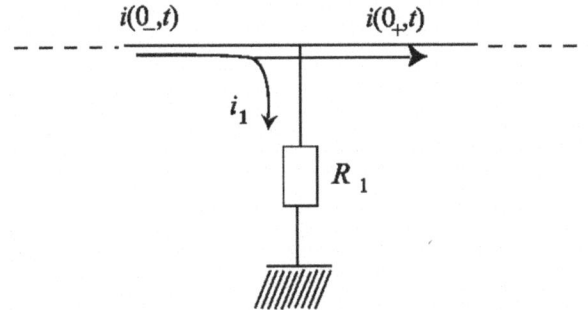

FIGURE 11.11.5
Schematic diagram of an infinite transmission line grounded at $x = 0$ through a concentrated resistivity.

Consequently, the value of the current jump there is determined by equality

$$R_1 [i(-0, t) - i(+0, t)] = v(0, t) . \qquad (24)$$

So, we need to find the voltage $v(x,t)$ and current $i(x,t)$ for $t > 0$ satisfying (3a) and (3b) ($CR = LG$) for ($x < 0$) and ($x > 0$), respectively, taking into account the jump condition (24) and the fact that for $x < 0$ and $t < 0$, there existed a prior wave of the shape prescribed above.

To find the desired solution, let us represent voltage and current in the left half of the transmission line in the form

$$v(x, t) = e^{-\frac{R}{L}t} [f(x - at) + \varphi(x + at)] ,$$
$$i(x, t) = \sqrt{\frac{C}{L}} e^{-\frac{R}{L}t} [f(x - at) - \varphi(x + at)] , \qquad (25)$$

and in the right half by

$$v(x, t) = e^{-\frac{R}{L}t} g(x - at) ,$$
$$i(x, t) = \sqrt{\frac{C}{L}} e^{-\frac{R}{L}t} g(x - at) . \qquad (26)$$

The absence here of terms containing the expression $x + at$ follows from the obvious condition that in the right half of the transmission line there are no waves incident on the grounded point (the so-called *radiation condition*).

Our problem will be solved if we find a function $\varphi(x)$ for $x > 0$ and a function $g(x)$ for $x < 0$. Substituting (25) and (26) in (24), we arrive at the following equation for φ and g:

$$\rho[f(-at) - \varphi(at) - g(-at)] = g(-at),$$

where

$$\rho = R_1 \sqrt{\frac{C}{L}}$$

is the nondimensionalized grounding resistivity (see also Exercise 11.12.5). The second equation,

$$f(-at) + \varphi(at) = g(-at),$$

results from the continuity condition $v(-0,t) = v(+0,t)$ for the voltage function at the grounding point.

Substituting $z = -at < 0$ in the last two equations, we get

$$\begin{aligned} \rho[f(z) - \varphi(-z) - g(z)] &= g(z), \\ g(z) &= f(z) + \varphi(-z). \end{aligned} \tag{27}$$

Solving (27) for φ and g, we obtain

$$g(z) = f(z)\frac{2\rho}{1+2\rho}, \quad \varphi(-z) = -f(z)\frac{1}{1+2\rho}.$$

Finally, replacing z by x in the first formula and z by $-x$ in the second one, we obtain

$$g(x) = f(x)\frac{2\rho}{1+2\rho} \quad (x < 0),$$

$$\varphi(x) = -f(-x)\frac{1}{1+2\rho} \quad (x > 0).$$

The physical consequences of the shape of the solution obtained above can be better understood if we plot (see Fig. 11.11.6) the dependence of the reflection coefficient K and the transmission coefficient T on the nondimensionalized grounding resistance ρ:

$$K(\rho) = -\frac{1}{1+2\rho}, \quad T(\rho) = \frac{2\rho}{1+2\rho}.$$

11.11. Telegrapher's Equation

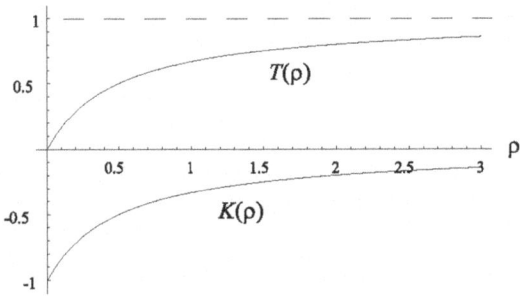

FIGURE 11.11.6
Dependence of the reflection coefficient K and the transmission coefficient T on the nondimensionalized grounding resistance ρ, for an infinite transmission line grounded in the middle through a concentrated resistivity.

The graphs imply that if the transmission line is short-circuited, i.e., $R_1 = 0$, $T = 0$, so that $g(x) \equiv 0$, then voltage and current do not propagate into the right half of the line. If $R_1 = \infty$, i.e., if the line is not grounded, then $K = 0 \iff \varphi(x) \equiv 0$, and the reflected wave disappears.

11.11.4 A Moving Boundary Problem: "Slithering" Short Circuit

Consider a semi-infinite $(x > 0)$ distortionless transmission line in which for $t < 0$, the voltage wave

$$v(x,t) = e^{-\frac{R}{L}t} f(x + at)$$

was present, and where $f(x) = 0$ for $x \leq 0$. We shall determine the distribution of the voltage in this line for $t > 0$, given that from the moment $t = 0$ on, the line is short-circuited to the ground at a moving point of contact $x = \psi(t)$ $(\psi(0) = 0)$.

In mathematical terms, the problem can be formulated as follows: We seek a solution of (3a) $(CR = LG)$ in the domain $\psi(t) < x < +\infty$, $0 < t < +\infty$, satisfying the *moving boundary* condition

$$v(\psi(t), t) = 0 \qquad (\psi(0) = 0)$$

and the initial condition described above.

To find the solution, we shall try to find the voltage function in the form

$$v(x,t) = e^{\frac{R}{L}t}\left[f(x+at) - \varphi(at-x)\right], \qquad t > 0,$$

where the unknown function $\varphi(z)$ equals zero for $z < 0$, and for $z > 0$, it is determined by the condition that voltage at the point of short circuit is zero:

$$f(\psi(t) + at) = \varphi(at - \psi(t)).$$

Thus

$$\varphi(z) = f(y(z)), \qquad z > 0, \qquad (28)$$

where y is a function of z given parametrically:

$$y = at + \psi(t), \quad z = at - \psi(t). \qquad (29)$$

Replacing z by $at - x$, we finally obtain that for $t > 0$ and $x > \psi(t)$, the voltage in a long transmission line evolves as follows:

$$v(x,t) = e^{-\frac{R}{L}t}\left[f(x+at) - f(y(x-at))\right]. \qquad (30)$$

Physically, the above solution makes sense only when the velocity of motion of the contact point does not exceed the propagation velocity of the wave in the transmission line, i.e., when $|\dot\psi(t)| < a$. In this case,

$$y'(z) = \frac{\dot y}{\dot z} = \frac{a + \dot\psi(t)}{a - \dot\psi(t)} > 0,$$

i.e., the function $y(z)$ must be monotonically increasing, and the reflected wave will satisfy the following causality principle: it should arrive at an arbitrary point $x > \psi(t)$ in the same time order in which the incident wave propagated through x.

To better understand the phenomenon of reflection from a moving short-circuited point, it is worthwhile to analyze it in a few typical cases of the regime of motion.

Example 1. Boundary moving with constant velocity. Consider a contact point sliding along the transmission line with constant velocity, that is, $\psi(t) = ct$. In this case, it follows from (29) that

$$y = \frac{a+c}{a-c} z. \qquad (31)$$

11.11. Telegrapher's Equation

Now assume that the contact point moving with constant velocity is overtaken by an incident sine wave,

$$e^{-\frac{R}{L}t}\chi(x+at)\sin(kx+\omega t),$$

where ω is the wave frequency, $k = \omega/a$ is its wave number, and $\chi(x)$ is the Heaviside unit step function (equal to 1 for $x > 0$, and 0 for $x \leq 0$).

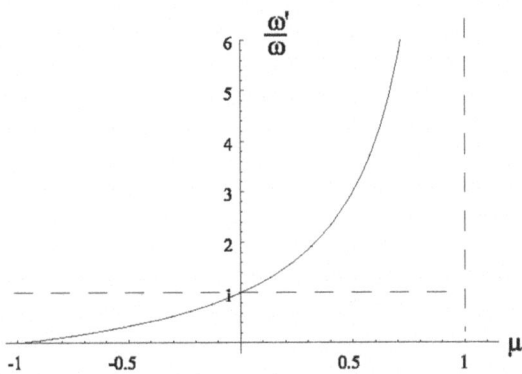

FIGURE 11.11.7
Dependence of the frequency of the reflected wave on the nondimensional parameter $\mu = c/a$, describing the ratio of the velocity of the boundary to the propagation velocity of the wave.

In this situation, it follows from (30) and (31) that

$$v_{\text{refl}}(x,t) = -e^{-\frac{R}{L}t}\chi(at-x)\sin(\omega' t - k' x),$$

where

$$\omega' = \omega\frac{a+c}{a-c}, \qquad k' = k\frac{a+c}{a-c}. \tag{32}$$

Equalities (32) represent the familiar *Doppler effect*: a wave reflected from a moving boundary has frequency ω' and wave number k' different from the frequency ω and the wave number k of the incident monochromatic wave. Figure 11.11.7 shows the dependence of the frequency of the reflected wave on the nondimensional parameter $\mu = c/a$.

Observe that the Doppler effect results in compression of the reflected wave profile (as compared to the incident wave) if the boundary is moving toward the wave ($c > 0$), and in dilation of the reflected wave profile if the boundary is moving away from the wave ($c < 0$).

Example 2. Harmonically oscillating boundary. As our second example, let us consider the situation in which the contact point itself undergoes a harmonic vibration $\psi(t) = H \sin \Omega t$, with frequency Ω and amplitude H. In this case, the profile of the reflected wave undergoes cyclical compressions and dilations with their magnitude determined by the function $y'(z)$.

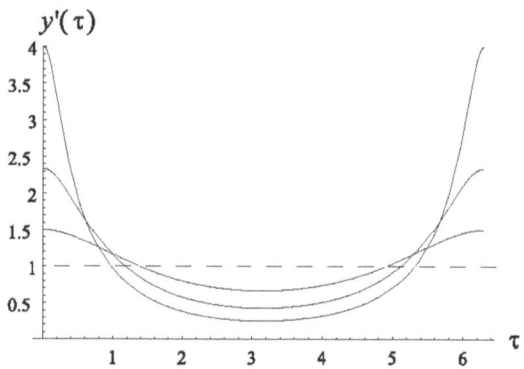

FIGURE 11.11.8
The magnitude $y'(\tau)$ of the compression of the wave reflected from a vibrating boundary as a function of the nondimensional variable $\tau = \Omega z/a$. Graphs are shown for three different values of the nondimensional parameter $\mu = c/a$, representing the ratio of the maximal velocity $c = \Omega H$ of the moving boundary point to the velocity of the propagating wave.

Let us take a close look at this phenomenon by changing to a nondimensional variable $\tau = \Omega z/a$. The above discussion gives
$$\frac{dy(\tau)}{d\tau} = \frac{1 + \mu \cos \rho}{1 - \mu \cos \rho}, \qquad \tau = \rho - \mu \sin \rho.$$
The only nondimensional parameter $\mu = c/a$ entering in the above formula is the ratio of the maximal velocity $c = \Omega H$ of the reflection point to the velocity of the propagating wave. Several graphs of the function $y'(\tau)$ for different values of the parameter μ are shown in Fig. 11.11.8.

11.11.5 Transmission Line Grounded at One End and Insulated at the Other

In this example we shall find the voltage $v(x,t)$ in a homogeneous transmission line, $0 \leq x \leq l$, with left endpoint $x = 0$ grounded and right endpoint $x = l$ insulated. We shall assume that the initial current and voltage are zero,

11.11. Telegrapher's Equation

but that at the initial moment of time, $t = 0$, concentrated electric charge Q has been placed at the point $x = x_0$. For reasons that will become clear later on, we shall concentrate on the situation in which

$$\frac{\pi}{l\sqrt{CL}} > \left|\frac{R}{L} - \frac{G}{C}\right|. \tag{33}$$

Let us begin with a mathematical formulation of the boundary conditions. Grounding of the left endpoint yields

$$v(0, t) = 0 \quad (0 < t < +\infty).$$

Insulation of the right endpoint means that $i(l, t) = 0$, but also that $i_t(l, t) = 0$. Thus, taking (1) into account, the voltage at the right boundary satisfies the condition

$$v_x(l, t) = 0 \quad (0 < t < +\infty).$$

Let us reformulate the initial conditions and consider a more general situation, which, as sometimes happens, leads to a simpler analysis. So assume that the initial electric charge at time $t = 0$ has a linear charge density $\rho(x)$. Consider a small segment $[x, x + dx]$ having length dx (see Fig. 11.11.1). The charge of this segment is $dQ = \rho(x)dx$, and its capacitance is $dC = Cdx$. Accordingly, the potential of the above segment is $v(x, 0) = dQ/dC = (1/C)\rho(x)$. Now, if we substitute here the density of the concentrated charge $\rho(x) = Q\delta(x - x_0)$, we shall obtain the first initial condition,

$$v(x, 0) = \frac{Q}{C}\delta(x - x_0).$$

The second initial condition is derived with the help of (2). Since at the initial time, $i(x, 0) = 0$ and $i_x(x, 0) = 0$, we have $v_t(x, 0) = -(G/C)v(x, 0)$, or taking into account the first initial condition,

$$v_t(x, 0) = -\frac{QG}{C^2}\delta(x - x_0).$$

To solve this problem it is convenient to pass from the voltage function $v(x, t)$ to an auxiliary function $u(x, t)$, connected with the voltage by the equality

$$v(x, t) = e^{-\mu t} u(x, t), \tag{34}$$

and satisfying the equation

$$u_{tt} = a^2 u_{xx} + b^2 u. \tag{35}$$

From (34) and (17), it follows that the boundary and initial conditions for $u(x,t)$ corresponding to the above boundary and initial conditions for $v(x,t)$ are of the form

$$u(0,t) = 0, \quad u_x(l,t) = 0, \tag{36a}$$

$$v(x,0) = \frac{Q}{C}\delta(x-x_0), \quad v_t(x,0) = -\frac{QG}{C^2}\delta(x-x_0). \tag{36b}$$

Let us solve the problem by the method of *separation of variables*, representing $u(x,t)$ as a product

$$u(x,t) = X(x)T(t),$$

which we substitute into (35), thus obtaining the equations

$$\frac{X''}{X} = \frac{\ddot{T} - b^2 T}{a^2 T} = -\lambda.$$

Here λ is still an arbitrary constant. The above relation implies that $X(x)$ satisfies the equation

$$X'' + \lambda X = 0 \tag{37a}$$

and the boundary conditions

$$X(0) = 0, \quad X'(l) = 0, \tag{37b}$$

resulting from (36). The function $T(t)$ must satisfy the equation

$$\ddot{T} + (\lambda a^2 - b^2)T = 0. \tag{38}$$

Nontrivial (i.e., not identically zero) solutions of (35) satisfying the condition (36) exist only if

$$\lambda = \lambda_n = \left(\frac{\pi(2n+1)}{2l}\right)^2, \quad n = 0, 1, 2, \ldots,$$

and are then of the form

$$\tilde{X}_n(x) = \sqrt{\frac{2}{l}} \sin\sqrt{\lambda_n}x = \sqrt{\frac{2}{l}} \sin\left(\frac{\pi(2n+1)}{2l}x\right).$$

The tilde emphasizes the fact that the functions $\tilde{X}_n(x)$ are *normalized* on the segment $[0,l]$, i.e.,

$$\left(\tilde{X}_n(x), \tilde{X}_n(x)\right) = \int_0^l \tilde{X}_n^2(x)\,dx = 1.$$

11.11. Telegrapher's Equation

As always, $(f(x), g(x))$ denotes the *inner product* of functions $f(x)$ and $g(x)$ on the interval $[0, l]$.

The family of functions $\{\tilde{X}_n(x)\}$, $n = 0, 1, 2, \ldots$, forms an orthonormal system of the *eigenfunctions* of the boundary value problem (37). The corresponding numbers λ_n are called the *eigenvalues*. The graphs of the first three eigenfunctions (with $l = 1$) are shown in Fig. 11.11.9.

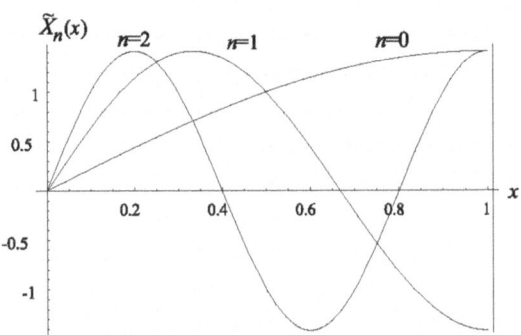

FIGURE 11.11.9
The first three eigenfunctions, $\tilde{X}_n(x)$, $n = 0, 1, 2, \ldots$, of the boundary value problem (37) ($l = 1$).

Substituting the above λ_n in (39), we obtain an equation for the corresponding functions $T_n(t)$,

$$\ddot{T}_n + (\lambda_n a^2 - b^2) T_n = 0 \ .$$

It is easy to see that under condition (33), $\lambda_n a^2 > b^2$ for all $n = 0, 1, 2, \ldots$, the general solution of the equation for $T_n(t)$ is of the form

$$T_n(t) = a_n \cos \omega_n t + b_n \sin \omega_n t \ ,$$

where

$$\omega_n = \sqrt{\lambda_n a^2 - b^2} \ .$$

Hence, the general solution of (35) satisfying boundary conditions (36a) is

$$u(x, t) = \sum_{n=0}^{\infty} (a_n \cos \omega_n t + b_n \sin \omega_n t) \tilde{X}_n(x) \ .$$

The coefficients a_n and b_n can be found from the initial conditions (36b):

$$u(x,0) = \frac{Q}{C}\delta(x-x_0) = \sum_{n=0}^{\infty} a_n \tilde{X}_n(x),$$

$$u_t(x,0) = b\frac{Q}{C}\delta(x-x_0) = \sum_{n=0}^{\infty} b_n \omega_n \tilde{X}_n(x).$$

The last two equalities give a series expansion of the functions u and u_t with respect to the orthonormal system of functions $\tilde{X}_n(x)$, with coefficients

$$a_n = \left(\frac{Q}{C}\delta(x-x_0), \tilde{X}_n(x)\right) = \frac{Q}{C}\tilde{X}_n(x_0),$$

$$b_n \omega_n = \left(b\frac{Q}{C}\delta(x-x_0), \tilde{X}_n(x)\right) = b\frac{Q}{C}\tilde{X}_n(x_0).$$

Consequently, the sought function $u(x,t)$ is given by

$$u(x,t) = \frac{Q}{C}\sum_{n=0}^{\infty}\left(\cos\omega_n t + \frac{b}{\omega_n}\sin\omega_n t\right)\tilde{X}_n(x_0)\tilde{X}_n(x).$$

Recalling that

$$\cos\omega_n t + \frac{b}{\omega_n}\sin\omega_n t = A_n \sin(\omega_n t - \varphi_n),$$

where

$$\tan\varphi_n = \frac{b}{\omega_n}, \quad A_n = \frac{\sqrt{\omega_n^2 + b^2}}{\omega_n},$$

we get another representation of the solution:

$$u(x,t) = \frac{Q}{C}\sum_{n=0}^{\infty} A_n \sin(\omega_n t - \varphi_n)\tilde{X}_n(x_0)\tilde{X}_n(x).$$

Finally, using formula (34), we can return from the function $u(x,t)$ to the voltage function $v(x,t)$ to obtain

$$v(x,t) = e^{-\mu t}\frac{2Q}{lC}\sum_{n=0}^{\infty} A_n \sin(\omega_n t - \varphi_n)\sin\frac{\pi(2n+1)}{2l}x_0 \sin\frac{\pi(2n+1)}{2l}x.$$

To discuss the physical meaning of the above solution we shall introduce, as usual, nondimensional variables and parameters. Mathematicians almost

11.11. Telegrapher's Equation

always make a tacit assumption that all quantities under consideration are nondimensional, but in physics, the choice of which quantities to express as nondimensional depends on the nature of the given problem and should be carefully considered. In our case, first of all, notice that without any loss of generality it is possible to take $l = 1$ and $a = 1$. This choice of spatial and temporal scale is adequate and most suitable for our analysis. Thus b becomes the only nondimensional parameter of the problem; any change in its value will affect the properties of the solutions. Moreover, to simplify our analysis we shall suppress the inessential factor in front of the sum. The damping effect caused by the exponential factor $e^{-\mu t}$ will be easy to reestablish at the very end. For the sake of concreteness, we shall also assume that $x_0 = 1/2$; in other words, the concentrated charge has been placed in the middle of the transmission line.

So at this point, all that remains is a discussion of the behavior of the auxiliary function

$$U(x,t) = \sum_{n=0}^{N} A_n \sin(\omega_n t - \varphi_n) \sin\frac{\pi}{4}(2n+1) \sin\frac{\pi}{2}(2n+1)x, \qquad (40)$$

where the nondimensional values are

$$\omega_n = \sqrt{\frac{\pi^2}{4}(2n+1)^2 - b^2} \qquad b < \frac{\pi}{2},$$

$$A_n = \sqrt{1 + \frac{b^2}{\omega_n^2}}, \qquad \varphi_n = \arctan\left(\frac{b}{\omega_n}\right).$$

In the expression for $U(x,t)$, only the first $N+1$ terms are retained, since it is our intention to study the formula numerically; Fig. 11.11.10 shows the function (40) for $N = 50$, $b = 1$, and $t = 0.3$. The bold arrow in the middle symbolizes the Dirac delta initial charge placement.

Note that the above graph does not really contradict our intuition, which would suggest that the initial Dirac delta impulse should generate narrow impulses propagating in both directions with velocity $a = 1$. The high peaks present in the figure are located in the vicinity of the points $x = 0.2$ and $x = 0.8$, whereas heuristically, one would expect them at $t = 0.3$. More challenging to explain, however, are the small oscillating "splashes" of amplitude an order of magnitude smaller than the main peaks. Their presence can be explained by the existence of dispersion in any long transmission line with $b \neq 0$. This dispersion distorts the profile of the propagating wave.

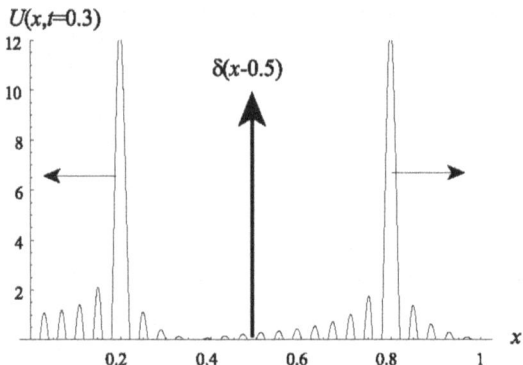

FIGURE 11.11.10
The auxiliary function $U(x,t)$ for $N = 50$, $b = 1$, and $t = 0.3$.

A further analysis shows that the small "splashes" have no physical meaning and are only an artifact of the cutoff procedure for the infinite Fourier series (i.e., its replacement by a finite sum). Indeed, the scale of the "splashes" is close to the period of the first neglected eigenfunction, which is of magnitude $4/103 \approx 0.039$. Such undesirable artifacts can reduce the effectiveness of the simple cutoff procedures for Fourier series, but can be repaired by the use of different *summation methods*; some of these were discussed in Volume 1.

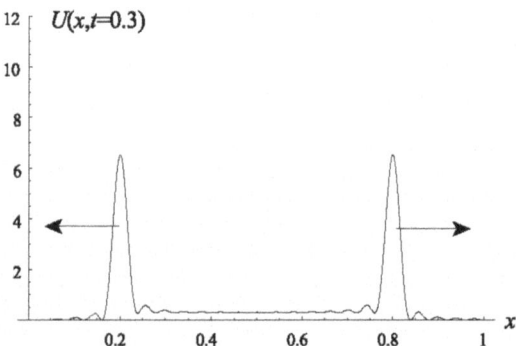

FIGURE 11.11.11
The result of Cesàro summation of the sum representing the auxiliary function $U(x,t)$ for $t = 0.3$, $b = 1$, and $N = 50$.

As it turns out, our series is divergent, and here we shall use the *Cesàro method*, which was introduced in Chap. 8, Volume 1. This method allows the summation of some divergent (in the classical sense) series and the acceleration of the convergence of some already convergent series; for example, the

11.11. Telegrapher's Equation

Cesàro sum of the divergent series $1 - 1 + 1 - 1 + 1 - \cdots$ turns out to be $1/2$. In our case, an application of the *Cesàro method* requires that we multiply the nth term of the finite sum $U(x,t)$ by $(N - n)/N$. The result of Cesàro summation, for $N = 50$, is shown in Fig. 11.11.11.

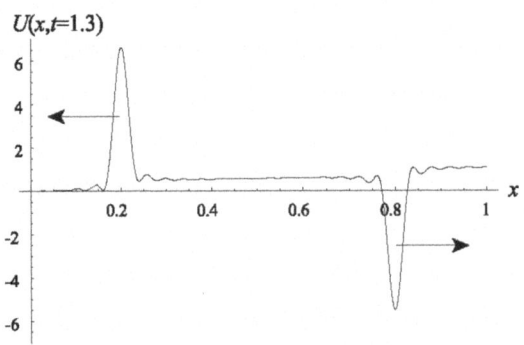

FIGURE 11.11.12
The result of Cesàro summation of the sum representing the auxiliary function $U(x,t)$ for $t = 1.3$, $b = 1$, and $N = 50$.

Cesàro summation thus washes out small "splashes," and in addition, eliminates the "forerunners" of the impulses present in Fig. 11.11.10 to the left of the leftward peak and to the right of the rightward peak. Physically, they are not supposed to be there, since the signal cannot propagate in the transmission line with a velocity greater than a.

Figure 11.11.12 shows the auxiliary function $U(x,t)$ at a later time, $t = 1.3$, obtained again via the Cesàro method, for $N = 50$. At first glance, it seems that the pulses did not move, but in reality, they have already had enough time to be reflected from the boundaries and exchange positions. Moreover, the right impulse changed its polarity after being reflected from the left boundary.

Until now we have investigated the spatial form of the wave at a fixed time instant t. Next, we shall take a look at its time dependence. For this purpose let us fix $x = 0.8$ and find the graph of $U(x = 0.8, t)$ as a function of t. Figure 11.11.13 shows this dependence in the case of the transmission line without distortion ($b = 0$); the graph is constructed using the Cesàro method applied to the sum containing 100 terms.

The shape of the graph is easily explained. Recall that in a distortionless ($b = 0$) transmission line, the Dirac delta impulse propagates from the middle, without changing its form, with velocity a. It changes its polarity as it

140 Chapter 11. Waves and Hyperbolic Equations

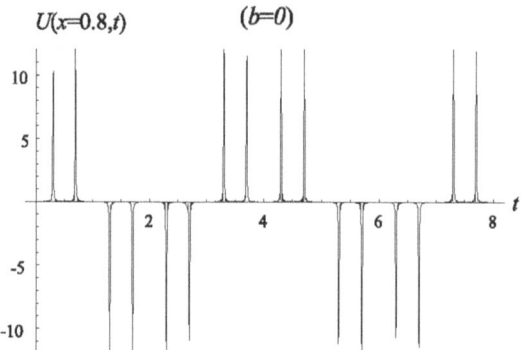

FIGURE 11.11.13
The time dependence of the auxiliary function $U(0.8, t)$ obtained via the Cesàro summation method applied to the sum containing 100 terms; here, the distortion parameter b is equal to zero.

reflects from the left (grounded) boundary, but the polarity does not change when the signal is reflected from the right (insulated) boundary.

Finally, Fig. 11.11.14 shows the analogous picture in the realistic case of a distortion parameter $b = 1$. In this case, dispersion is present but does not destroy the pulses themselves; it just adds trailing "echoes" to them, such as were also present in Figs. 11.11.11 and 11.11.12.

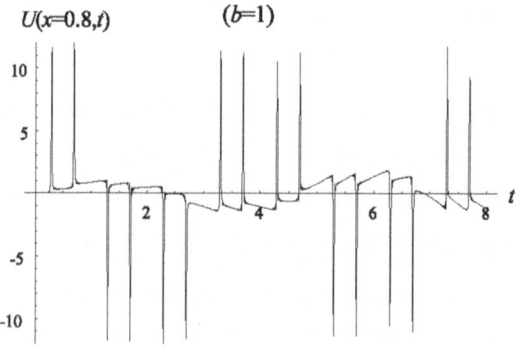

FIGURE 11.11.14
The time dependence of the auxiliary function $U(0.8, t)$ obtained via the Cesàro summation method applied to the sum containing 100 terms; here, the distortion parameter b is equal to 1.

11.12 Exercises

1. Recall that the wave packets (11.1.10) possess (11.3.3) as an invariant. Prove that they also have an infinite set of other invariants of the form

$$S = \int u^*(x,t) v(x,t)\, dx = const, \qquad (1)$$

where $v(x,t)$ is a linear functional of $u(x,t)$ determined by the formula

$$v(x,t) = \int h(s,\tau) u(x-s, t-\tau)\, ds\, d\tau,$$

and where $h(x,t)$ is a "response function."

2. The equation

$$\frac{\partial^2 u}{\partial t^2} + \omega_0^2 u = a^2 \frac{\partial^2 u}{\partial x^2} \qquad (2)$$

arises in various physical and engineering applications. For large-scale waves ($ak \ll \omega_0$), it is close to the harmonic oscillator equation, while for small-scale waves ($ak \gg \omega_0$), it is close to the wave equation. Find the phase and group velocities of the harmonic wave satisfying this equation.

3. Assume that the profile of a wave packet $u(x,t)$ propagating to the right has, at $t=0$, a jump of size $\lfloor f \rfloor$ at the point x_0. Assume that $u(x,t)$ is a solution of (2). What is the time evolution of the jump?

4. Determine voltage and current in an infinitely long distortionless transmission line ($CR = LG$) at an arbitrary time $t > 0$, given the initial distribution of the voltage and current in the line.

5. Find, for $t > 0$, current and voltage in a semi-infinite distortionless line if for $t < 0$, the line carried a wave of the form

$$v(x,t) = e^{-\frac{R}{L}t} f(x + at), \quad i(x,t) = -\sqrt{\frac{C}{L}} e^{-\frac{R}{L}t} f(x+at),$$

with $f(x) = 0$ for $x < 0$, and given the information that at $x = 0$, the line is grounded through a concentrated resistance R_0.

6. Solve the previous problem for a transmission line grounded through a concentrated inductance L_0.

7. A voltage source $E(t)$ is attached at the endpoint $x = 0$ of a semi-infinite $(x \geq 0)$ distortionless transmission line. Calculate the voltage distribution in the line for $t > 0$.

8. At the endpoint $x = 0$ of a semi-infinite distortionless transmission line, a voltage source E was applied long enough to establish the stationary distribution of voltage and current in the line. Then, at time $t = 0$, the endpoint was grounded through a concentrated resistance R_0. Find voltage and current in the line for $t > 0$.

9. Find the voltage distribution for $t > 0$ in a homogeneous transmission line, $0 < x < l$, if the initial voltage and current are zero, the left endpoint is insulated, and the right endpoint, $x = l$, is at time $t = 0$ connected to the voltage source, $E(t) = E_0 \sin \omega t$, $0 < t < +\infty$. Find the main asymptotics of the solution for $t \to +\infty$. As in Subsection 11.11.5, assume that condition (11.11.33) is satisfied.

Part IV

Nonlinear Partial Differential Equations

Chapter 12
First-Order Nonlinear PDEs and Conservation Laws

Linear partial differential equations discussed in Part III often offer only a very simplified description of physical phenomena. To get a deeper understanding of some of them, it is necessary to move beyond the linear "universe" and consider nonlinear models, which in the case of continuous media, means nonlinear partial differential equations. Even today, their theory is far from complete and is the subject of intense study. On closer inspection, almost all physical phenomena in continuous media—from growing molecular interfaces at atomic scales to the structure of the distribution of matter in the universe at intergalactic scales—are nonlinear. The variety of nonlinear physical phenomena necessitates the use of various mathematical models and techniques to study them. In this part we shall restrict our attention to nonlinear waves of hydrodynamic type in media with weak or no dispersion. Since weak dispersion has little influence on the development of many nonlinear effects, we shall have a chance to observe typical behavior of these systems in strongly nonlinear regimes. The basic features of strongly nonlinear fields and waves are already evident in solutions of first-order nonlinear partial differential equations, and we take them as our starting point.

12.1 Riemann Equation

We have already encountered the Riemann equation on several occasions: in discussing flows of noninteracting particles, analyzing optical waves behind a phase screen, and studying wave packets in dispersive media. Here, we return to this equation once again and develop its theory in a systematic fashion.

12.1.1 The Canonical Form of the Riemann Equation

The first-order partial differential equation

$$\frac{\partial u}{\partial t} + C(u)\frac{\partial u}{\partial x} = 0, \qquad u = u(x,t), \tag{1}$$

is called the *Riemann equation*. We shall assume that the function $C(u)$ is continuously differentiable as a function of u. Multiplying both terms of the equation (1) by $C'(u)$, applying the chain rule, and making the substitution $v(x,t) = C(u(x,t))$, we can reduce equation (1) to the following equation for an unknown function v,

$$\frac{\partial v}{\partial t} + v\frac{\partial v}{\partial x} = 0, \tag{2}$$

which we shall call the *canonical Riemann equation*. Thus, without any loss of generality, our further discussion of the Riemann equation can be restricted to the equation (2). Typically, we shall look for solutions satisfying the initial condition

$$v(x, t=0) = v_0(x), \tag{2a}$$

where $v_0(x)$ is a given function of the variable x.

To better understand the construction of solutions of the Riemann equation and its typical features, we shall recall some key hydrodynamic concepts such as the Lagrangian and Eulerian coordinates (see Volume 1, Chap. 2). Providing the following vivid mechanical interpretation of the equation should also help the reader to develop correct intuitions for the model.

12.1.2 Uniform Flow of Particles

Consider a flow of particles, each moving with a uniform velocity along the x-axis. Assume that at the initial time instant $t = 0$, a particle with coordinate $x = y$ moves with velocity $v_0(y)$. Then the subsequent particle motion is described by the equations

$$X(y,t) = y + v_0(y)t, \qquad V(y,t) = v_0(y). \tag{3}$$

Varying y, we shall obtain the laws of motion of all the particles in the flow. Note that in addition to the time variable t, there appears here another variable, y, which represents the particle coordinate at the initial time instant $t = 0$. This coordinate, rigidly connected to each particle of the flow, is called the *Lagrangian coordinate* of the particle.

On the other hand, in many physical applications, the observer is more likely to describe the flow by measuring its velocity at a certain selected

12.1. Riemann Equation

point in space with a given coordinate, say x, which we shall call the *Eulerian coordinate*. The transformation of Lagrangian coordinates into Eulerian coordinates is described by the equality

$$x = X(y,t). \tag{4}$$

In the case of uniform particle motion discussed here,

$$x = y + v_0(y)t. \tag{5}$$

Suppose that the particle velocity field $v(x,t)$ is known as a function of the Eulerian coordinate x and the time t. If in addition, we know the relationship (4) which determines the transformation of Lagrangian coordinates into Eulerian coordinates, then we can express the velocity field in terms of the Lagrangian coordinates via the formula

$$V(y,t) = v(X(y,t),t). \tag{6}$$

In what follows, the fields describing the behavior of particles in the Lagrangian coordinate system will be called *Lagrangian fields*, and the fields describing the motion of particles in the Eulerian coordinate system will be called *Eulerian fields*. So, $v(x,t)$ is the Eulerian field of the particles' velocity, but $X(y,t)$ is the Lagrangian field of the Eulerian coordinates of the particles. Since the motion of particles is assumed to be uniform, the velocity $V(y,t)$ of a particle with a given Lagrangian coordinate y does not depend on the time, and is thus described by the simple differential equation

$$\frac{dV}{dt} = 0, \tag{7}$$

while its coordinate satisfies an equally simple equation

$$\frac{dX}{dt} = V. \tag{8}$$

Equations (7) and (8) are immediately recognizable as characteristic equations for the first-order partial differential equation (2), and solutions of the Riemann equation can be obtained from solutions of the equations (7) and (8) as soon as we find the inverse function to the function (4),

$$y = y(x,t),$$

displaying the Lagrangian coordinates in terms of the Eulerian coordinates. In this case, the Lagrangian laws of motion (3) yield the following solution of the Riemann equation:

$$v(x,t) = V(y(x,t),t) = v_0(y(x,t)). \tag{9}$$

We emphasize that the existence of an unambiguous inverse function $y(x,t)$, and thus the possibility of a construction via the formula (9) (for a smooth initial field $v_0(x)$) for a classical solution of the Riemann equation (2) depends on the transformation (4)–(5) from the Lagrangian to the Eulerian coordinates being a strictly increasing function of y, mapping **R** onto **R**; we shall make this assumption throughout this chapter. In the following chapter we shall analyze a more complex situation when the function $X(y,t)$ is no longer monotonically increasing.

12.1.3 Classical Solutions of the Riemann Equation

In this subsection we shall describe typical features of the classical solutions $v(x,t)$ of the Riemann equation as functions of time t and the Eulerian spatial coordinate x. Let us begin with listing several basic ways of writing out solutions of the Riemann equation that are used in different application areas.

Replacing the variable y in equation (5) (describing the uniform particle motion) by $y(x,t)$,

$$y(x,t) = x - v_0(y(x,t))t \quad \Rightarrow \quad y(x,t) = x - v(x,t)t, \tag{10}$$

and substituting the right-hand side of the resulting expression into (9), we arrive at the solution of the Riemann equation in an implicit form,

$$v(x,t) = v_0(x - v(x,t)t). \tag{11}$$

On the other hand, the second equality in (10) gives the solution in the form

$$v(x,t) = \frac{x - y(x,t)}{t}. \tag{12}$$

The physical meaning of (12) is clear: the velocity v of a particle in uniform motion is equal to the distance $x - y$ traveled during time t, divided by that time. In what follows, we shall find a deeper physical and geometric meaning of (12).

If one needs to plot a solution of the Riemann equation, it is most convenient to rely on the Lagrangian field and construct $v(x,t)$ as a parametric family of curves given by the equations

$$x = y + v_0(y)t, \quad v = v_0(y). \tag{13}$$

12.1. Riemann Equation

Figure 12.1.1 shows a plot of the solution of the Riemann equation in the case that the initial profile (dashed curve) of the velocity field is represented by a Gaussian function

$$v_0(x) = V_0 \exp\left(-\frac{x^2}{2\ell^2}\right). \qquad (14)$$

The graph uses nondimensional variables

$$z = \frac{x}{\ell}, \qquad \tau = \frac{V_0}{\ell} t, \qquad (15)$$

and the solid curve represents the solution at time $\tau = 1$. The arrows show displacement of particles up to time τ. One can observe that a greater particle velocity causes a larger displacement from the original position. As a result, the right front of the field $v(x, t)$ steepens as particles bunch up in trying to overtake the particles positioned to their right. The left front flattens out, and the particles located there spread out.

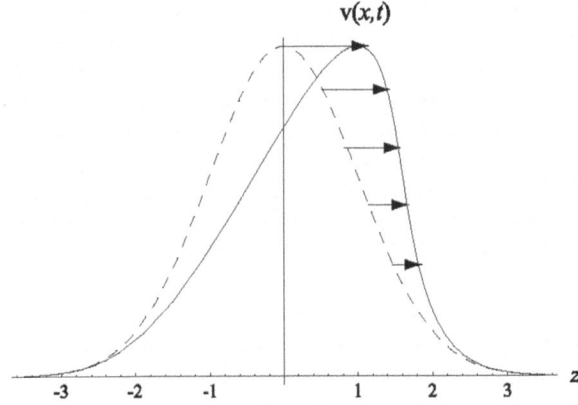

FIGURE 12.1.1
A solution of the Riemann equation for a Gaussian initial condition.

12.1.4 Compression and Rarefaction of Particle Flows

Observe that the steepening of the right-hand side of the velocity profile $v(x, t)$ in Fig. 12.1.1 is accompanied by an increase of the particle density in the flow, an important physical effect. Indeed, as we observed before, in this portion of the flow the particles positioned to the left have a greater velocity than the particles to the right, and the former catch up with the latter in the course of time. By contrast, the decreasing slope of the left portion of the velocity profile $v(x,t)$ corresponds to the rarefaction in the particle density. Quantitatively, the magnitude of the compression and rarefaction effects is measured by the Jacobian of the transformation that maps Lagrangian to Eulerian coordinates,

$$J(y,t) = \frac{\partial X(y,t)}{\partial y}. \tag{16}$$

For a uniform flow of particles (5), the Jacobian is given by

$$J(y,t) = 1 + v_0'(y)t. \tag{17}$$

For a given Lagrangian coordinate y, the larger the Jacobian is, the larger are the rarefaction effects seen in a neighborhood of y. For this reason, the field $J(y,t)$ is often called the *divergence field* of the particle flow. A graph of the flow $X(y,t)$ and its divergence field corresponding to the flow of particles with Eulerian velocity field satisfying the Riemann equation (2), with initial condition (14), is shown in Fig. 12.1.2.

The field $J(y,t)$ described by (16) is a Lagrangian divergence field. The corresponding Eulerian divergence field is obviously

$$j(x,t) = J(y(x,t),t) \iff J(y,t) = j(X(y,t),t). \tag{18}$$

If the mapping $y(x,t)$ of Eulerian coordinates into Lagrangian coordinates is known, then the divergence field can be determined via the following, geometrically obvious, relationship:

$$\frac{\partial y(x,t)}{\partial x} = \frac{1}{j(x,t)}. \tag{19}$$

12.2 Continuity Equation

12.2.1 Evolution of the Particle Density

If we stay with the hydrodynamic interpretation of solutions of the Riemann equation as the velocity field of the flow of uniformly moving particles, then it is natural to ask about the temporal evolution of other features of the flow such as the particle density $\rho(x,t)$. Principles of hydrodynamics (see also Volume 1, Chap. 1) tell us that the density field for any type of particle flow must satisfy the continuity equation that expresses the physical law of mass conservation. Here we derive the continuity equation once more, utilizing a

12.2. Continuity Equation

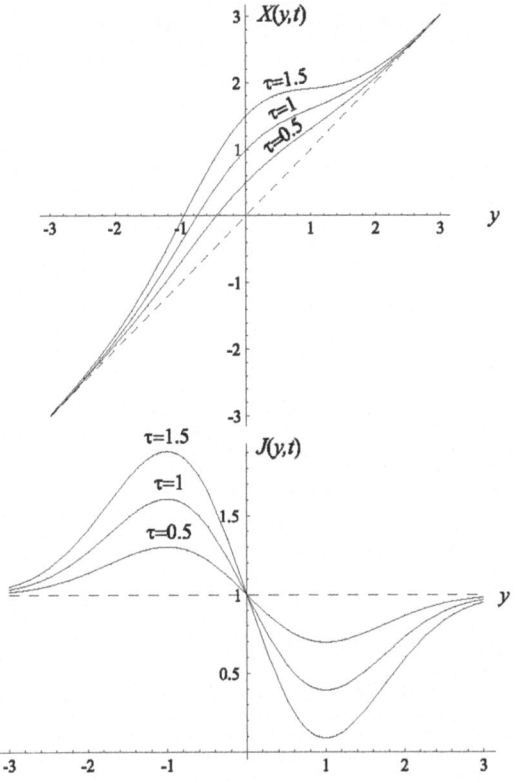

FIGURE 12.1.2
Graph of the flow $X(y,t)$ (*top*), and its divergence field (*bottom*), corresponding to a flow of uniformly moving particles, with Eulerian velocity field satisfying the Riemann equation (2), with initial condition (14). For values of $J(y,t)$ greater (resp. smaller) than 1, the particle density decreases (resp. increases).

method that will permit a deeper understanding of the analysis of nonlinear partial differential equations to be carried out in what follows.

Assume that the initial density distribution $\rho_0(x)$ is such that the *cumulative mass field*

$$m_0(x) = \int_{-\infty}^{x} \rho_0(z)\,dz$$

of particles located to the left of an arbitrary point x is finite. The current mass

$$m(x,t) = \int_{-\infty}^{x} \rho(z,t)\,dz$$

of particles located to the left of the point x at time t is an Eulerian field. If the law of motion $X(y,t)$ is known, then the Lagrangian mass evolution is easily expressed via the Eulerian field

$$M(y,t) = m(X(y,t), t).$$

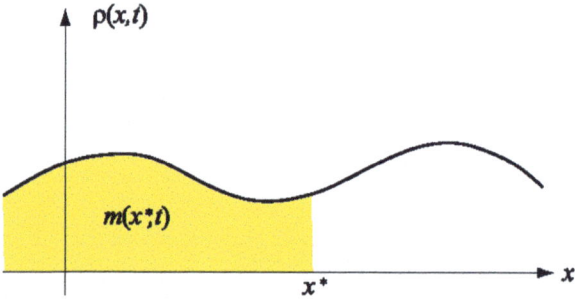

FIGURE 12.2.1
The Eulerian field $m(x^*, t)$ representing the mass of particles of the flow to the left of the point x^*.

The latter can be easily found from elementary physical considerations (Fig. 12.2.1). Indeed, if at time t, particles did not change their order, the mass to the left of the point with Lagrangian coordinate y is

$$M(y,t) = \int_{-\infty}^{y} \rho_0(z) \, dz = m_0(y), \tag{1}$$

and it does not vary in time. In other words, the Lagrangian field of the mass on the left satisfies the equation

$$\frac{dM}{dt} = 0.$$

The equivalent equation for the corresponding Eulerian field $m(x,t)$ is

$$\frac{\partial m}{\partial t} + v \frac{\partial m}{\partial x} = 0. \tag{2}$$

Now we are ready to define the particle density. In the 1-D case under consideration, the density field is simply the derivative of the mass on the left with respect to x,

$$\rho(x,t) = \frac{\partial m(x,t)}{\partial x}. \tag{3}$$

12.2. Continuity Equation

Consequently, differentiating the equation (3) termwise with respect to x, we arrive at the sought 1-D version of the continuity equation,

$$\frac{\partial \rho}{\partial t} + \frac{\partial}{\partial x}(v\rho) = 0. \tag{4}$$

Remark 1. Note that in contrast to the Riemann equation for the velocity field of uniformly moving particles, our derivation of the continuity equation never made use of the fact that the motion was uniform. So, the continuity equation expresses a universal law, valid for any particle motion.

12.2.2 Construction of the Density Field

To find a solution of the continuity equation (4), let us initially write down the Eulerian cumulative mass field. It follows from (12.1.1) and the relationship between the Eulerian and Lagrangian coordinates that

$$M(y,t) = m_0(y) \quad \Longleftrightarrow \quad m(x,t) = m_0(y(x,t)). \tag{5}$$

Differentiating the last equality with respect to x, we obtain

$$\rho(x,t) = \rho_0(y(x,t)) \frac{\partial y(x,t)}{\partial x}, \tag{6}$$

or taking into account (12.1.19),

$$\rho(x,t) = \frac{\rho_0(y(x,t))}{j(x,t)} \quad \Longleftrightarrow \quad R(y,t) = \frac{\rho_0(y)}{J(y,t)}. \tag{7}$$

The last relationship has an obvious geometric meaning: the flow's density at a given point is equal to the initial density in the vicinity of the particle's initial location at $t = 0$, divided by the degree of compression (or rarefaction) of the particle flow.

Separately, consider the density field for the flow of uniformly moving particles in which the velocity field $v(x,t)$ satisfies the Riemann equation, and $y(x,t)$ is expressed by the relation (12.1.10). In this case, in view of (6) and (12.1.10), the flow's density is expressed in terms of the solution of the Riemann equation as follows:

$$\rho(x,t) = \rho_0\left(x - v(x,t)t\right)\left(1 - \frac{\partial v(x,t)}{\partial x} t\right). \tag{8}$$

In particular, under the assumption of uniform initial flow density $\rho_0 = const$, which does not depend on x, the density is described by the expression

$$\rho(x,t) = \rho_0 \left(1 - \frac{\partial v(x,t)}{\partial x} t\right), \qquad (9)$$

which directly demonstrates the close relationship between the flow's density and the steepening of its velocity profile.

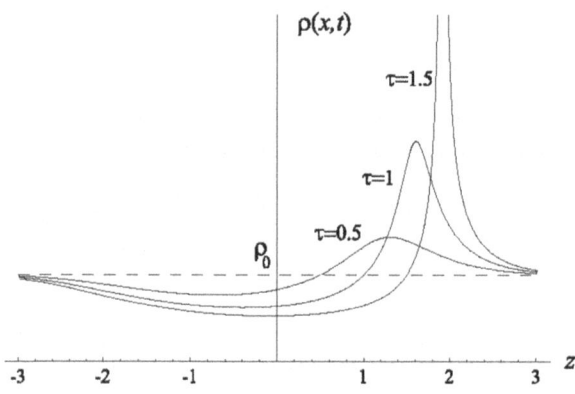

FIGURE 12.2.2
Graphs of the evolving density in the flow of uniformly moving particles for a Gaussian initial velocity field (12.1.14) and constant initial density field $\rho(x, t = 0) = \rho_0 = const$.

If we need to draw a graph of the density field $\rho(x,t)$ of the flow of uniformly moving particles at time t, then, as in the case of the velocity field, it is convenient to describe the curve $\rho(x,t)$ parametrically, utilizing the fact that the Lagrangian laws of flow evolution are given explicitly via the formulas

$$x = y + v_0(y)t, \qquad \rho = \frac{\rho_0(y)}{1 + v_0'(y)t}. \qquad (10)$$

Graphs of the evolving density field for a Gaussian initial velocity field (12.1.14) and constant initial density field $\rho_0(x) = \rho_0 = const$ are shown in Fig. 12.2.2.

12.2.3 Momentum Conservation Law

In addition to the obvious mass conservation law, the flow of uniformly moving particles obeys an infinite number of other conservation laws. However,

12.2. Continuity Equation

only some of them, such as the energy and momentum conservation laws, have a clearcut physical significance; in the present subsection we shall briefly discuss the latter. Recall that the cumulative momentum function at time t is determined by the formula

$$p(x,t) = \int_{-\infty}^{x} v(x,t)\rho(x,t)\, dx.$$

Substituting here the expressions (12.1.9) and (6) for the velocity and density in the flow of uniformly moving particles and then passing to the Lagrangian coordinates in the integral, we obtain that the cumulative momentum is given by

$$p(x,t) = \int_{-\infty}^{x} v_0(y(x,t))\rho_0(x(y,t))\frac{\partial y(x,t)}{\partial x}dx = \int_{-\infty}^{y(x,t)} v_0(y)\rho_0(y)\, dy.$$

It follows from the above chain of relations that

$$P(y,t) = p_0(y) = \int_{-\infty}^{y} \rho_0(y)v_0(y)\, dy,$$

that is, in Lagrangian coordinates, the cumulative momentum remains constant in time. Consequently, the Eulerian cumulative momentum satisfies the equation

$$\frac{\partial p}{\partial t} + v\frac{\partial p}{\partial x} = 0,$$

while the momentum's density

$$g(x,t) = \rho(x,t)v(x,t) = \frac{\partial p(x,t)}{\partial x}$$

satisfies the continuity equation

$$\frac{\partial g}{\partial t} + \frac{\partial}{\partial x}(vg) = 0.$$

Remark 2. The last equation could be derived directly as a consequence of the Riemann, (12.1.2), and the continuity, (4), *differential* equations. However, we deliberately selected a more circuitous, "integral" route, since later on, it will permit us to construct generalized solutions of these equations.

Remark 3. Since the momentum density satisfies the same equation as the mass density function $\rho(x,t)$, substituting in (7) the initial momentum in place of the initial density, we immediately obtain expressions for the Eulerian and Lagrangian momentum density fields,

$$g(x,t) = \frac{\rho_0(y(x,t))v_0(y(x,t))}{j(x,t)} \quad \Longleftrightarrow \quad G(y,t) = \frac{\rho_0(y)v_0(y)}{J(y,t)}. \tag{11}$$

12.2.4 Fourier Images of Density and Velocity Fields

In many physical and engineering applications, the important objects are not the fields themselves but their spectra. For this reason, we shall derive explicit expressions for the spatial Fourier images of the velocity field $v(x,t)$ and the density field $\rho(x,t)$ that are convenient for computation. We shall begin with calculation of the Fourier image

$$\tilde{\rho}(\kappa,t) = \frac{1}{2\pi} \int e^{-i\kappa x} \rho(x,t)\,dx \tag{12}$$

of the density field. For this purpose, let us substitute in (12) the solution of the continuity equation (6) to obtain

$$\tilde{\rho}(\kappa,t) = \frac{1}{2\pi} \int e^{-i\kappa x} \rho_0(y(x,t))\,dy(x,t)\,.$$

Switching to integration with respect to the Lagrangian coordinates, we finally arrive at the formula

$$\tilde{\rho}(\kappa,t) = \frac{1}{2\pi} \int e^{-i\kappa X(y,t)} \rho_0(y)\,dy\,. \tag{13}$$

Similar but slightly more complex calculations yield a formula for the Fourier image of the velocity field,

$$\tilde{v}(\kappa,t) = \frac{i}{2\pi kt} \int \left[e^{-ikX(y,t)} - e^{-iky} \right] dy\,. \tag{14}$$

Example 1. Harmonic initial velocity field. Formulas (13) and (14) are remarkable because they express Fourier images of implicitly defined (for example, by equality (12.1.11)) functions $\rho(x,t)$ and $v(x,t)$ through integrals of explicitly given integrands. This fact gives us an opportunity to find an explicit expression for the density field $\rho(x,t)$ in the case that the initial

12.2. Continuity Equation

velocity field is given by a simple harmonic function and the initial density field is constant,

$$v_0(x) = a\sin(kx), \qquad \rho_0(x) = \rho_0 = const. \tag{15}$$

In our calculations we shall have need of the following well-known formula from the theory of Bessel functions:

$$e^{iw\sin z} = \sum_{n=-\infty}^{\infty} J_n(w)\, e^{inz}. \tag{16}$$

In the case under discussion, the mapping of Lagrangian into Eulerian coordinates is given by the formula

$$x = X(y,t) = y + at\,\sin(ky). \tag{17}$$

Substituting it into the formula (13) for the Fourier image of the density field gives

$$\tilde{\rho}(\kappa,t) = \frac{\rho_0}{2\pi k}\int e^{-i\mu z - i\mu\tau\sin z}\,dz,$$

where we introduced the nondimensional variables of integration $z = ky$, time $\tau = kat$, and spatial frequency $\mu = \kappa/k$. An application of the formula (16) gives

$$\tilde{\rho}(\kappa,t) = \frac{\rho_0}{2\pi k} \sum_{n=-\infty}^{\infty} J_n(-\mu\tau) \int e^{-i(\mu-n)z}\,dz.$$

In view of the distributional formula (3), of Sect. 3.3, Volume 1, we have

$$\frac{1}{2\pi k}\int e^{-i(\mu-n)z}\,dz = \frac{1}{k}\delta(\mu - n) = \delta(\kappa - kn).$$

So, the generalized Fourier image is given by

$$\tilde{\rho}(\kappa,t) = \rho_0 \sum_{n=-\infty}^{\infty} J_n(-n\tau)\delta(\kappa - kn).$$

Substituting this expression into the inverse Fourier integral

$$\rho(x,t) = \int \tilde{\rho}(\kappa,t)\,e^{i\kappa x}\,d\kappa$$

and taking into account the symmetry properties of Bessel functions

$$J_n(-w) = (-1)^n J_n(w),\; J_{-n}(w) = (-1)^n J_n(w) \Rightarrow J_{-n}(-w) = J_n(w), \tag{18}$$

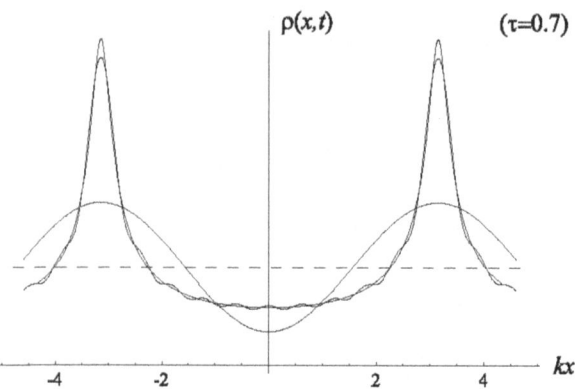

FIGURE 12.2.3
A comparison of the sum of the first two and the first eleven terms of the series (19) with the exact expression for the density field obtained parametrically using the formula (10). Time $\tau = akt$ equals 0.7, the initial velocity is assumed to be harmonic (15), and the initial density, constant.

we arrive at the formula for the density field expressed as a Fourier series:

$$\rho(x,t) = \rho_0 + 2\rho_0 \sum_{n=1}^{\infty} (-1)^n J_n(n\tau) \cos(kx). \qquad (19)$$

A comparison of the sum of the first few terms of (19) with the exact form of the density field obtained parametrically using formula (10) is given in Fig. 12.2.3. It shows that even for moderate values of τ, the agreement is quite good.

12.3 Interface Growth Equation

In this section we shall discuss the time evolution of a growing interface between two spatial regions. The interface itself can have many concrete physical interpretations such as the surface of a growing thin semiconductor film in a chemical vapor deposition reactor, the surface of a shock wave generated by a supersonic aircraft, or even the boundary line of a spreading forest fire; it is the latter case that will be considered in some detail in the next subsection. Evolution of these curves and surfaces will be described via nonlinear partial differential equations.

12.3.1 Spreading Forest Fires

To describe mathematically the spreading of a forest fire, let us consider a planar forest surface with the Cartesian coordinate system (x, z). The z-axis will be considered the main direction of advance of the fire front, with the moving front itself described by the function (see Fig. 12.3.1)

$$z = h(x, t). \tag{1}$$

A natural assumption is that locally, the fire spreads in a direction perpendicular to the fire front $h(x, t)$, say at speed c. The last statement means that if we consider a point $\{y, h(y, t = 0)\}$ located at the fire front at time $t = 0$ and track its motion on the paths perpendicular to the fire front, then its velocity will be c. Let us write a parametric description of the position of such a point at time t in the form $(X(t), Z(t))$. Its trajectory $\{(X(t), Z(t)), t > 0\}$ will be called a *ray*.

The above discussion leads to the conclusion that the ray coordinates must satisfy the equations

$$\frac{dX}{dt} = c \sin \theta, \qquad \frac{dZ}{dt} = c \cos \theta, \tag{2}$$

where θ is the angle between the ray and the z-axis.

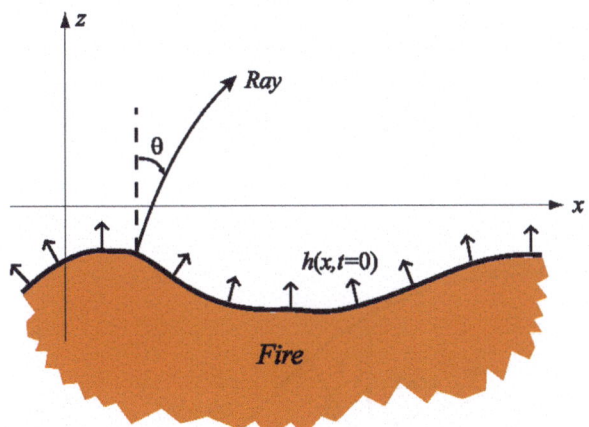

FIGURE 12.3.1
A schematic illustration of a moving forest fire front.

Furthermore, observe that the coordinate $Z(t)$ can be expressed via the previously introduced fire front function (1) as follows:

$$Z(t) = h(X(t), t). \tag{3}$$

Substituting this equality into the second equation in (2), we get

$$\frac{\partial h}{\partial t} + \frac{dX}{dt}\frac{\partial h}{\partial x} = c\cos\theta.$$

Taking into account the first equation in (2), we arrive at the following partial differential equation for the unknown function $h(x,t)$:

$$\frac{\partial h}{\partial t} + c\sin\theta\,\frac{\partial h}{\partial x} = c\cos\theta. \tag{4}$$

At first sight, this equation is not closed, because it also contains a second unknown function $\theta(x,t)$. However, it is not difficult to close this equation using the obvious geometric relationship between the fire front function $h(x,t)$ and the angle θ:

$$\frac{\partial h}{\partial x} = -\tan\theta. \tag{5}$$

Taking (5) into account, equation (4) can now be rewritten in the form

$$\frac{\partial h}{\partial t} = \frac{c}{\cos\theta}. \tag{6}$$

Finally, since $\cos\theta = 1/\sqrt{1+\tan^2\theta}$, we arrive at the desired nonlinear first-order partial differential equation,

$$\frac{\partial h}{\partial t} = c\sqrt{1 + \left(\frac{\partial h}{\partial x}\right)^2}. \tag{7}$$

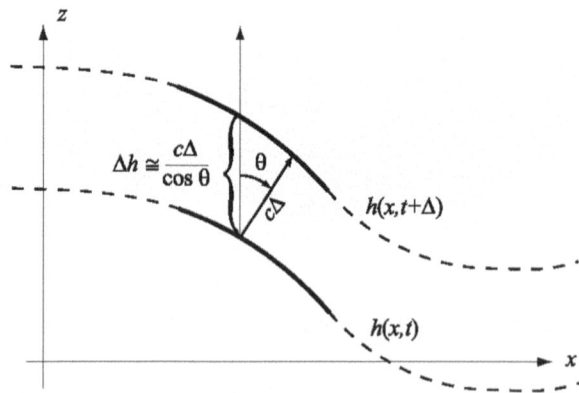

FIGURE 12.3.2
A geometric illustration of the validity of equation (7).

12.3. Interface Growth Equation

Remark 1. Equation (6) seems to violate a common geometric sense. Indeed, intuitively, the greater the deviation of the normal to the curve $h(x,t)$ from the z-axis, that is, the greater the angle between the z-axis and the direction of front's growth, the more slowly the front should grow in the direction of the z-axis. But equation (6) implies that for larger θ, the front line grows faster, and for $\theta = \pi/2$, the growth' speed is infinite. Nevertheless, a more detailed geometric analysis shows that equation (6) is valid. The geometric construction shown in Fig. 12.3.2 displays a fragment of the front line $h(x,t)$ at two time instants that differ by a small Δ. The illustration clearly indicates that the increment

$$\Delta h = h(x, t+\Delta) - h(x,t) \approx \frac{c\Delta}{\cos \theta}$$

of the elevation of the front line at an arbitrary point x is proportional to the inverse of $\cos\theta$. Perhaps, for some readers, this geometric picture will be more convincing than the preceding rigorous derivation of the equations (6)–(7).

Remark 2. Let us also note that if the velocity in equation (7) is negative ($c < 0$), then the equation describes a receding front such as, for instance, the surface of a melting ice cube in a glass of Coke, or the corroding surface of ship's hull.

Remark 3. The parametric curve $\{(X(t), Z(t)), t > 0\}$ perpendicular to the moving front line $h(x,t)$ was called a ray for good reason. The wave front of an optical wave propagates according to the same principle as a forest fire: the motion is perpendicular to the front line, and the velocity is constant. In that case, the lines remaining perpendicular to the optical wave front surfaces are called the *optical rays*. So, equation (7) represents a 2-D version of the equation describing the evolution of the optical wave front.

Remark 4. Optical waves usually propagate in a certain preferred direction, or at angles that deviate from that direction by a small amount. If we select the z-axis as the main propagation direction, then the angles between the rays and the z-axis are small ($\theta \ll 1$), and

$$-\frac{\partial h}{\partial x} = \tan\theta \approx \theta, \qquad \sqrt{1+\tan^2\theta} \approx 1 + \frac{1}{2}\left(\frac{\partial h}{\partial x}\right)^2.$$

In this case, instead of equation (7), one can use a simpler approximate equation,

$$\frac{\partial h}{\partial t} = c + \frac{c}{2}\left(\frac{\partial h}{\partial x}\right)^2.$$

For a planar wave propagating along the z-axis, this equation has a very simple solution, $h = ct$. If we are interested only in the form of the wave front but not in its absolute position, then we can introduce the "comoving" coordinate system

$$w(x,t) = h(x,t) - ct. \tag{8}$$

The new function $w(x,t)$ satisfies a more elegant equation,

$$\frac{\partial w}{\partial t} = \frac{c}{2}\left(\frac{\partial w}{\partial x}\right)^2. \tag{9}$$

It is a special case of the so-called *Hamilton–Jacobi* equation.

12.3.2 Anisotropic Surface Growth

The subsequent discussion will be more transparent if we introduce the *slope (gradient) field*

$$u(x,t) = -\frac{\partial h(x,t)}{\partial x}. \tag{10}$$

Obviously, $u = \tan\theta$. Sometimes the evolving surface grows with different speeds in different directions. This is the case, for example, for optical waves in an anisotropic medium, where the propagation speed depends on the direction. Another example is provided by the process of melting of a mountain snow cover, where the rate of melting depends on the angle of the snow-covered surface to the incoming sun rays. To take these types of anisotropic effects into account, we will assume that the velocity of front propagation depends on the gradient u, and replace equation (6) by the more general equation

$$\frac{\partial h}{\partial t} = \Phi(u), \qquad h(x,t=0) = h_0(x), \tag{11}$$

where

$$\Phi(u) = c(u)\sqrt{1+u^2}. \tag{12}$$

In the remainder of this subsection we shall discuss two examples of anisotropically growing surfaces.

Example 1. Snowfall in absence of wind. Consider snow vertically falling to the ground in windless weather. An elementary infinitesimal surface area

12.3. Interface Growth Equation

ds positioned at an angle θ to the zenith receives an amount of snowfall proportional to $\cos\theta\, ds$. Consequently, the velocity of the growing snow surface in the direction θ is determined by the equality

$$c(u) = c\cos\theta = \frac{c}{\sqrt{1+u^2}}. \tag{13}$$

Substituting this expression into (12), we obtain $\Phi = c = const$, so that the equation (11) of the growing surface has the trivial solution

$$h(x, z) = h_0(x) + ct, \tag{14}$$

and the topography of the growing snow cover does not change in time; only its elevation increases uniformly in time.

Example 2. Adsorption of particles with a nontrivial distribution of incident angles. Next, let us consider the more general situation of a surface growing by adsorption of particles hitting it from different directions. Let $D(\theta)$ represent the angle-dependent distribution of the intensity of a stream of incident particles, with θ, as usual, the angle between the particle's direction and the z-axis. Then the full intensity of particles falling from any direction on a unit patch of the growing surface is

$$c(\theta) = \int_{\theta_-}^{\theta_+} D(\theta')\cos(\theta - \theta')\, d\theta'. \tag{15}$$

In particular, the case of vertically falling particles considered in Example 1 corresponds to the singular angle-dependent intensity diagram

$$D(\theta) = c\delta(\theta).$$

Let us note that the limits in the integral (15) have to take into account effects of shadowing of the incident stream of particles by the growing interface $h(x,t)$ and that they depend on the concrete topography of the interface $h(x,t)$. As a concrete example, consider a segment of the interface that forms an angle $\theta > 0$ with the z-axis. In this case, clearly (see Fig. 12.3.3),

$$\min\theta_- = \theta - \frac{\pi}{2}, \quad \max\theta_+ = \frac{\pi}{2}.$$

The asymmetry of the minimal and maximal angles for the directional diagram is due to the fact that the local geometry ($\theta > 0$) of the interface

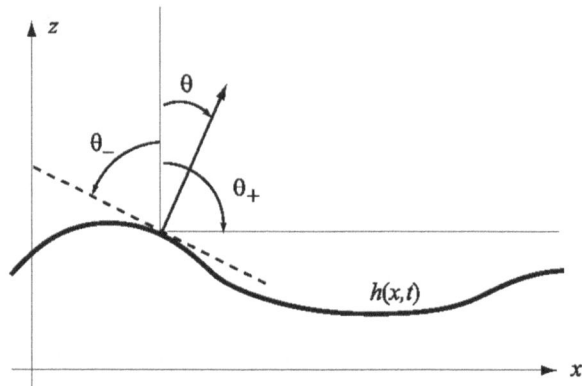

FIGURE 12.3.3
A schematic illustration of the process of determining limits in the integral (15).

segment makes it impossible for the particle to be deposited on the interface at an angle smaller than $\min \theta_-$. A similar argument can be made for $\max \theta_+$.

In the case of an isotropic particle flow $D = const$, taking into account the shadowing effects of the interface $h(x,t)$ produces the following formula for the velocity of interface growth:

$$c(\theta) = D \int_{\theta-\pi/2}^{\pi/2} \cos(\theta' - \theta)\, d\theta' = D(1 + \cos\theta). \tag{16}$$

Obviously, the function $c(\theta)$ is even. This means that the formula (16) remains valid for both $\theta > 0$ and $\theta < 0$.

12.3.3 Solution of the Interface Growth Equation

In this subsection we shall solve the equation (11) via the method of characteristics described in Sect. 2.6, Volume 1. To begin, let us find the derivative of (11) with respect to x. As a result, we obtain a closed Riemann equation (10) for the gradient field $u(x,t)$,

$$\frac{\partial u}{\partial t} + C(u)\frac{\partial u}{\partial x} = 0, \tag{17}$$

where

$$C(u) = \frac{d\Phi(u)}{du} = \frac{d}{du}\left[c(u)\sqrt{1+u^2}\right]. \tag{18}$$

12.3. Interface Growth Equation

Now, for the system of equations (11) and (17), the corresponding characteristic equations are

$$\frac{d\tilde{X}}{dt} = C(U), \quad \frac{dU}{dt} = 0, \quad \frac{dH}{dt} = \Lambda(U), \tag{19}$$

where the fields along the characteristic curves are denoted by capital letters, and

$$\Lambda(u) = \Phi(u) - uC(u) = -u^2 \frac{d}{du}\left(\frac{\Phi(u)}{u}\right). \tag{20}$$

The auxiliary function $\tilde{X}(y,t)$ introduced above should not be confused with the function $X(t)$, which must satisfy the first equation in (2),

$$\frac{dX}{dt} = V(u), \quad V(u) = c\sin\theta = \frac{u\,c(u)}{\sqrt{1+u^2}}, \tag{21}$$

and which, in the case of optical wave fronts, has a clearcut geometric meaning: it is the horizontal coordinate of the ray, always perpendicular to the wave front propagating in an anisotropic (if $c = c(u)$) medium.

To distinguish $X(t)$ from $\tilde{X}(y,t)$, we shall call the latter function (together with $\tilde{Z} = h(\tilde{X},t)$) the *isocline trajectory* of the growing interface $h(x,t)$.

The solutions of (19) are of the form

$$U(y,t) = u_0(y), \quad \tilde{X}(y,t) = y + C(u_0(y))t,$$
$$H(y,t) = h_0(y) + \Lambda(u_0(y))t. \tag{22}$$

In particular, it follows that the isoclines are straight lines, which is not always true for rays, which in an anisotropic medium could be curved.

As in the case of the Riemann equation, the sought fields $h(x,t)$ and $u(x,t)$ are obtained from (22) by substituting in $H(y,t)$ and $U(y,t)$ the function $y = \tilde{y}(x,t)$, which is the inverse function to $x = \tilde{X}(y,t)$.

In the remainder of this section we shall provide two concrete examples of the interface evolution discussed above.

Example 3. Isotropic velocity of a growing surface. Suppose that the growth rate of an evolving interface is independent of its gradient, that is, $c(u) = c = const$. In the language of optical waves, it corresponds to the case of wave propagation in an isotropic medium. In such a medium, the speeds of wave propagation along rays and isoclines are the same,

$$C(u) = V(u) = \frac{cu}{\sqrt{1+u^2}} = c\sin\theta.$$

This means in particular that rays and isoclines coincide and are represented by straight lines perpendicular to the wave fronts. In this case, the interface evolution can be described parametrically as follows:

$$x = y + \frac{u_0(y)}{\sqrt{1 + u_0^2(y)}} ct, \qquad h = h_0(y) + \frac{ct}{\sqrt{1 + u_0^2(y)}}. \tag{23}$$

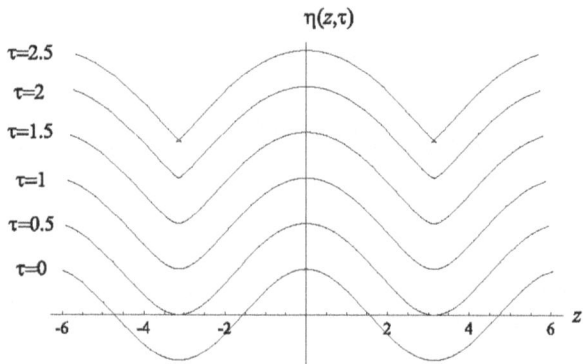

FIGURE 12.3.4
Time evolution of the interface described by the field $h(x,t)$ (24) in the case of positive growth rate ($c > 0$). Physically, it may be interpreted as evolution of an optical wave front that started out as a sine curve. The medium is assumed to be isotropic. In the course of time, the crests become flatter, but the troughs become sharper.

Now consider the initial sine profile

$$h_0(x) = h_0 \cos(kx) \quad \Rightarrow \quad u_0(x) = h_0 k \sin(kx).$$

In this case, we obtain the following parametric representation for $h(x,t)$:

$$z = \mu + \frac{\varepsilon \tau \sin \mu}{\sqrt{1 + \varepsilon^2 \sin^2 \mu}}, \qquad \eta = \varepsilon \cos \mu + \frac{\tau}{\sqrt{1 + \varepsilon^2 \sin^2 \mu}}, \tag{24}$$

where the nondimensional variables

$$kx = z, \quad ky = \mu, \quad ckt = \tau, \quad kh = \eta, \tag{25}$$

have been introduced, with the parameter

$$\varepsilon = kh_0. \tag{26}$$

Graphs of the evolving interface $h(x,t)$ are shown in Fig. 12.3.4 for different τ and $\varepsilon = 1/2$.

12.4. Exercises

Example 4. Melting surface. In this example, we will assume that the parametric equations (12.3.3) describing the evolution of the interface are the same as in Example 3, but the growth velocity is negative ($c < 0$). Physically, this corresponds to the model of a surface melting with speed c.

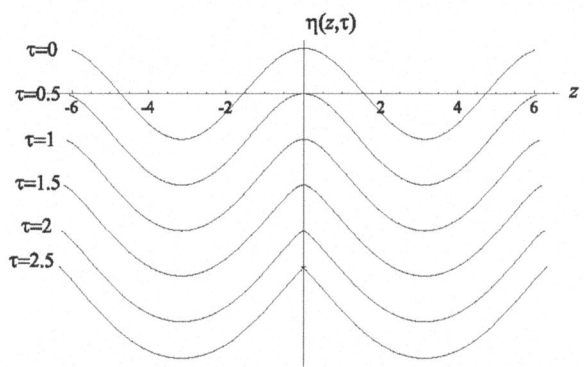

FIGURE 12.3.5
Time evolution of a melting interface $h(x,t)$ from Example 4. In contrast to the growing interface from Example 3 (Fig. 12.3.4), in the course of time, the troughs of this interface become smoother while the crests become sharper. The reader may have observed this phenomenon while tasting the ice cubes melting in a glass of Coke.

In this case, the profiles of the evolving interface are obtained by changing in (12.3.3) the signs in front of τ from plus to minus. The corresponding graphs are shown in Fig. 12.3.5.

12.4 Exercises

1. In the 1980s, Ya.B. Zeldovich, one of the fathers of the Russian hydrogen bomb, developed the theory of nonlinear gravitational instability, which describes the evolution of the large-scale mass distribution in the universe (at the scale of galaxy clusters). In its simplified form, his theory reduces to an analysis of solutions of the Riemann equation (12.1.2), and the continuity equation (12.2.4), with initial conditions

$$v(x, t=0) = u_0(x) + Hx, \qquad \rho(x, t=0) = \varrho_0(x), \qquad (1)$$

where H is the Hubble constant (which reflects the expansion rate of the universe), and $u_0(x)$ and $\varrho_0(x)$ describe fluctuations of the velocity and density of matter, respectively, at the initial time instant.

Show that the solution of the above problem can be expressed in terms of the velocity field $u(x,t)$ and the density field $\varrho(x,t)$ in the "expansionless" universe, solving the Cauchy problem

$$\frac{\partial u}{\partial t} + u\frac{\partial u}{\partial x} = 0, \quad \frac{\partial \varrho}{\partial t} + \frac{\partial}{\partial x}(u\varrho) = 0,$$
$$u(x, t=0) = u_0(x), \qquad \varrho(x, t=0) = \varrho_0(x).$$

2. Suppose that particles move in a medium, such as a viscous fluid, that offers resistance to their motion. As a result, the particles' velocity decreases in time, and the velocity field is described by the equation

$$\frac{\partial v}{\partial t} + v\frac{\partial v}{\partial x} + \frac{1}{\tau}v = 0, \quad v(x, t=0) = v_0(x),$$

where τ is a constant representing the characteristic velocity "dissipation" time. Find the evolution of the velocity field $v(x,t)$ and determine its asymptotic behavior as $t \to \infty$.

3. Prove the formula (12.2.14) for the Fourier image of the solution of the Riemann equation. Assume that $v(x,t)$ and $y(x,t)$ are smooth functions of x such that $v(x,t)$ decays to zero rapidly as $x \to \pm\infty$, while $y(x,t)$ is strictly increasing and maps the x-axis onto the whole y-axis.

4. Physicists and engineers often see the generation of higher harmonics in a field as an indicator of its nonlinearity. Study the process of generation of higher harmonics in a field $v(x,t)$ satisfying the Riemann equation (2) if at $t=0$, the field was purely harmonic: $v_0(x) = a\sin(kx)$.

5. The 1-D motion of cold plasma is described by the equations

$$\frac{\partial v}{\partial t} + v\frac{\partial v}{\partial x} = -\frac{e}{m}E, \quad \frac{\partial \rho}{\partial t} + \frac{\partial}{\partial t}(v\rho) = 0, \qquad (2)$$
$$v(x, t=0) = v_0(x), \quad \rho(x, t=0) = \rho_0(x),$$

describing the electron velocity $v(x,t)$ and their density $\rho(x,t)$, and the equation

$$\frac{\partial E}{\partial x} = -4\pi e(\rho - \rho_0)$$

12.4. Exercises

describing the longitudinal electric field $E(x,t)$. Here e and m are respectively the electron charge and mass, and ρ_0 is the density of ions that are assumed to be at rest. Solve the above equations by the method of characteristics and discuss their solutions. Assume that as $x \to -\infty$, the electron velocity field decays to zero, and so does the cumulative mass function of the electrons,

$$m(x,t) = \int_{-\infty}^{x} [\rho(x',t) - \rho_0] \, dx'.$$

Provide a detailed analysis in the case of uniform initial density $\rho(x, t = 0) = \rho_0$. *Hint:* To reduce the equations to a form convenient for analysis, use the obvious relationship $E(x,t) = -4\pi\, e\, m(x,t)$.

6. *Oblique snowfall.* Consider snow falling in the presence of a horizontal wind. This means that the snowflakes are deposited at a certain angle $\theta_0 \neq 0$ to the vertical direction. Additionally, assume that the initial profile $h_0(x)$ of the snow surface is such that the incidence angle θ always satisfies the inequality

$$|\theta - \theta_0| < \frac{\pi}{2}. \tag{3}$$

Find the time evolution of the snow elevation $h(x,t)$.

7. *More oblique snowfall.* Extend the solution of the preceding exercise to the case of an arbitrary directional diagram $D(\theta')$ that vanishes outside the interval $[\theta_1, \theta_2]$ and for the initial profile $h_0(x)$ with angles θ that satisfy the inequalities

$$\{\theta - \theta_1,\; \theta_2 - \theta\} < \frac{\pi}{2}.$$

8. Suppose that a directional diagram is given by the formula

$$D(\theta) = c \cos^2 \theta.$$

Study the evolution of the interface $h(x,t)$. While constructing the corresponding equation, use the small angle approximation and assume that the inequalities $|\theta| \ll 1$ and $|u| \ll 1$ are satisfied. Conduct a detailed investigation of the case that the initial interface profile is given by $h_0(x) = h\cos(kx)$ $(kh \gg 1)$.

9. Investigate the interfacial growth in the case that the growth velocity depends on the angle of incidence θ as follows:
$$c(\theta) = c\cos^2\theta.$$
Use the small angle approximation. Construct graphs of the solutions for different time instants t and initial profile
$$h_0(x) = -h\cos(kx) \qquad (h > 0). \tag{4}$$

10. Assuming the cosine initial condition (4), expand the solution of the previous exercise in a Fourier series. For $\tau = 10$, compare the graph in Fig. 2 shown in the Appendix A: Answers and Solutions with the graph of the partial sum of the first ten terms of the Fourier series.

Chapter 13

Generalized Solutions of First-Order Nonlinear PDEs

Most of the equations of mathematical physics, and in particular nonlinear first-order partial differential equations, are a result of idealizing and simplifying assumptions. This approach promotes the effectiveness and elegance of mathematical models that adequately reflect some important qualitative features of the physical phenomena under consideration. However, sooner or later, one has to pay the price for the simplifying assumptions. The influence of factors not taken into account sometimes is gradual, and does not affect the qualitative picture of the physical phenomenon, but sometimes it is abrupt, and the simplified model is unable to describe the real course of events.

If such difficulties are encountered, a natural remedy is to consider more complex models that are able to describe more adequately the phenomena being studied. However, in certain situations one can achieve the same goal by the construction of *generalized (or distributional) solutions* of the original equations. Here, the main point is that many differential equations of mathematical physics are consequences of more general integral laws. If a distributional solution that does not satisfy the original equation in the classical sense satisfies the corresponding integral laws, it can be justifiably used for descriptions of the physical phenomena under study even in cases in which the classical solutions of the original differential equations do not exist.

In this chapter we shall discuss the basic principles of construction of generalized solutions of nonlinear first-order partial differential equations that were introduced in Chap. 12.

13.1 Master Equations

Our goal here is to illustrate the methods of construction of generalized, distributional solutions for first-order nonlinear partial differential equations. To achieve this goal, we concentrate only on the simplest, but also critical from the viewpoint of physical applications, examples. Most of them were encountered in the previous chapter. For the reader's benefit, we shall review them briefly in the following subsections.

13.1.1 Equations of Particle Flows

Our first example is the familiar Riemann equation

$$\frac{\partial v}{\partial t} + v\frac{\partial v}{\partial x} = 0, \qquad v(x, t = 0) = v_0(x), \tag{1}$$

for the velocity field $v(x,t)$ in the 1-D flow of uniformly moving particles and the continuity equation

$$\frac{\partial \rho}{\partial t} + \frac{\partial}{\partial x}(v\rho) = 0, \qquad \rho(x, t = 0) = \rho_0(x), \tag{2}$$

for the particle density $\rho(x,t)$. In addition, let us introduce the velocity potential

$$s(x,t) = \int^{x} v(x', t)\, dx'. \tag{3}$$

The lower limit in the integral has been deliberately omitted, indicating the presence of an arbitrary constant; physical potentials are always defined only up to a constant. The potential of the velocity field satisfies the following nonlinear equation:

$$\frac{\partial s}{\partial t} + \frac{1}{2}\left(\frac{\partial s}{\partial x}\right)^2 = 0, \qquad s(x, t = 0) = s_0(x). \tag{4}$$

The reader can verify the validity of the above equation by differentiating it with respect to x to obtain again the Riemann equation (1) for the velocity field.

13.1.2 Interface Growth Equation in Small Angle Approximation

Our construction of generalized solutions of the velocity field and its potential will be geometrically more transparent if we present them as generalized

13.1. Master Equations

solutions of the interface growth equation written in the small angle approximation. Recall (see Chap. 12) that the forest fire boundary line $h(x,t)$, or the boundary of a 2-D wave front, propagating with the velocity c perpendicularly to the interface $h(x,t)$ satisfies, in the small angle approximation, the equation

$$\frac{\partial h}{\partial t} = \frac{c}{2}\left(\frac{\partial h}{\partial x}\right)^2, \qquad h(x, t=0) = h_0(x), \tag{5}$$

and the gradient field

$$u(x,t) = -\frac{\partial h}{\partial x} \tag{6}$$

of the moving interface satisfies the related Riemann equation

$$\frac{\partial u}{\partial t} + cu\frac{\partial u}{\partial x} = 0, \qquad u(x, t=0) = u_0(x). \tag{7}$$

Moreover, if one introduces the change of variables

$$v = cu, \qquad s = -ch, \tag{8}$$

then the particle flow and the interface equations coincide.

13.1.3 Equation of Nonlinear Acoustics

As our third example, we will quote an equation describing propagation of nonlinear acoustic waves. Consider fluctuations $P(r,t)$ of the atmospheric pressure caused by the presence of an intensive acoustic wave, such as jet engine noise or thunder, that satisfy the equation

$$\frac{\partial P}{\partial r} - \frac{1}{c}\frac{\partial P}{\partial t} - \beta P\frac{\partial P}{\partial t} + \frac{n}{2r}P = 0 \tag{9}$$

containing the parameter n. For $n = 2$, the equation (9) accurately describes the propagation of spherical waves (cylindrical waves for $n = 1$ and flat waves for $n = 0$). The constant c stands for the sound velocity in the medium, and β is a parameter that quantifies the nonlinearity of the medium.

Usually, one reduces equation (9) to a more convenient form by introducing the local time coordinate

$$\theta = t - \frac{r}{c},$$

which "delays" the wave by the time needed for the wave to propagate from the origin $r = 0$ to the point r under consideration. This substitution eliminates one term of the equation (9), so that we arrive at the equation

$$\frac{\partial P}{\partial r} - \beta P \frac{\partial P}{\partial \theta} + \frac{n}{2r} P = 0.$$

The last term takes into account the wave attenuation due to its geometric divergence, but it can also be eliminated via a change of variables. Indeed, for example, in the case of a spherical wave ($n = 2$), one can introduce the new variables

$$p = \frac{r}{r_0} \beta P, \qquad z = r_0 \ln\left(\frac{r}{r_0}\right),$$

which transform the equation of nonlinear acoustics into the canonical Riemann equation for the field $p(z, \theta)$,

$$\frac{\partial p}{\partial z} + p \frac{\partial p}{\partial \theta} = 0. \tag{10}$$

Usually, r_0 is taken to be the radius of the spherical source radiating the acoustic wave. Then the boundary condition in equation (10), which describes vibrations of the surface of the radiator, is

$$p(z = 0, \theta) = p_0(\theta). \tag{11}$$

The first-order nonlinear partial differential equations described above will help us illustrate the basic ideas and methods used in the construction of generalized solutions of general equations of hydrodynamic type.

13.2 Multistream Solutions

It is convenient to begin our discussion of generalized solutions of first-order nonlinear partial differential equations by an analysis of the so-called *multistream solutions*.

13.2.1 The Interval of Single Stream Motion

In what follows, our construction of generalized solutions will be based on the formula

$$x = X(y, t) = y + v_0(y) t \tag{1}$$

13.2. Multistream Solutions

connecting the Lagrangian coordinate y with the Eulerian coordinate x. We shall assume that $v_0(x)$ is continuously differentiable everywhere, and that its derivative is bounded from below by $-u_{\min}$,

$$v_0'(x) \geq -u_{\min}. \tag{2}$$

Then, in the time interval

$$0 < t < t_n, \qquad \text{where} \quad t_n = 1/u_{\min}, \tag{3}$$

there exists a monotonically increasing and continuously differentiable function

$$y = y(x, t) \tag{4}$$

inverse to the function (1) that maps $\mathbf{R}$ onto $\mathbf{R}$ and is such that the classical solution of the Riemann equation (1) can be written in the form

$$v(x, t) = \frac{x - y(x, t)}{t}. \tag{5}$$

Recall that this solution has a transparent geometric meaning: the flow velocity at time t at the Eulerian coordinate x is equal to the velocity of a uniformly moving particle that finds itself at x at time t. The time interval $(0, t_n)$ will be called the *single stream motion interval*.

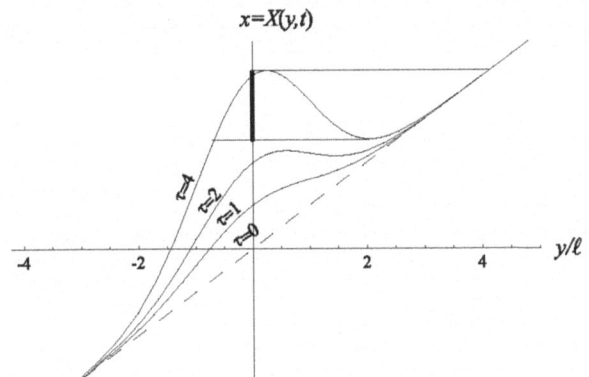

FIGURE 13.2.1
Lagrangian-to-Eulerian coordinate mappings for the flow of uniformly moving particles at different instants of nondimensional time $\tau = 1, 2, 4$. The thick-line interval on the vertical x-axis indicates the presence of a multistream motion there, at $\tau = 4$.

13.2.2 Appearance of the Multistream Regime

The time instant t_n will be called the *wave turnover time*. For $t > t_n$, the Lagrangian-to-Eulerian mapping (1) is no longer strictly increasing everywhere. In the language of particle flow, this means that some particles begin to overtake other particles so that at the same point of the x-axis there may appear several particles with different Lagrangian (initial) coordinates y. In this case, we shall talk about the appearance of *multistream motions* on portions of the x-axis.

The appearance of the multistream regime is easiest to demonstrate with the graph of the Lagrangian-to-Eulerian mapping $x = X(y,t)$ (1) (see Fig. 13.2.1). It was generated for the initial Gaussian velocity profile (12.1.14) at several instants of nondimensional time $\tau = V_0 t/\ell$. An easy calculation shows that for the initial Gaussian velocity field, the multistream regime appears for $\tau \geq \sqrt{e} \approx 1.65$.

The lack of monotonicity of the mapping $x = X(y,t)$ in the multistream regime (for $t > t_n$) implies nonuniqueness of the inverse function $x = y(x,t)$. In other words, for $t > t_n$, there are segments of the x-axis on which the function $y(x,t)$ has not one, but $n > 1$ values,

$$\{y_1(x,t), y_2(x,t), \ldots, y_n(x,t)\}. \tag{6}$$

In other words, the function is *multivalued* there. A graph of the function $y(x,t)$ for a Gaussian initial profile (12.1.14) and time $\tau = 4$ is shown in Fig. 13.2.2. Clearly, there exists an interval of the x-axis at every point of which one can find three ($n = 3$) particles with different initial (Lagrangian) coordinates.

Let (x_1, x_2) be the interval where the mapping $y(x,t)$ is not unique. Substituting different branches of the multivalued, in the interval $[x_1, x_2]$, functions $y(x,t)$ (6) in the expression (5) for the velocity field, we obtain a multistream field $v(x,t)$, which inside the nonuniqueness interval, takes n values,

$$v(x,t) = \{v_1(x,t), v_2(x,t), \ldots, v_n(x,t)\}, \qquad v_i(x,t) = \frac{x - y_i(x,t)}{t}. \tag{7}$$

A graph of this multivalued field $v(x,t)$ is shown in Fig. 13.2.3.

In addition to the velocity field, for $t > t_n$, other fields related to the particle flow also become multivalued. A good example here is the density field, each branch of which,

$$\rho_i(x,t) = (-1)^{i-1} \rho_0(y_i(x,t)) \frac{\partial y_i(x,t)}{\partial x}, \qquad (i = 1, 2, \ldots, n) \tag{8}$$

13.2. Multistream Solutions

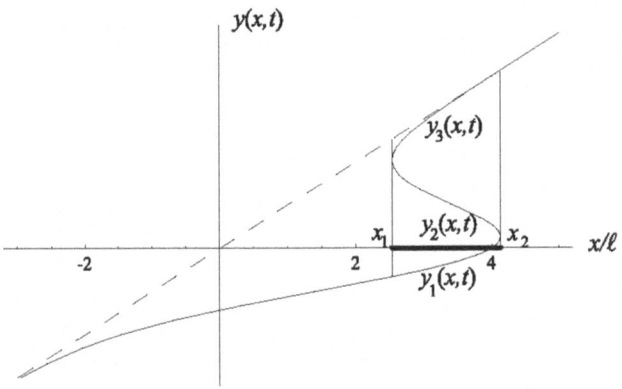

FIGURE 13.2.2
Eulerian-to-Lagrangian mapping at the nondimensional time $\tau = 4$. The thick-line interval on the x-axis indicates points where the multistream regime is present. Three branches of the function $y(x,t)$ are visible, and are numbered in the order of increase of the corresponding Lagrangian coordinates.

describes the density evolution in the corresponding stream. The factors $(-1)^{i-1}$ take into account the alternating signs of the derivatives $\partial y_i(x,t)/\partial x$ in different streams corresponding to the change of ordering of particles as compared with their original positions. They guarantee that the flow density remains a positive field—an obvious physical requirement.

13.2.3 The Gradient Catastrophe

The multistream solutions of the Riemann equation described above appear natural if one discusses them in the context of flows of uniformly moving particles. However, the Riemann equation also arises in the analysis of other physical phenomena. For example, as we already mentioned before, the Riemann equation (13.1.10) describes the fluctuations $p(x,\theta)$ of atmospheric pressure due to propagation of an intensive acoustic wave. In this case, naturally, the field $p(z,\theta)$ cannot be multivalued. For that reason, mathematicians made an effort to produce a general treatment of the above equations that would be independent of their narrower physical interpretation.

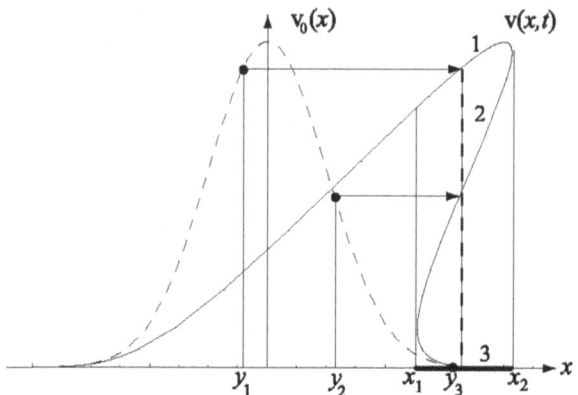

FIGURE 13.2.3
The velocity field $v(x,t)$ at time $\tau = 4$. The dashed line indicates the initial Gaussian velocity profile (12.1.14) on which three particles with different Lagrangian coordinates y_1, y_2, y_3 are marked. They all end up, at time $\tau = 4$, at the same point in space. It is clear that particles y_1, y_2, initially positioned to the left have a greater velocity than y_3, and in the course of time, they catch up with the particle with Lagrangian coordinate y_3. The points x_1 and x_2 bound the multistream segment under consideration.

Judged from the perspective of classical analysis, the multivalued functions introduced above cannot be solutions of differential equations. The first problem is that the onset of the multistream regime is preceded by the *gradient catastrophe*, a characteristic explosion of the spatial derivative,

$$q(x,t) = \frac{\partial v(x,t)}{\partial x}, \tag{9}$$

of the solutions of the Riemann equation in the neighborhood $(t \to t_n - 0)$ where the multistream regime appears. We shall carefully trace a sequence of events that lead to it.

It follows from (5) that

$$q(x,t) = \frac{1}{t}\left(1 - \frac{\partial y(x,t)}{\partial x}\right). \tag{10}$$

The expression contains the derivative of the mapping $y = y(x,t)$, and is best studied by investigating a "nicer," everywhere differentiable, inverse function $x = X(y,t)$, see (1), and its derivative

13.2. Multistream Solutions

$$J(y,t) = \frac{\partial X(y,t)}{\partial y} = 1 + v_0'(y)\, t \tag{11}$$

with respect to y. At the turnover time t_n, the mapping $X(y,t)$ ceases to be strictly monotone, and the minimum value of its derivative (11), attained at a certain point y_*, becomes zero:

$$J(y_*, t_n)) = 1 + v_0'(y_*)\, t_n = 0\,.$$

Since the point y_* is also the point minimizing values of the function $J(y, t_n)$, not only does the function $J(y, t_n)$ itself vanish there, but its derivative vanishes as well,

$$\left.\frac{\partial J(y,t)}{\partial y}\right|_{y=y_*} = v_0''(y_*)\, t_n = 0 \quad \Rightarrow \quad v_0''(y_*) = 0\,.$$

Consequently, the mapping $X(y,t)$, see (1), has, in a vicinity of the point y_*, cubic asymptotics,

$$x = X(y,t) \sim x_* + b(y - y_*)^3\,, \qquad b = -\frac{v_0'''(y_*)}{v_0'(y_*)}\,, \tag{12}$$

where the *turnover point* x_* is the Eulerian coordinate of the singular point y_* discussed above.

In view of (12), the function $y(x,t)$, inverse to $X(y,t)$, and the velocity field $v(x,t)$, see (5), have, in the vicinity of the turnover point x_*, the following asymptotics:

$$y(x,t) \sim y_* + \sqrt[3]{\frac{|x - x_*|}{b}}\, \mathrm{sign}\,(x - x_*) \quad \Rightarrow$$

$$v(x,t) \sim v_* - \sqrt[3]{\frac{|x - x_*|}{d}}\, \mathrm{sign}\,(x - x_*)\,,$$

$$v_* = \frac{x - x_*}{t}\,. \qquad d = \frac{v_0'''(y_*)}{[v_0'(y_*)]^4}\,.$$

Consequently, at the turnover time, the derivative $q(x,t)$, see (10), of the field $v(x,t)$ has an infinite singularity of the following type:

$$q(x,t) \sim -\frac{1}{3\sqrt[3]{d}}\, \frac{1}{|x - x_*|^{2/3}}\,.$$

This asymptotic expression provides a mathematical formulation of the concept of the gradient catastrophe. It implies that at the turnover time, the field $v(x,t)$ is nondifferentiable at least at one point, and thus cannot be a classical solution of a differential equation.

Recall that the density field $\rho(x,t)$, see (8), is proportional to the derivative of the mapping $y(x,t)$ with respect to x. It is this derivative that determines the character of the gradient catastrophe. In the language of particle flows, the gradient catastrophe is caused by an infinite compression of the flow at turnover time. The density at this point (for $\rho_0(y_*) \neq 0$) becomes infinite. The nature of the density collapse as the time approaches the turnover time t_n is well illustrated in Fig. 12.2.2, where the last time considered, $\tau = 1.5$, is close to the turnover time $\tau = \sqrt{e} \approx 1.65$. The sharp peak in the density field at that time does not fit in the figure.

In concluding this section, it is important to reemphasize that because of the gradient catastrophe, the classical, everywhere differentiable solutions of the Riemann equation exist only for $t \in (0, t_n)$, the latter being the interval of the single stream regime. However, multistream solutions constructed for $t > t_n$, via mappings (1), have a well-defined physical interpretation and must be taken seriously. For that reason, we will include multivalued solutions as natural generalized solutions of the Riemann equation.

And the moral of the above story is this: if the family of classical solutions of an equation is not rich enough to describe the physical phenomena under discussion, one should explore the possibility of the existence of generalized solutions that might be sufficient for the purpose.

13.3 Summing over Streams in Multistream Regimes

In the previous section we observed that a first-order nonlinear partial differential equation can have a number of different generalized solutions satisfying the same initial and boundary conditions. To distinguish among them, and to make them unique (in a certain sense), it may be necessary to impose on them additional restrictions. One such requirement could be that the solution is single-valued. Then, of course, the multistream solutions discussed above are a forbidden fruit. However, different pieces of different streams can be used in constructions of various single-valued generalized solutions as long as the latter fit various physical constraints of the problem under consideration.

We begin by discussing properties of single-valued fields obtained simply by the algebraic summation of the values of different streams in multistream flows.

13.3.1 Total Particle Density

Let us consider a multistream density field and define the single-valued density field $\rho(x,t)$ as the sum of densities (13.2.8) that correspond to a given point x in space and time instant t:

$$\rho(x,t) = \sum_{i=1}^{n} \rho_i(x,t) = \sum_{i=1}^{n} (-1)^{i-1} \rho_0(y_i(x,t)) \frac{\partial y_i(x,t)}{\partial x}. \tag{1}$$

The physical argument in favor of this construction is obvious: the density of the gas of particles in an infinitesimal interval dx is unambiguous and corresponds to the full count of particles inside this interval divided by its length.

Furthermore, observe that using Dirac delta notation, we can write out the total density field (1) in a convenient way:

$$\rho(x,t) = \int \rho_0(y) \delta(X(y,t) - x)\, dy. \tag{2}$$

The validity of the above formula can be verified using standard properties of the Dirac delta distributions and in particular, the formula for a nonmonotone change of variables; see Volume 1.

13.3.2 Summing over Streams, and the Inverse Fourier Transform

Now let us turn our attention to the fact that the right-hand side of the expression (12.2.13) for the Fourier image

$$\tilde{\rho}(\kappa,t) = \frac{1}{2\pi} \int e^{-i\kappa X(y,t)} \rho_0(y)\, dy \tag{3}$$

of the density field is well defined even in the case of nonmonotone mappings $x = X(y,t)$. The inverse Fourier transform of the right-hand side of (3),

$$\rho(x,t) = \int \tilde{\rho}(\kappa,t) e^{i\kappa x}\, d\kappa, \tag{4}$$

is a single-valued function, which can be considered a generalized solution of the continuity equation (13.1.2). By a detailed analysis of the transition from the Eulerian variable of integration x in the integral (12.2.12) to the Lagrangian y in the final integral (3), it is not difficult to see that the inverse

Fourier transform (4) is exactly the sum (1) of the multistream densities in the flow of uniformly moving particles.

Although construction of the density field via the inverse Fourier transform may seem somewhat artificial, the technique permits us to prove several integral properties of the total density (1). For example, the theory of the Fourier transform gives us the following equality:

$$\int \rho(x,t)\,dx = 2\pi \tilde{\rho}(\kappa = 0, t)\,.$$

For $\kappa = 0$, this fact and equality (3) imply that the total density (1) satisfies the integral mass conservation law

$$\int \rho(x,t)\,dx = \int \rho_0(y)\,dy = \text{const}\,. \tag{5}$$

Similarly, considering the inverse Fourier transform of the expression (12.2.14), one can construct another single-valued field, $v_\pm(x,t)$. As in the previous case, one can demonstrate that this field is the signed sum of the velocity fields corresponding to different streams of the multistream solution of the Riemann equation

$$v_\pm(x,t) = \sum_{i=1}^{n}(-1)^{i-1}\,v_i(x,t) = \sum_{i=1}^{n}(-1)^{i-1}\,v_0(y_i(x,t))\,. \tag{6}$$

In contrast to the total density fields (1) and (4), the above function (6) does not have any particular physical interpretation when applied to particle flows. This does not preclude the possibility that such an interpretation might be found in the future for other physical, economic, etc., phenomena described by the Riemann equation. Notice that the field (6) has an integral invariant. Indeed,

$$\int v_\pm(x,t)\,dx = \int v_0(y)\,dy = \text{const}\,. \tag{7}$$

This can be easily proved by calculation of the limit, as $\kappa \to 0$, of the right-hand side of (12.2.14).

Figure 13.3.1 shows a graph of the field (6) corresponding to the sine initial condition

$$v_0(x) = a\sin(kx)\,. \tag{8}$$

13.3. Summing over Streams

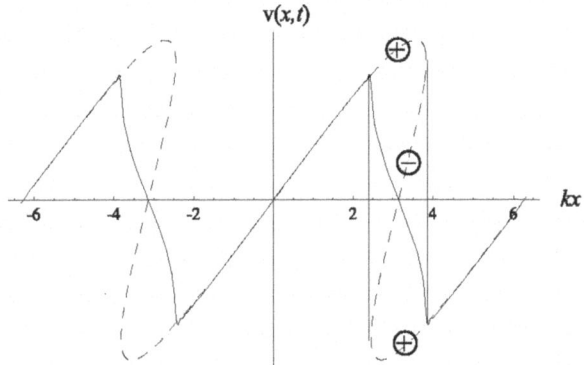

FIGURE 13.3.1
The field $v_\pm(x,t)$, found by summation of the first 50 terms of the series (9), at time $\tau = kat = 2$ (which is larger than the time $\tau_n = 1$ when the multistream regime appears for the first time). The *dashed line* shows the multistream velocity field for the flow of uniformly moving particles. The *vertical lines* mark one of the multistream intervals. The *pluses and minuses* indicate which sign was assigned for which stream. The *sharp cusps* reflect the lack of differentiability for branches of the multistream velocity field at the points corresponding to the onset of different regimes.

The graph is constructed using the series expansion

$$v_\pm(x,t) = 2a \sum_{n=1}^{\infty} (-1)^{n+1} \frac{J_n(n\tau)}{n\tau} \sin(nkx), \tag{9}$$

which was found in one of the exercises in Chap. 12 by taking the inverse Fourier transform of the Fourier image (12.2.14) of the velocity field $v_\pm(x,t)$. A comparison of the graph of the partial sum of the series with the corresponding multistream solution shows the equivalence of the inverse Fourier transform of the Fourier image (12.2.14) and the signed sum (6).

13.3.3 Density of a "Warm" Particle Flow

Let us take another look at the behavior of the above total density field (1) in the flow of uniformly moving particles, for $t > t_n$. This time, to avoid singularities created by particles overtaking each other, we shall include the effect of a *smoothing heat bath*. To help the reader better understand what we have in mind, let us recall the relevant physical facts.

In addition to the global hydrodynamic velocity field v, particles of gas are now subject to thermal velocity fluctuations u. In the model of the flow of uniformly moving particles, this "heat" component of the velocity can be taken into account by replacing $v_0(y)$ by $v_0(y) + u$ in (13.2.1), which yields the equality

$$x = X(y, u, t) = y + [v_0(y) + u]\,t = X(y, t) + ut. \tag{10}$$

Here, $X(y, t)$ represents the "hydrodynamic" component of the velocity field; recall that it is given by the expression (13.2.1).

Now the total flow of particles is described with the help of the initial density $\rho_0(x, u)$, depending not only on x, but also on u. Physically, the dependence of $\rho_0(x, u)$ on u shows the distribution of particles over different velocities. The initial density $\rho_0(x, u)$ must satisfy the consistency condition

$$\int \rho_0(x, u)\,du = \rho_0(x),$$

where $\rho_0(x)$ is the ordinary initial particle density at the point x.

This suggests the following obvious generalization of the density field (2) to the case of "warm" flows of particles:

$$\rho(x, t) = \iint \rho_0(y, u)\,\delta\left(X(y, u, t) - x\right)\,du\,dy. \tag{11}$$

Let us take a look at the most illustrative special case,

$$\rho_0(y, u) = \rho_0(y)\,f(u), \tag{12}$$

where $f(u)$ is a distribution of "thermal particle" velocities for a uniformly "heated" flow. Furthermore, assume that the density $f(u)$ is Gaussian (physicists often call it the Maxwell density), that is,

$$f(u) = \frac{1}{\sqrt{2\pi\varepsilon}}\exp\left(-\frac{u^2}{2\varepsilon}\right), \tag{13}$$

where ε is the "temperature" of the flow. Substituting (13) together with (12) into (11), we obtain the elegant convolution formula

$$\rho_\varepsilon(x, t) = \int \rho(x - ut, t)\,f(u)\,du. \tag{14}$$

The field $\rho(x - ut, t)$ is the original "cold" density field (2) shifted along the x-axis by ut. The subscript ε in (14) points to the dependence of the density field $\rho_\varepsilon(x, t)$ on the "temperature" ε.

13.3. Summing over Streams

The formula (14) means that the thermal scattering of particle velocities leads to the spatial averaging of the hydrodynamic density field $\rho(x,t)$. Such averaging eliminates singularities of the original density field. We shall illustrate this fact with the example of the uniform original density $\rho_0(x) = \rho_0 = $ const and the sinusoidal initial velocity field (8).

Substituting the Fourier expansion (12.2.19) for $\rho(x,t)$ into (14), we obtain

$$\rho(x,t) = \rho_0 + 2\rho_0 \sum_{n=1}^{\infty} (-1)^n J_n(n\tau) \varphi_n(x,t), \qquad (15)$$

where

$$\varphi_n(x,t) = \int \cos(knx - kunt) f(u)\, du.$$

In the case of the Gaussian distribution (13), this gives

$$\varphi_n(x,t) = \exp\left[-\frac{\varepsilon}{2}(knt)^2\right] \cos(nkx).$$

Substituting this expression into (15), we finally get

$$\rho(x,t) = \rho_0 + 2\rho_0 \sum_{1}^{\infty} (-1)^n J_n(n\tau) \exp\left(-\frac{\delta}{2}\tau^2 n^2\right). \qquad (16)$$

Here and in (15), we used the nondimensional time $\tau = kat$ as well as the parameter $\delta = \varepsilon/a^2$, which characterizes the relative contribution of thermal velocity fluctuations to the flow's behavior. For small δ, for a long stretch of time, the field $\rho(x,t)$ is formed by hydrodynamic compression and rarefaction of the particle flow. For large δ, very quickly, namely at the time

$$\tau_* = \sqrt{\frac{2}{\delta}},$$

we observe that the thermal effects gradually eliminate the lack of uniformity of the density. The remaining small density fluctuations are asymptotically described by the first two terms of the series (16):

$$\rho(x,t) \sim \rho_0 \left[1 - 2J_1(\tau) e^{-\delta\tau^2/2} \cos(kx)\right].$$

The graphs of the density (16), with $\delta = 0.002$, are shown in Fig. 13.3.2 at different moments of time.

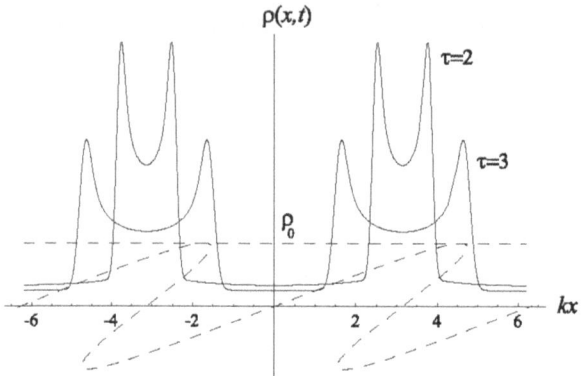

FIGURE 13.3.2
The density field (16) for $\delta = 0.002$ and the time instants $\tau = 2$ and 3. The *dashed line* depicts the multistream velocity field at the time $\tau = 3$. The value of the parameter δ is small, and as a result, all the structural elements of the "hydrodynamic" density field are clearly visible. In particular, the multistream intervals and the characteristic density peaks near the interval boundaries can be seen. The peaks' heights diminish with time. The presence of even weak thermal fluctuations of particle velocities eliminates the infinite singularities.

13.4 Weak Solutions of First-Order Nonlinear Equations

In the mathematical literature, *weak solutions* of differential equations are formally defined as solutions of the integrated (against some class of test functions) versions of these equations. However, in this section we shall take a different, more elementary course, but one that is also more intuitive physically, thinking of weak solutions as solutions that satisfy the equations, and the associated initial and boundary conditions, everywhere where the required derivatives exist, but which can have discontinuities (including discontinuities of the derivatives) along certain curves $x^*(t)$ in the (x,t)-plane. The shape of these curves can be found from the integral conservation laws and other general principles that must be obeyed by the solution fields. Sometimes, to determine unique weak solutions, we will be able to use simple intuitive geometric arguments. This is the case for the familiar example of forest fires.

13.4.1 Forest Fire Revisited

Recall that if a forest fire propagates predominantly along the z-axis, the small angle approximation is applicable, and the fire front $h(x,t)$ satisfies the equation (13.1.5). To kill two birds with one stone, let us carry out our analysis using the substitutions (13.1.8), so that it applies to the study of the particle flows as well. Recall that those substitutions replaced the fire front function $h(x,t)$ and its gradient $u(x,t)$ by the fields $s(x,t)$ and $v(x,t)$. The latter have a natural interpretation in the language of the flows of uniformly moving particles. In the context of forest fires, the transition from h to s means that the fire runs in the direction of the negative z-axis. In other words, by time t, the area above the line $s(x,t)$ has burned down.

By now, we already have at least three physical interpretations of the function $s(x,t)$. It could be the velocity field's potential (13.1.3), the fire front, or the optical wavefront; in this section we shall make use of the last two.

Let us begin by finding the fire front function $s(x,t)$, propagating downward, by solving equations (13.1.1) and (13.1.4) by the method of characteristics. These equations can be rewritten in a more uniform fashion as follows:

$$\frac{\partial s}{\partial t} + v \frac{\partial s}{\partial x} = \frac{1}{2} v^2, \qquad s(x, t=0) = s_0(x),$$
$$\frac{\partial v}{\partial t} + v \frac{\partial v}{\partial x} = 0, \qquad v(x, t=0) = v_0(x). \qquad (1)$$

The corresponding characteristic equations are of the form

$$\frac{dX}{dt} = V, \qquad \frac{dV}{dt} = 0, \qquad \frac{dS}{dt} = \frac{1}{2} V^2.$$

These equations have the familiar solutions

$$X(y,t) = y + v_0(y)\, t, \qquad V(y,t) = v_0(y),$$
$$S(y,t) = s_0(y) + \frac{1}{2} v_0^2(y)\, t. \qquad (2)$$

The corresponding Eulerian fields $s(x,t)$ and $v(x,t)$ are conveniently expressed in terms of the mapping $y(x,t)$, inverse to the function $x = X(y,t)$:

$$s(x,t) = s_0(y(x,t)) + \frac{(y(x,t)-x)^2}{2t}, \qquad v(x,t) = \frac{x - y(x,t)}{t}. \qquad (3)$$

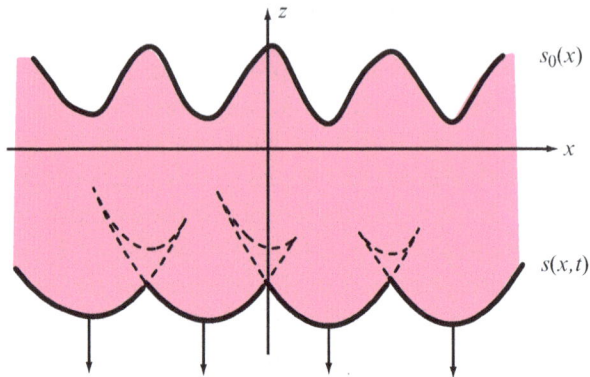

FIGURE 13.4.1
The initial location of the fire front $s_0(x)$ (the upper curve) and the fire front function $s(x,t)$ at a certain time $t > 0$ (the lower curve). The two graphs are dilated along the z-axis to emphasize the direction (opposite to the direction of the z-axis) of propagation of the fire front. The *shaded area* indicates the part of the forest that has burned down during the time interval $(0,t)$. The generalized weak solution of the fire front equation is obtained by rejecting the dashed portions of the multivalued solutions.

A typical fire front function $s(x,t)$ constructed parametrically with the help of the corresponding Lagrangian fields

$$x = X(y,t) = y + v_0(y)\,t, \qquad s = S(y,t) = s_0(y) + \frac{1}{2}v_0^2(y)\,t, \qquad (4)$$

is shown in Fig. 13.4.1. The graph of $s(x,t)$ is presented at the moment of time when the function $s(x,t)$ already has become multivalued.

In the context of the optical wavefronts, the multivaluedness of the function $s(x,t)$ is not a contradiction. The changes in ordering of the pieces of the wavefront simply mean that those wavefront pieces passed through each other.

If $s(x,t)$ is interpreted as a forest fire front, it must be single-valued. Indeed, the fire front can sweep through any forest area only once, because the combustible material is exhausted on the first pass.

13.4. Weak Solutions of First-Order Equations

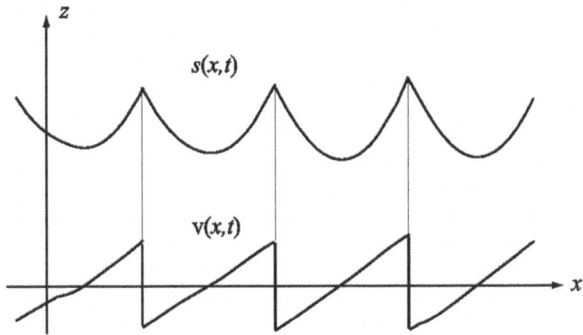

FIGURE 13.4.2
Typical sawtooth weak solutions $v(x,t)$, see (6), of the Riemann equation, and the corresponding piecewise smooth weak solutions $s(x,t)$, see (5), of the equation (13.1.4).

Consequently, if we want to recover the description of the real fire front by solving the first-order nonlinear partial differential equation (13.1.4), it is necessary to define the corresponding generalized (weak) solution via the formula

$$s_w(x,t) = \min_i \left\{ s_0(y_i(x,t)) + \frac{(y_i(x,t) - x)^2}{2t} \right\}, \tag{5}$$

where the minimum is sought over all values of the multivalued function $y(x,t)$ at a given point x and time t. The corresponding discontinuous weak solution of the Riemann equation is defined by the formula

$$v_w(x,t) = \frac{x - y_w(x,t)}{t}, \tag{6}$$

where $y_w(x,t)$ is the value of the multivalued function $y(x,t)$ for which the function $s(x,t)$ (3) assumes its minimum value. The function $v_w(x,t)$ in (6) describes the slope of the real fire front with respect to the x-axis. The graphs of typical weak solutions $s(x,t)$ and $v(x,t)$ are provided in Fig. 13.4.2.

13.4.2 Oleinik–Lax Global Minimum Principle

The weak solutions $s_w(x,t)$ and $v_w(x,t)$ introduced above, while quite satisfactory to a firefighter, have an essential defect from the viewpoint of a mathematician: they force us to manipulate the implicitly given multivalued function $y(x,t)$. So it is desirable to find another, more satisfying for a mathematician, procedure to construct the weak solutions $v_w(x,t)$ and $s_w(x,t)$ that relies on explicitly defined functions.

This can be done, but there is a price to pay for the alternative approach. To achieve our goal, it will be necessary to expand the number of variables of the function under investigation. Namely, we shall take our familiar mapping $x = y + v_0(y) t$ of Lagrangian into Eulerian coordinates, which depends on two variables y and t, and use it to construct a function of three variables,

$$\mathcal{R}(y; x, t) = X(y, t) - x = v_0(y) t + (y - x). \tag{7}$$

Its graph, as a function of the variable y, for a fixed x and t, passes through the zero level at certain points $\{y_i(x, t)\}$, which in general, are values of the multivalued mapping

$$y = y(x, t) \tag{8}$$

of Eulerian into Lagrangian coordinates.

At this point, let us introduce another auxiliary function,

$$\mathcal{G}(y; x, t) = \int^y \mathcal{R}(z, x, t) \, dz = \int^y (X(z, t) - x) \, dz, \tag{9}$$

which, up to an arbitrary constant, is equal to

$$\mathcal{G}(y; x, t) = s_0(y) t + \frac{1}{2}(y - x)^2, \tag{10}$$

where $s_0(x)$ is the familiar initial potential (13.1.3) of the field $v(x, t)$.

Now we can interpret $\mathcal{G}(y; x, t)$ as a (continuous) function of the variable y, depending on x and t as parameters. Note the following remarkable property of this function, which follows directly from its definition: the coordinate $y_i(x, t)$ of each of its extrema, including the minimum, coincides with one of the values of the mapping (8). The global minimum of $\mathcal{G}(y; x, t)$ represents the actual fire front (5); this follows immediately by a comparison of (10) with (5).

The above analysis leads us to the construction of the weak solutions $v_w(x, t)$ and $s_w(x, t)$ via the following formulas:

$$s_w(x, t) = \frac{1}{t}\mathcal{G}(y_w(x, t), x, t) = s_0(y_w(x, t)) + \frac{(y_w(x, t) - x)^2}{2t}, \tag{11}$$

$$v_w(x, t) = \frac{1}{t}[x - y_w(x, t)], \tag{12}$$

where $y_w(x, t)$ is the coordinate of the global minimum of the function $\mathcal{G}(y; x, t)$; see (10). This is the *Oleinik–Lax global minimum principle* mentioned in the title of the present subsection. In essence, the principle defines a discontinuous mapping $y = y_w(x, t)$ of Eulerian into Lagrangian coordinates whose the substitution into the right-hand side of the equalities (11) and (12) gives the desired weak solutions.

13.4. Weak Solutions of First-Order Equations

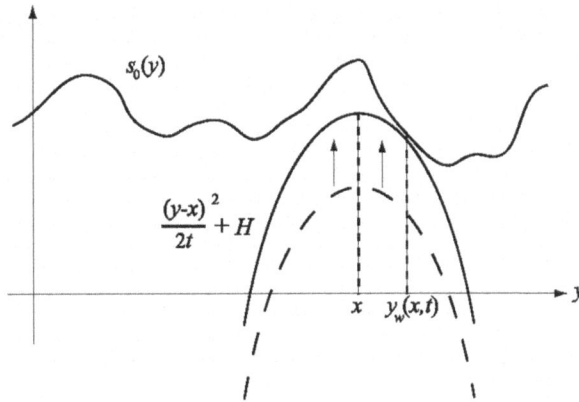

FIGURE 13.4.3
Search of the coordinate of the absolute minimum of the auxiliary function $\mathcal{G}(y; x, t)$ via a geometric algorithm. Raising the parabola $\mathcal{P}$, centered at the point x, we find the coordinate $y_w(x, t)$ where the parabola first touches the graph of the initial potential $s_0(y)$.

13.4.3 A Geometric Construction of the Weak Solutions

The Oleinik–Lax global minimum principle has a striking geometric interpretation. According to this principle, to find the coordinate $y_w(x, t)$ of the global minimum of the function $\mathcal{G}(y; x, t)$, it is necessary to lift, starting from $H = -\infty$, the graph of the parabola

$$\mathcal{P}(y; x, t) = -\frac{(y-x)^2}{2t} + H \tag{13}$$

until it touches the graph of the initial potential $s_0(y)$. The coordinate $y_w(x, t)$ of the first point of tangency will be the desired coordinate of the global minimum of the function $\mathcal{G}(y; x, t)$; see (10). The fields (11) and (12) found using this global minimum will be the desired weak solutions. This geometric algorithm is depicted in Fig. 13.4.3.

Note that the Eulerian coordinate x serves as the parabola's center and that the point of tangency of the parabola with the graph of the initial potential $s_0(y)$ defines the corresponding Lagrangian coordinate $y_w(x, t)$. Thus, the pair of points $\{x, y_w(x, t)\}$ geometrically illustrate the relationship between Lagrangian and Eulerian coordinates.

Another observation is that the parabola (13) needs to be raised to the point of tangency only once; afterward, changing x, we can slide it along the curve $s_0(y)$ (hence the name *osculating parabola*), keeping track of the coordinate $y_w(x,t)$ of the point of tangency.

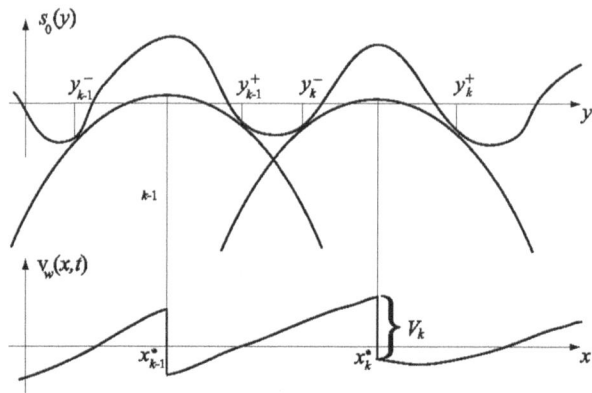

FIGURE 13.4.4
Top: **The initial potential $s_0(y)$ and two critical parabolas.** *Bottom:* **Their centers determine locations of the discontinuities (shocks) of the weak solutions of the Riemann equation.**

The qualitative character of the osculation of the parabola along the initial potential curve depends in an essential way on the magnitude of t. For $0 < t < t_n$, the branches of the parabola drop off rapidly, forming a narrow, pencil-like shape, and the parabola itself has, for every x, a single (but different for different x's) point of tangency with the graph of $s_0(y)$.

As t becomes larger, the parabola becomes more spread out. As a result, for $t > t_n$, there are values $x = x_k^*$ for which the parabola has two points of tangency, say with coordinates $y_k^-(x,t)$ and $y_k^+(x,t)$ ($y_k^+ > y_k^-$), with the curve $s_0(y)$. An infinitesimal increase of x from $x_k^* - 0$ to $x_k^* + 0$ brings about a jump of the function $y_w(x,t)$. Consequently, the field $v_w(x,t)$, see (12), has a jump as well.

Thus the point of double tangency of the parabola $\mathcal{P}$ (13) and the initial potential $s_0(y)$ determines the position of the discontinuity of the generalized (weak) solution $v(x,t)$ of the Riemann equation; the distance between the coordinates of the double tangency points gives the jump's magnitude:

$$V_k = v_w(x_k^* - 0, t) - v_w(x_k^* + 0, t) = \frac{y_k^+ - y_k^-}{t}. \tag{14}$$

An example of *critical parabolas* determining locations and magnitudes of discontinuities (*shocks*) of the weak solutions of the Riemann equation is shown in Fig. 13.4.4.

13.4. Weak Solutions of First-Order Equations

13.4.4 The Convex Hull

The global minimum principle discussed above can be reduced to a geometric procedure for finding the generalized Eulerian-to-Lagrangian mapping $y_w(x,t)$. In general, the mapping is discontinuous. For this reason, it is sometimes convenient to work with the inverse function $X_w(y,t)$, which is continuous everywhere. It can also be constructed via an elegant geometric algorithm. To describe the latter, let us resort once more to the auxiliary function $\mathcal{G}(y;x,t)$ (10):

$$\mathcal{G}(y;x,t) = s_0(y) + \frac{y^2}{2} - xy + \frac{x^2}{2}. \tag{15}$$

As before, we are looking for the coordinate of the global minimum of $\mathcal{G}(y;x,t)$ as a function of y. The first, immediate, observation is that the last term in (15) plays no role in finding the desired location. Thus it can be dropped. Moreover, if we introduce the new notation

$$\varphi(y,t) = \frac{y^2}{2} + s_0(y)\, t, \tag{16}$$

then our problem is reduced to finding the coordinate of the global minimum of the function

$$\varphi(y,t) - xy. \tag{17}$$

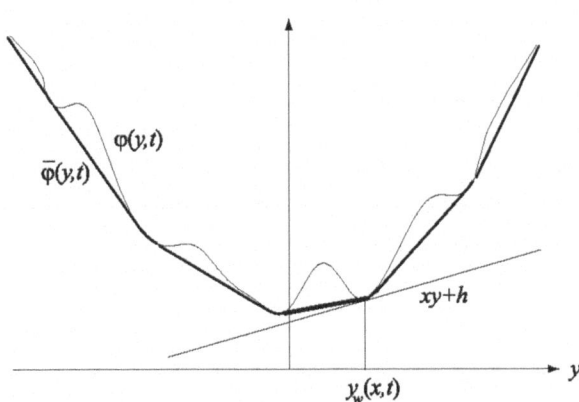

FIGURE 13.4.5
The function $\varphi(y,t)$ and its convex envelope $\bar{\varphi}(y,t)$. Note that the convex envelope is always located above any of the tangent lines to $\varphi(y,t)$.

To find it, we shall "push" (from below) the line

$$xy + h \tag{18}$$

against the graph of the function $\varphi(y,t)$, see (16), thus selecting a particular value of h. The coordinate of the tangency point of the line (18) and the function $\varphi(y,t)$ is the desired coordinate $y_w(x,t)$. Changing the slope x of the line (18), we can find the values of $y_w(x,t)$ for all x.

The key observation is that the outcome of the above-described algorithm will not change if the function $\varphi(y,t)$ is replaced by its *convex envelope* $\bar{\varphi}(y,t)$, and we search for the coordinate of the global minimum of the function

$$\bar{\varphi}(y,t) - xy \tag{19}$$

instead of that of the function (17).

Let us briefly recall the notion of convex envelope. Figuratively speaking, the convex envelope of a function $\varphi(y)$ growing sufficiently fast for $y \to \pm\infty$ has the form of a rubber band stretched on the curve from below. An example of the graph of the convex envelope $\bar{\varphi}(y,t)$ of a function $\varphi(y,t)$ is shown in Fig. 13.4.5. An important property of the convex envelope of a function is that it is always located above all of the function's tangent lines (support hyperplanes in higher dimensions).

Having constructed the convex envelope of $\varphi(y,t)$, we are now ready to look at the problem of finding the location of the minimum of the functions (17) and (19) from a different perspective. Instead of trying to press the straight line (18) against the graph of the curve $\varphi(y,t)$ from below, we can just try to find the slope of the convex envelope $\bar{\varphi}(y,t)$ at a given point y. Obviously, this slope is $x = X_w(y,t)$, the value of the Lagrangian-to-Eulerian mapping for that y. In other words, the mapping $x = X_w(y,t)$ inverse to $y = y_w(x,t)$ is defined by the equality

$$X_w(y,t) = \frac{\partial \bar{\varphi}(y,t)}{\partial y}. \tag{20}$$

The above method of construction of the generalized mapping $X_w(y,t)$ guarantees that it is a continuous function of the variable y, which may be constant on some intervals. The boundaries of the intervals of constancy are formed by the left and right limit values of the inverse mapping at the jump points.

13.4.5 Integral Conservation Laws and Maxwell's Rule

Until now, our strategy of finding weak solutions $v_w(x,t)$ of the Riemann equation has been to look at it as a function of the slope, with respect to the x-axis, of the actual piecewise differentiable "fire front" curve $s_w(x,t)$. However, in numerous applications, for example in nonlinear acoustics, the Riemann equation and its generalized solutions have other interpretations, and in this context, let us introduce another, alternative, definition of weak solutions of the Riemann equation.

The differential form (13.1.1) of the Riemann equation expresses an integral conservation law. Indeed, let us rewrite it in the form

$$\frac{\partial v}{\partial t} + \frac{\partial}{\partial x}\left(\frac{v^2}{2}\right) = 0.$$

Integrating this equation term by term over the whole x-axis and assuming that $v(x,t)$ decays rapidly to zero as $x \to \pm\infty$, we arrive at the equality

$$\frac{d}{dt}\int v(x,t)\,dx = 0.$$

This indicates the existence of the invariant

$$\mathcal{I} = \int v(x,t)\,dx = \int v_0(x)\,dx = \text{const}. \tag{21}$$

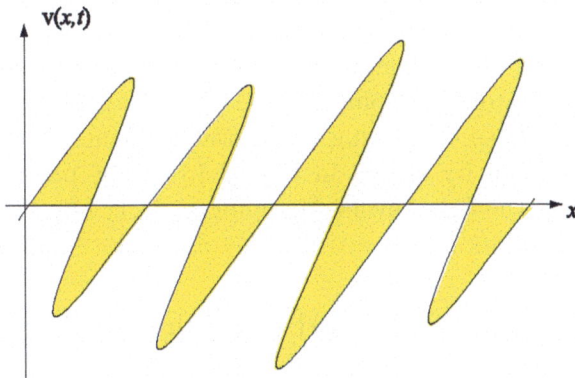

FIGURE 13.4.6
Geometric interpretation of the invariants (21) and (23) in the case of a multistream field $v(x,t)$. The sum of the areas of the shaded domain is equal to the invariant $\mathcal{I}$.

Of course, there are infinitely many integral invariants for the Riemann equation, but in nonlinear acoustic applications, the weak solutions $v_w(x,t)$ are physically meaningful as long as they satisfy the relation (21). We shall now construct solutions of this type (Fig. 13.4.6).

For that purpose, let us substitute in (21) the right-hand side of the second equality in (3),

$$\mathcal{I} = \frac{1}{t} \int [x - y(x,t)]\, dx. \tag{22}$$

A simple geometric argument shows that our invariant can be expressed via the mapping $x = X(y,t)$ inverse to $y = y(x,t)$:

$$\mathcal{I} = \frac{1}{t} \int [X(y,t) - y]\, dy. \tag{23}$$

The last integral makes sense even for the multistream velocity field $v(x,t)$ of uniformly moving particles. In this case, the right-hand side of (3) is equal to the area contained between the graph of the multivalued function $v(x,t)$ and the x-axis. From (23) and from the explicit expression (13.2.1) for $X(y,t)$, equality (21) follows. This confirms that the invariant (21) is also present in multistream flows.

Usually, construction of weak solutions of the Riemann equation that possess the invariant (21) proceeds as follows: In the intervals where the multistream field $v(x,t)$ is multivalued, we draw vertical segments. Their position is chosen so that the cut-off areas of the multivalued field $v(x,t)$, shown in Fig. 13.4.7 (left picture), are equal.

Representation of the invariant (21) in the integral form (23) implies that the above *rule of equal areas* will be satisfied if one uses *Maxwell's rule*, which is well known in mechanics. According to this principle, one has to replace the nonmonotone mapping $X(y,t) = y + v_0(y)\,t$ by the piecewise constant mapping $X_w(y,t)$, placing horizontal segments so that the cut-off pieces of nonmonotonicity shown in Fig. 13.4.7 (right picture) have equal areas. Substituting then the inverse mapping $y_w(x,t)$ into (12), we will get the required weak solution that satisfies the invariance condition (21).

13.4.6 Maxwell's Rule and the Oleinik–Lax Principle

Analytically, Maxwell's rule, which demands that the nonmonotone piece of the mapping $X(y,t)$ be replaced by a horizontal segment at the level $x = x_k^*$, can be written as the condition

13.4. Weak Solutions of First-Order Equations

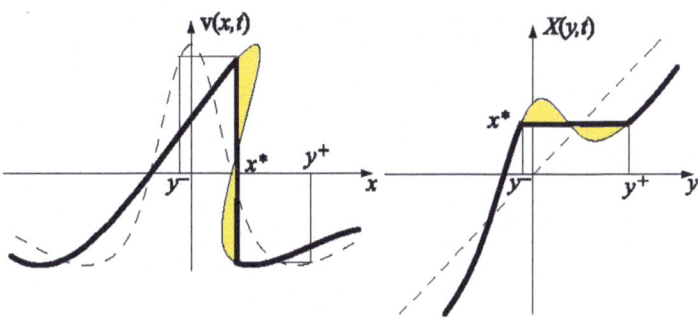

FIGURE 13.4.7
Illustration of the rule of equal areas (*left*) and the equivalent Maxwell's rule (*right*). The former states that to obtain a weak solution (*thick line*) of the Riemann equation, it is necessary to cut off the multivalued branches of the multistream field while keeping the left and right cut-off areas equal. The latter prescribes how to remove the intervals of nonmonotonicity of the Lagrangian-to-Eulerian mapping $x = y + v_0(y)\, t$. Again the upper and lower cut-off areas must be equal. The *dashed lines* indicate the initial conditions.

$$\int_{y_k^-}^{y_k^+} [X(z,t) - x_k^*]\, dy = 0, \tag{24}$$

where $[y_k^-, y_k^+]$ is the interval of the y-axis that is a projection of the horizontal segment described above.

In this subsection we shall establish the equivalence of Maxwell's rule with the Oleinik–Lax global minimum principle described in Sect. 13.4.2.

For this purpose, let us observe that if the mapping $y_w(x,t)$ jumps from y_k^- to y_k^+ as we pass through the point x_k^* from left to right, then for $x = x_k^*$, the function $\mathcal{G}(y; x_k^*, t)$ has two identical global minima attained at y_k^- and y_k^+. In other words, at every discontinuity point x_k^* of the weak solution of the Riemann equation, we have

$$\mathcal{G}(y_k^-, x_k^*, t) = \mathcal{G}(y_k^+, x_k^*, t). \tag{25}$$

In view of the definition (9) of the function $\mathcal{G}(y; x_k^*, t)$, the above condition of equality of the two minima can be written in the form

$$\mathcal{G}(y_k^+, x_k^*, t) - \mathcal{G}(y_k^-, x_k^*, t) = \int_{y_k^-}^{y_k^+} [X(z,t) - x_k^*]\, dy = 0,$$

which, obviously, is equivalent to Maxwell's rule (24).

13.5 E–Rykov–Sinai Principle

In this section we shall present yet another class of weak solutions of the Riemann equation based on a construction of the discontinuous velocity field for the 1-D flow of inelastically colliding (sticky) particles.

13.5.1 Flow of Sticky Particles

To explain our goal, let us begin with an elementary physics problem.

Consider n material points moving uniformly on the x-axis. Their initial (at $t = 0$) coordinates, velocities, and masses are equal, respectively, to $\{x_k, v_k, m_k\}$ ($k = 1, \ldots, n$). On running into each other, the particles stick, following the law of completely inelastic collision. This means that the total mass and momentum of the colliding particles are preserved. At a finite time T, all the particles stick together to form a single *macroparticle*. The problem is to find the coordinate and the velocity of this macroparticle for an arbitrary time instant $t > T$.

The solution to this problem is trivial and is based on Newton's laws of motion: in the absence of external forces, the center of mass of the particles (let us denote it by $x^*(t)$) is subject to uniform motion

$$x^*(t) = x_c + v^* t, \qquad (1)$$

where

$$x_c = \frac{1}{m^*} \sum_{k=1}^{n} m_k x_k \qquad (2)$$

is the initial center of mass, and

$$v^* = \frac{p^*}{m^*} \qquad (3)$$

is the velocity calculated from the law of conservation of momentum with

$$m^* = \sum_{k=1}^{n} m_k, \qquad p^* = \sum_{k=1}^{n} m_k v_k. \qquad (4)$$

Obviously, once all the particles stick together, the position of the thus formed macroparticle coincides with its center of mass, and the macroparticle itself moves according to (1). Observe that to arrive at the solution to the above elementary problem it was not necessary to know the entire history of individual particles. In particular, information about where and in what order particles stuck together was not needed.

13.5. E–Rykov–Sinai Principle

Now let us formulate the more challenging problem: Consider a hydrodynamic flow of microparticles moving uniformly along the x-axis. The dependence of the initial particle velocity on x is given by the function $v_0(x)$, and their initial density is $\rho_0(x)$. The goal is to find the generalized fields of velocity $v(x,t)$ and density $\rho(x,t)$ describing the flow at an arbitrary time $t > 0$, provided that the microparticles stick together on collisions that preserve their total momentum.

The above problem has an elegant solution in the form of the *E–Rykov–Sinai (ERS) principle*. This solution has one feature in common with the solution of the elementary problem discussed at the beginning of this subsection: to find fields $v(x,t)$ and $\rho(x,t)$ at $t > 0$, it is not necessary to know the particles' history in the time interval $[0,t]$. A detailed description of the ERS principle follows.

13.5.2 Inelastic Collisions of Particles

Let us start with an analysis of the velocity field of 1-D flow of inelastically colliding particles. Up to the moment of the first collision, that is, up to the time t_n, see (13.2.3), the velocity and density fields satisfy, in the classical sense, the Riemann equation (13.1.1) and the continuity equation (13.1.2) respectively, and the motion of the particles is completely described by the Lagrangian-to-Eulerian mapping (13.2.1). For times $t > t_n$, the above mapping $x = X(y,t)$ begins to display intervals of nonmonotonicity, indicating the fact that some pairs of particles interchanged their order.

A helpful mental picture of this "interchanging" regime is that of highway traffic.[1] In traffic on a multilane expressway, cars (particles) moving with different velocities can pass each other without colliding. However, if traffic moves in a single lane, then passing is impossible, and cars moving with different velocities eventually bunch up. In our idealized case, we assume that the collisions are totally inelastic, so that after a collision, cars move together as a single "macrocar" satisfying the law of conservation of momentum.

Mathematically, particle sticking can be taken into account by an approach similar to that used in the description of a moving fire front or of nonlinear acoustic waves. The pieces where the mapping (13.2.1) is nonmonotone are cut off by horizontal segments, which in the present case, have a clear-cut mechanical sense: All the particles that initially were located inside the interval $[y^+(t), y^-(t)]$, whose boundaries are projections onto the y-axis

---

[1] In fact, theoretical traffic studies often employ this model.

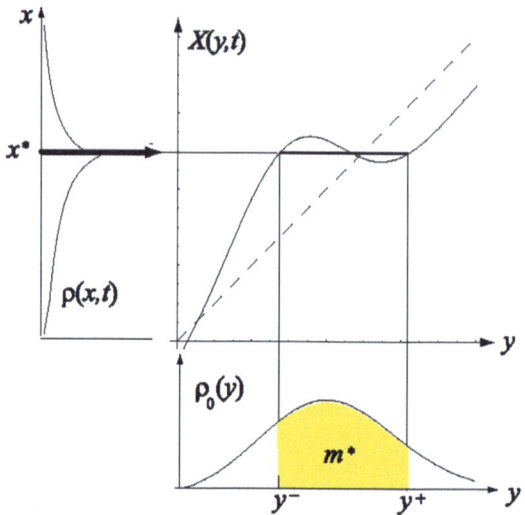

FIGURE 13.5.1
The Lagrangian-to-Eulerian mapping $X(y,t)$ of the original and current density of the particles in a flow of inelastically colliding particles. The area of the shaded domain is equal to the mass m^* of the macroparticle located at the point x^*. The discrete component of the macroparticle density is schematically pictured as the thick arrow on the left.

of the endpoints of the cut-off segments, stick together to form a macroparticle located at the point with coordinate $x^*(t)$, which is a projection of the cut-off segment on the x-axis; see Fig. 13.5.1.

The velocity field is then pieced together from the smooth (in the case of smooth initial condition $v_0(x)$) segments describing uniform motion of the particles that have not yet collided and the discontinuity points created by the formation of macroparticles.

Now the key issue is the determination of the law of motion, $x^*(t)$, of the resulting macroparticles. Recall that we already have a recipe for finding the location of the cut-off nonmonotonicity segments based on the global minimum principle, or the equivalent Maxwell's principle. However, these recipes, although useful in the context of fire fronts and shock creation in nonlinear acoustic waves, have serious shortcomings from the perspective of the physics of particle flows: they place the discontinuities without paying attention to the density distribution of the particle flow.

13.5. E–Rykov–Sinai Principle

In our present context, the initial distribution of particles should essentially influence the motion of macroparticles. Indeed, according to the law of conservation of momentum for colliding particles, the velocity

$$v^*(t) = \frac{dx^*(t)}{dt} \tag{5}$$

of a macroparticle should be equal to

$$v^*(t) = \frac{p^*(t)}{m^*(t)}, \tag{6}$$

where

$$m^*(t) = \int_{y^-(t)}^{y^+(t)} \rho_0(y)\, dy \tag{7}$$

is a macroparticle's mass and

$$p^*(t) = \int_{y^-(t)}^{y^+(t)} \rho_0(y)\, v_0(y)\, dy \tag{8}$$

is its total momentum. Changing the initial density distribution $\rho_0(x)$, we unavoidably affect a macroparticle's velocity and as a result, its position on the x-axis as well.

The E–Rykov–Sinai (ERS) principle, akin to the global minimum principle, will permit us to find the generalized velocity field $v(x,t)$ and the density field $\rho(x,t)$ in inelastically colliding particle flow via a simple geometric construction.

13.5.3 Formulation of the ERS Principle

Like the global minimum principle, the ERS principle can be reduced to finding the coordinate of the global minimum of the function

$$\mathcal{S}(y; x, t) = \int_{y_0}^{y} [X(z,t) - x]\, \rho_0(z)\, dz \tag{9}$$

as a function of the variable y, which also depends on x and t as parameters. Here $X(y,t)$ is the Lagrangian-to-Eulerian mapping (13.2.1) for the flow of noninteracting, uniformly moving particles, and y_0 is an arbitrary constant. To discuss the physical meaning of the ERS principle, we will set $y_0 = -\infty$. Moreover, we shall assume that $\mathcal{S}(y; x, t)$ is a continuous function of its arguments.

Note that the single but crucial difference between the function $\mathcal{S}(y;x,t)$ and the function $\mathcal{G}(y;x,t)$, see (13.4.9), appearing in the absolute minimum principle consists in the presence of the initial density $\rho_0(x)$ under the integral (9). The two functions have a common feature: the coordinates of the extrema in both cases are equal to the Lagrangian coordinate $y_i(x,t)$ of noninteracting particles that are located at x at time t.

The ERS principle can be formulated as follows: *the value $y_w(x,t)$ of the mapping appearing in the weak solution*

$$v_w(x,t) = \frac{x - y_w(x,t)}{t} \tag{10}$$

of the Riemann equation is equal to the coordinate of the global minimum in the variable y of the function $\mathcal{S}(y;x,t)$ (9).

13.5.4 Mechanical Interpretation of the ERS Principle

The mechanical interpretation of the ERS principle is best understood if we rely on physical concepts and arguments. We will explain it by substituting expression (13.2.1) into the integral (9) and rewriting the function $\mathcal{S}(y;x,t)$ in the form

$$\mathcal{S}(y;x,t) = P(y)\,t - x\,M(y) + x_c(y)\,M(y), \tag{11}$$

where

$$M(y) = \int_{-\infty}^{y} \rho_0(y)\,dy \tag{12}$$

is the cumulative mass of the particles located to the left of y. Formula (11) also contains the cumulative momentum

$$P(y) = \int_{-\infty}^{y} \rho_0(y)\,v_0(y)\,dy \tag{13}$$

as well as the initial center of mass of all the particles located to the left of y:

$$x_c(y) = \frac{N(y)}{M(y)}, \qquad N(y) = \int_{-\infty}^{y} z\,\rho_0(z)\,dz\,.$$

Furthermore, observe that if for an arbitrary fixed time t and jump from $x^*(t) - 0$ to $x^*(t) + 0$ of the parameter x, the coordinate $y(x,t)$ of the global minimum of $\mathcal{S}(y;x,t)$, see (9), jumps from $y^-(t)$ to $y^+(t)$ ($y^+ > y^-$), then $\mathcal{S}(y;x^*,t)$ has, because of its continuity with respect to x, two identical global

13.5. E–Rykov–Sinai Principle

minima, located at $y^-(t)$ and $y^+(t)$. This, in turn, implies the validity of the following equality:

$$\mathcal{S}(y^+; x^*, t) - \mathcal{S}(y^-; x^*, t) = \int_{y^-(t)}^{y^+(t)} [x^*(t) - y - v_0(y)\, t]\, \rho_0(y)\, dy = 0. \quad (14)$$

Separation of the terms of the integral and term-by-term integration give

$$x^*(t) = x_c(t) + v^*(t)\, t, \quad (15)$$

where $v^*(t)$ is described by the expression (6) with

$$p^*(t) = P(y^+(t)) - P(y^-(t)), \quad \text{and} \quad m^*(t) = M(y^+(t)) - M(y^-(t)), \quad (16)$$

respectively the momentum and the mass of particles that at $t = 0$ were located within the interval $[y^-(t),\, y^+(t)]$, and

$$x_c(t) = \frac{1}{m^*(t)} \int_{y^-(t)}^{y^+(t)} y\, \rho_0(y)\, dy \quad (17)$$

is their center of mass.

The relationship (15) has the same clear-cut mechanical interpretation as the solution (1) of the elementary problem mentioned at the beginning of the present section. It describes the motion of the center of mass of particles that originally (at $t = 0$) were located in the interval $[y^-(t),\, y^+(t)]$. If by time t, these particles become amalgamated into a single macroparticle, then the equality (15) describes the motion of that macroparticle.

The ERS principle avoids the problem of nonmonotonicity of the mapping $X(y, t)$, and places discontinuities in the velocity field (10) in such a way that they move according to the laws governing the flow of inelastically sticking particles.

13.5.5 The Admissibility Criterion

For a definitive confirmation of the ERS principle as a description of the physical process of inelastic collisions and the subsequent motion of the macroparticles thus created, it is necessary to verify two requirements imposed on the system by the mechanics of inelastic collisions.

First of all, one has to show that the law of macroparticle motion implied by the ERS principle is consistent. In other words, we need to convince ourselves that the velocity $v^*(t)$ of the macroparticle defined in (6) and appearing in (15) is related to the macroparticle's position $x^*(t)$, see (15), through the obvious kinematic relation (5).

Let us demonstrate this by differentiating the jump condition (14) with respect to t:

$$\frac{d}{dt}\int_{y^-(t)}^{y^+(t)}[x^*(t) - y - v_0(y)\,t]\,\rho_0(y)\,dy = 0. \tag{18}$$

Note that the derivatives with respect to the integral limits vanish in view of the obvious equalities

$$x^* = X(y^+, t) = y^+ + v_0(y^+)\,t \quad \text{and} \quad x^* = X(y^-, t) = y^- + v_0(y^-)\,t, \tag{19}$$

which mean that at the boundary of the interval of integration, the integrand in (18) is equal to zero. Differentiation of the integrand leads to the required relationship

$$v^*(t) = \frac{d}{dt}x^*(t) = \frac{p^*(t)}{m^*(t)},$$

where $p^*(t)$ and $m^*(t)$ are given by the formulas (16).

The second requirement, which we shall call the *absorption rule*, says that in the course of time, macroparticles should absorb the surrounding particles without releasing those that already had been absorbed previously.

Since requirements like the one above play a key role in the theory of weak solutions of nonlinear partial differential equations, a comment about general *admissibility criteria* is in order. They all reflect certain physical realities of the system under consideration. For example, in studying the multivalued fire front function $h(x,t)$, it is clear that once the branches of the function fall behind the front (see Fig. 13.4.1), they will never again influence its future evolution.

In all of the above applications of weak solutions of the Riemann equation (fire fronts, nonlinear acoustic waves, flows of inelastically sticking particles), the admissibility criterion reduces to the inequalities

$$v^-(t) \geq v^*(t) \geq v^+(t), \tag{20}$$

where

$$v^-(t) = v_0(y^-) = v(x^* - 0, t) \quad \text{and} \quad v^+(t) = v_0(y^+) = v(x^* + 0, t)$$

are values of the field $v_w(x,t)$ immediately to the left and, respectively, right of the jump point. In the mathematical literature, such criteria are also called *entropy conditions*.[2]

---

[2]See, for example, J. Smoller, *Shock Waves and Reaction–Diffusion Equations*, Springer-Verlag, 1994, Chapter 16.

13.5. E–Rykov–Sinai Principle

When the admissibility criterion expresses the above-mentioned absorption rule, its physical meaning is clear: microparticles will stick to macroparticles (and not separate from each other) only if the velocity v^- on the left is larger than v^*, but the velocity v^+ on the right is smaller than that of the macroparticle.

Let us demonstrate that the ERS principle leads to weak solutions of the Riemann equation that satisfy the admissibility criterion (20). Indeed, first let us subtract the first equation in (19) from the second. Then elementary calculations give the following formula for half the jump size of the velocity field $v_w(x,t)$:

$$V = \frac{v^- - v^+}{2} = \frac{y^+ - y^-}{2t} > 0. \qquad (21)$$

Hence, it is clear that:

- The velocity v^- to the left of the jump point is always larger then the velocity v^+ to the right.

- The magnitude of the jump is directly expressed through the Lagrangian coordinate of a macroparticle that sticks to another macroparticle at time t.

It remains to show that the velocity v^* of the jump (the velocity of the macroparticle) is contained in the interval $[v^+, v^-]$. For this purpose, let us put together both parts of the equalities (19) to get that

$$2x^* = y^+ + y^- + (v^+ + v^-)t.$$

Substituting now x^* from (15) and calculating the macroparticle's velocity from the obtained expression, we get

$$v^* = U + W, \qquad (22)$$

where

$$U = \frac{v^+ + v^-}{2} \qquad (23)$$

is the average of the velocities of microparticles arriving at the jump from left and right, and

$$W = \frac{1}{t}\left[\frac{y^+ + y^-}{2} - x_c\right] \qquad (24)$$

is the difference between U and the velocity of the moving jump. We shall prove that v^*, see (22), satisfies the inequalities (20). For this purpose, we shall evaluate the quantity x_c that enters on the right-hand side of (24).

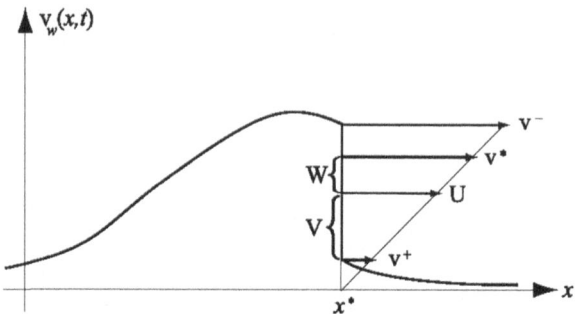

FIGURE 13.5.2
The graph of a discontinuous field $v_w(x,t)$. Shown are all the velocities used in the proof of the fact that the ERS principle generates weak solutions of the Riemann equation satisfying the admissibility criterion.

Using definition (17), one can prove rigorously that
$$y^- \leq x_c \leq y^+, \tag{25}$$
but for a physicist, the inequality is obvious and follows from the fact that the center of mass x_c of the matter contained in the interval $[y^-, y^+]$ lies necessarily inside this interval. Anyway, in view of the inequality (25), the expression (21) for the magnitude of the jump shows that
$$-\frac{V}{2} \leq W \leq \frac{V}{2} \quad \Rightarrow \quad v^+ < U + W < v^-.$$
Consequently, the macroparticle velocity (22) satisfies the admissibility criterion (20). A schematic illustration of all the velocities related to the motion of the jump is shown in Fig. 13.5.2.

Remark 1. The ERS principle is transformed into the global minimum principle if the initial flow density is constant everywhere, that is, $\rho_0(x) \equiv \rho_0 = $ const. Thereby, the weak solution of the Riemann equation, constructed via the Oleinik–Lax principle, can be treated as the velocity field of the flow of inelastically sticking particles, that originally were uniformly distributed.

Remark 2. It is also worthwhile to note that the Oleinik–Lax principle gives an elegant formula for the shock velocity. Indeed, substituting $\rho_0 = $ const into (7) and (17), we get
$$x_c = \frac{y^+ + y^-}{2}. \tag{25}$$

13.5. E–Rykov–Sinai Principle

Consequently, W in (24) is zero, and the shock moves with velocity U, see (23), which is the arithmetic mean of the values of the velocity field $v_w(x,t)$ immediately to the left and right of the discontinuity.

13.5.6 Geometric Construction Based on the ERS Principle

All the geometric constructions of the weak solutions of the Riemann equation based on the global minimum principle can be extended to the cases covered by the ERS principle. In this subsection we shall provide two examples of this type.

Recalling that the function $\mathcal{S}(y;x,t)$, see (9) and (11), is determined only up to an arbitrary constant, we shall rewrite it, in analogy to (13.4.17), in the form

$$\mathcal{S}(y;x,t) = \phi(y,t) - x\, M(y)\,, \tag{26}$$

where

$$\phi(y,t) = P(y)\,t + N(y)\,, \tag{27}$$

with

$$M(y) = \int_0^y \rho_0(z)\,dz\,, \quad P(y) = \int_0^y \rho_0(z)\,v_0(z)\,dz\,,$$
$$N(y) = \int_0^y y\,\rho_0(z)\,dz\,. \tag{28}$$

The mapping, $y_w(x,t)$ defining the weak solution (10) of the Riemann equation is a coordinate of the global minimum of the function (26). To find it, it is necessary to lift the graph of the function

$$M(y)x + h \tag{29}$$

from the minus infinity level until it touches the graph of the function $\phi(y,t)$ (27).

Example 1. A particle carried by a flow. Suppose that a flow of particles with constant density ρ moves uniformly along the x-axis with velocity v. At the moment $t = 0$, a particle of mass m is placed at $x = 0$. Let us find the law of motion of this particle.

To begin, we shall write out the explicit formulas for the functions (28):

$$M(y) = \rho\,y + m\,\chi(y)\,, \quad P(y) = \rho\,v\,y\,, \quad \text{and} \quad N(y) = \frac{\rho}{2}\,y^2\,.$$

Here, $\chi(z)$ is the Heaviside unit step function equal to 1 for $z > 0$ and 0 elsewhere. Also, notice that the particle has no impact on the initial momentum $P(y)$, since its momentum is zero.

Before we proceed to explain the solution of our problem, let us carry out a dimensional analysis of the parameters entering into the problem, something that is routine in physics and engineering. These parameters can be used to form the unique combinations that have the dimensionality of length and time:

$$\ell = \frac{m}{\rho}, \qquad \theta = \frac{m}{v\rho}.$$

Uniqueness of the above dimensional combinations means that ℓ and θ are typical spatial and temporal intervals over which most of the events that are important for the embedded particle are played out. For this reason, it makes sense to pass to the nondimensional variables

$$\eta = \frac{x}{\ell}, \qquad \zeta = \frac{y}{\ell} \quad \text{and} \quad \tau = \frac{t}{\theta},$$

measuring the spatial coordinate and the time in scales that are natural for the particle. Now multiply (27) and (29) by ρ/m^2, and replace $\phi(y,t)$ and $M(y)x$ by the nondimensional functions

$$\psi(\zeta, \tau) = \frac{\rho}{m^2} \phi(y, t)$$

and

$$\mu(\zeta, \eta) = \frac{\rho}{m^2} M(y) x = \eta[\zeta + \chi(\zeta)].$$

It is also convenient to take

$$\psi(\zeta, \tau) = \frac{(\zeta + \tau)^2}{2}. \tag{30}$$

Determination of the law of motion of the particle is equivalent to finding the value $\eta = \eta^*(\tau)$ for which the piecewise linear curve $\mu(\zeta, \eta)$ is tangent to the parabola (30) at two points. The graphics shown in Fig. 13.5.3 suggest that the desired η^* satisfies the quadratic equation

$$(\eta^* - \tau)^2 = 2\eta^*,$$

so that $\eta^* = \tau + 1 - \sqrt{2\tau + 1}$. The resulting law of motion for the macroparticle has the following asymptotics:

$$\eta^* \sim \frac{\tau^2}{2} \quad (\tau \to 0); \qquad \eta^* \sim \tau \quad (\tau \to \infty).$$

13.5. E–Rykov–Sinai Principle

In particular, it is clear that initially, the particle accelerates uniformly. Then, after the flow to the left sticks to it, the resulting particle of mass much bigger than the initial particle moves with a velocity that is asymptotically equal to the underlying flow velocity.

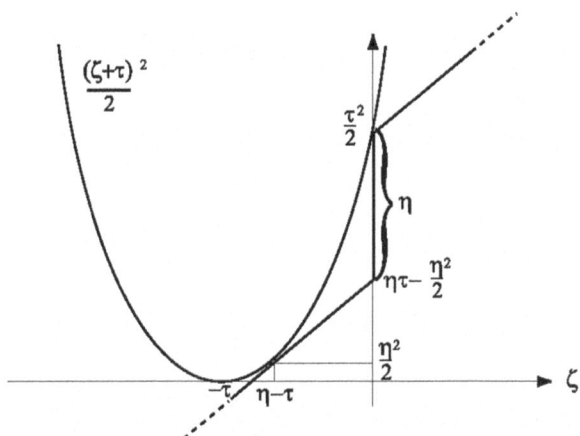

FIGURE 13.5.3
A particle placed in a flow of constant density. Two rectilinear segments ($\zeta - \tau$ for $\zeta < 0$ and $\zeta - \tau + \eta$ for $\zeta > 0$) touch the parabola (30) from below at two points.

The above-mentioned construction of the mapping $y = y_w(x,t)$ is reminiscent of the procedure of Sect. 13.4.4 that led us to the concept of the convex envelope. The latter permitted us to find, in a natural way, the inverse mapping $x = X_w(y,t)$. Let us try to extend the concept of the convex envelope to the case covered by the ERS principle. For this purpose, let us introduce a new coordinate

$$m = M(y) = \int_0^y \rho_0(z)\,dz. \tag{31}$$

If the initial flow density $\rho_0(x)$ is positive everywhere, then there exists a continuous inverse function $y = \mathcal{Y}(m)$. Substituting it in (26)–(29), we arrive at a construction of the weak solution of the Riemann equation that relies on the previously used formula (10), where now

$$y_w(x,t) = \mathcal{Y}(m(x,t)), \tag{32}$$

while $m(x,t)$ is the "coordinate" of the lower point of tangency of the line

$$xm + h \tag{33}$$

and the curve
$$\varphi(m,t) = \mathcal{P}(m)\,t + \mathcal{N}(m). \tag{34}$$
Here,
$$\mathcal{P}(m) = P\left(\mathcal{Y}(m)\right), \qquad \mathcal{N}(m) = N\left(\mathcal{Y}(m)\right). \tag{35}$$

Construction of the weak solution of the Riemann equation via the ERS principle gives the same result if the curve $\varphi(m,t)$ is replaced by its convex envelope $\bar{\varphi}(m,t)$. On the other hand, differentiating the convex envelope with respect to m, we obtain the mapping

$$x = X(m,t) = \frac{\partial}{\partial m}\bar{\varphi}(m,t),$$

inverse to $m(x,t)$.

If one is interested only in the dynamics of the macroparticles that were formed by the adhesion process, then one needs to know only the function $x = X(m,t)$. Indeed, the position $x^*(t)$ of the macroparticle coincides with the elevation of the plateau, the horizontal segment of the mapping $x = X(m,t)$. The mass of the macroparticle is the width $m^*(t) = m^+ - m^-$ of the plateau, and $[m^+, m^-]$ is the projection of the plateau on the m-axis.

13.5.7 Generalized Solutions of the Continuity Equation

Having completed a study of the velocity field of the 1-D flow of inelastically colliding particles, we can turn now to the corresponding generalized solutions of the continuity equation for the particle density in the flow. Since in the construction of the generalized density field one expects to employ the integral conservation laws, it is appropriate to analyze initially not the density, but its integrated version, which is the cumulative density function at the point with Eulerian coordinate x,

$$m(x,t) = \int_{-\infty}^{x} \rho(x',t)dx'. \tag{36}$$

For a time $t \in (0, t_n)$, while the motion remains single-streamed, the corresponding Lagrangian field $M(y,t)$ is independent of time, and is described by the expression (31). This means that the Eulerian field $m(x,t)$ satisfies the equation

$$\frac{\partial m}{\partial t} + v\frac{\partial m}{\partial x} = 0, \qquad m(x, t=0) = \int_{-\infty}^{x} \rho_0(x')\,dx' = M(x). \tag{37}$$

13.5. E–Rykov–Sinai Principle

Substituting for y in (31) the mapping $y_w(x,t)$ obtained via the ERS principle, we arrive at a generalized solution of equation (37),

$$m(x,t) = M(y_w(x,t)). \tag{38}$$

This expression fully agrees with the laws of physics. In particular, crossing through the discontinuity point x^* of the mapping $y_w(x,t)$, the cumulative mass function has a jump of size corresponding to the mass

$$m^* = \int_{y^-}^{y^+} \rho_0(z)\,dz \tag{39}$$

of the macroparticle created at this point. and as $x \to \infty$, when $y_w(x,t) \to \infty$, the generalized cumulative mass function $m(x,t)$, see (38), converges to the total mass of all the particles in the flow.

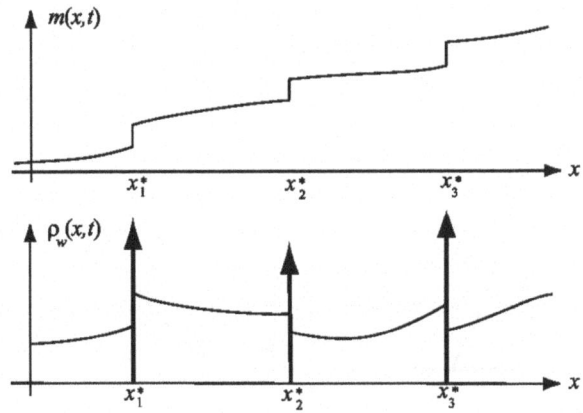

FIGURE 13.5.4
Schematic graphs of the generalized cumulative mass function (*top*), and the density (*bottom*) of the 1-D flow of inelastically colliding particles. The thick arrows in the bottom graph indicate the Dirac deltas describing the singular density of the macroparticles.

To obtain a generalized solution of the continuity equation (13.1.2), it suffices to differentiate the cumulative mass function (38) with respect to x. The sought derivative does not exist in the classical sense. However, in the generalized distributional sense, it is well defined and is equal to

$$\rho_w(x,t) = \sum_k m_k^*(t)\,\delta(x_k^*(t) - x) + \rho_c(x,t), \tag{40}$$

where the summation is taken over all the jump points $\{x_k^*\}$ of the mapping $y_w(x,t)$. The constants $\{m_k^*\}$ in front of the Dirac deltas are equal to the masses of the corresponding macroparticles, and the last term describes the bounded density of uniformly moving microparticles located in the intervals between the macroparticles. A schematic graph of the generalized density field in the flow of inelastically colliding particles is shown in Fig. 13.5.4.

Another method of determination of the generalized density of the flow of sticky particles is based on the representation of the density field with the help of the Dirac delta:

$$\rho_w(x,t) = \int \rho_0(y)\, \delta\left(X_w(y,t) - x\right)\, dy. \tag{41}$$

In the case that $X(y,t)$ is smooth and strictly monotone, the functional on the right-hand side is defined via the usual rules of distribution theory. However, if on some intervals of the y-axis the function $X(y,t)$ is constant, as is the case for the flow of sticky particles, then standard distribution theory fails to apply. But even in this case, the expression (41) can be endowed with a well-defined meaning relying on the concept of supersingular distributions discussed in Sect. 2.9.5 of Volume 1.

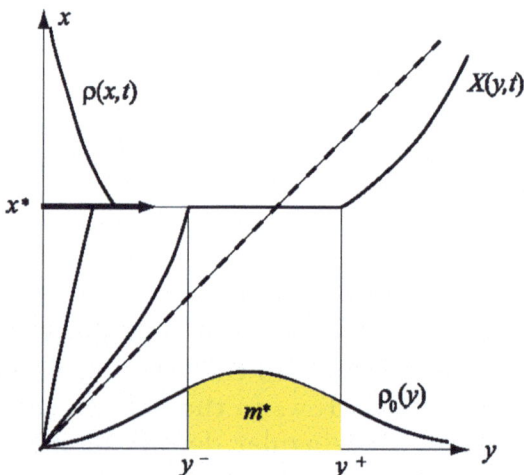

FIGURE 13.5.5
The generalized mapping $X_w(y,t)$ and the corresponding singular density $\rho_w(x,t)$. The initial density is Gaussian.

Operations on supersingular distributions applied to the expression (41) can be reduced to the following formal procedure: multiply the equality (41)

13.5. E–Rykov–Sinai Principle

by a test function $\phi(x)$, integrate both sides with respect to x over the whole real line, change the order of integration on the right-hand side, and finally, use the probing property of the Dirac delta. As a result, we get the equality

$$\int \phi(x)\rho_w(x,t)\,dx = \int \rho_0(y)\,\phi(X(y,t))\,dy, \tag{42}$$

where the left-hand side contains a functional of the generalized density $\rho_w(x,t)$, and the integral on the right discloses how this distribution acts on the test function $\phi(x)$.

Let us take a closer look at the right-hand side of (42), assuming for the sake of simplicity that $x = X(y,t)$ is a smooth function that maps the whole y-axis onto the whole x-axis and is strictly increasing, with the exception of one horizontal piece between the points y^- and y^+, where $X(y,t) = x^* = $ const (see, Fig. 13.5.5). In this situation, the integral on the right can be split into three components:

$$\int \phi(x)\rho_w(x,t)\,dx = \phi(x^*)m^* + \left[\int_{-\infty}^{y^-} + \int_{y^+}^{\infty}\right] \phi(X_w(y,t))\,\rho_0(y)\,dy, \tag{43}$$

where m^* is the mass of a macroparticle given by the expression (39).

Finally, let us take a look at the remaining integrals, taking advantage of the smoothness and monotonicity of the function $z = X(y,t)$ in both intervals of integration, $(-\infty, y^-)$ and (y^+, ∞). After a change of variables, both integrals can be represented as a single integral over the whole x-axis, and we obtain

$$\int \phi(x)\rho_w(x,t)\,dx = \phi(x^*)m* + \int \rho_0(y_w(x,t))\left\{\frac{\partial y_w(x,t)}{\partial x}\right\}\,dx. \tag{44}$$

The braces on the right-hand side indicate that values of the derivative are taken everywhere with the exception of the point x^*, where the derivative can be defined arbitrarily without affecting the value of the integral.

Equality (44) means that the distributional density of the flow of sticky particles is equal to

$$\rho_w(x,t) = m^*\delta(x-x^*) + \rho_0(y_w(x,t))\left\{\frac{\partial y_w(x,t)}{\partial x}\right\}. \tag{45}$$

It is in complete agreement with the equality (40), which was obtained via different reasoning.

13.6 Multidimensional Nonlinear PDEs

Until this section, the present chapter has studied only the properties of 1-D fields depending on the time t and on one spatial variable x. Such fields often arise as idealized and simplified descriptions of real physical processes that usually take place in 3-D Euclidean space or on 2-D surfaces. However, many of the ideas and methods developed for 1-D fields can be extended in a natural fashion to the much more complex—from the perspective of both geometry and mechanics—multidimensional case. So, to complete the picture, in this section we will take a look at multidimensional nonlinear first-order partial differential equations analogous to the 1-D equations considered above.

13.6.1 Basic Equations of 3-D Flows

We begin with the equation for the velocity $\boldsymbol{v}(\boldsymbol{x}, t)$ in a 3-D flow of uniformly moving particles:

$$\frac{\partial \boldsymbol{v}}{\partial t} + (\boldsymbol{v} \cdot \boldsymbol{\nabla})\boldsymbol{v} = 0, \qquad \boldsymbol{v}(\boldsymbol{x}, t=0) = \boldsymbol{v}_0(\boldsymbol{x}). \tag{1}$$

The *nabla operator*

$$\boldsymbol{\nabla} = \boldsymbol{j}_1 \frac{\partial}{\partial x_1} + \boldsymbol{j}_2 \frac{\partial}{\partial x_2} + \boldsymbol{j}_3 \frac{\partial}{\partial x_3}$$

is equal to the sum of the partial derivatives with respect to the three Cartesian coordinates x_1, x_2, x_3, each multiplied by the corresponding unit coordinate vector, $\boldsymbol{j}_1, \boldsymbol{j}_2, \boldsymbol{j}_3$.

If the velocity field is a *potential field*, that is, if there exists a scalar function $s(\boldsymbol{x}, t)$ such that

$$\boldsymbol{v}(\boldsymbol{x}, t) = \boldsymbol{\nabla} s(\boldsymbol{x}, t), \tag{3}$$

then $s(\boldsymbol{x}, t)$ satisfies the equation

$$\frac{\partial s}{\partial t} + \frac{1}{2}(\boldsymbol{\nabla} s)^2 = 0, \qquad s(\boldsymbol{x}, t=0) = s_0(\boldsymbol{x}). \tag{4}$$

In what follows, we shall always assume that the velocity field is a potential field and that the equality (3) is satisfied. In this case,

$$(\boldsymbol{\nabla} s)^2 = v^2 = (\boldsymbol{v} \cdot \boldsymbol{\nabla})s,$$

13.6. Multidimensional Nonlinear PDEs

and the previous equation can be rewritten in the form

$$\frac{\partial s}{\partial t} + (\boldsymbol{v} \cdot \boldsymbol{\nabla})s = \frac{v^2}{2} \tag{5}$$

which is more suitable for our analysis.

The geometric illustration of the behavior of 3-D fields requires construction of 4-D graphs. Since it is much easier to produce convincing 3-D illustrations, many of the results of the theory of nonlinear waves of hydrodynamic type will be here illustrated with the example of 2-D flows, with $\boldsymbol{x} = \{x_1, x_2\}$. Sometimes such fields are of independent interest. For example, the function $s(\boldsymbol{x}, t)$ of the 2-D vector $\boldsymbol{x} = \{x_1, x_2\}$ can have the following familiar interpretations: evolution of a fire front or a wavefront of an optical wave propagating along the third spatial coordinate $z = x_3$ (in the small angle approximation).

Rearrangement of the order of moving particles brings about a change in their density $\rho(\boldsymbol{x}, t)$, both in time and space. The latter satisfies the universal continuity equation

$$\frac{\partial \rho}{\partial t} + (\boldsymbol{\nabla} \cdot \rho \boldsymbol{v}) = 0, \qquad \rho(\boldsymbol{x}, t = 0) = \rho_0(\boldsymbol{x}). \tag{6}$$

Two other equations will also be useful in our exposition. Recall that in addition to the Eulerian coordinates $\boldsymbol{x} = \{x_1, x_2, x_3\}$, it is desirable to consider problems in the Lagrangian coordinates $\boldsymbol{y} = \{y_1, y_2, y_3\}$ "frozen" into the particle flow under consideration. If a certain quantity Q conserves its value in a neighborhood of an arbitrarily chosen particle of the flow, then its Lagrangian field $Q = Q(\boldsymbol{y})$ does not depend on time. The corresponding Eulerian field $q(\boldsymbol{x}, t)$ then satisfies the following partial differential equation:

$$\frac{\partial q}{\partial t} + (\boldsymbol{v} \cdot \boldsymbol{\nabla})q = 0, \qquad q(\boldsymbol{x}, t = 0) = q_0(\boldsymbol{x}). \tag{7}$$

This remains true not only for scalar but also for vector fields. Consequently, the vector mapping $\boldsymbol{y} = \boldsymbol{y}(\boldsymbol{x}, t)$ of Eulerian into Lagrangian coordinates satisfies the equation

$$\frac{\partial \boldsymbol{y}}{\partial t} + (\boldsymbol{v} \cdot \boldsymbol{\nabla})\boldsymbol{y} = 0, \qquad \boldsymbol{y}(\boldsymbol{x}, t = 0) = \boldsymbol{x}. \tag{8}$$

13.6.2 Lagrangian vs. Eulerian Description of 3-D Flows

In Lagrangian coordinates, the nonlinear partial differential equations of Sect. 13.6.1 are reduced to ordinary differential equations. Thus, for the Eulerian coordinates,

$$\frac{d\boldsymbol{X}}{dt} = \boldsymbol{V}, \qquad \boldsymbol{X}(\boldsymbol{y}, t=0) = \boldsymbol{y}, \tag{8}$$

for the Lagrangian velocity field $\boldsymbol{V}(\boldsymbol{y}, t)$,

$$\frac{d\boldsymbol{V}}{dt} = 0, \qquad \boldsymbol{V}(\boldsymbol{y}, t=0) = \boldsymbol{v}_0(\boldsymbol{y}), \tag{9}$$

and for the velocity potential $S(\boldsymbol{y}, t)$,

$$\frac{dS}{dt} = \frac{1}{2} V^2, \qquad S(\boldsymbol{y}, t=0) = s_0(\boldsymbol{y}). \tag{10}$$

The solution of the first equation gives the Lagrangian-to-Eulerian mapping

$$\boldsymbol{x} = \boldsymbol{X}(\boldsymbol{y}, t) = \boldsymbol{y} + \boldsymbol{v}_0(\boldsymbol{y})\, t, \tag{11}$$

and the solutions of the second and third equations provide, respectively, the Lagrangian velocity and the Lagrangian velocity potential fields:

$$\boldsymbol{V}(\boldsymbol{y}, t) = \boldsymbol{v}_0(\boldsymbol{y}), \qquad S(\boldsymbol{y}, t) = s_0(\boldsymbol{y}) + \frac{1}{2} v_0^2(\boldsymbol{y})\, t. \tag{12}$$

To determine the corresponding Eulerian fields it is necessary to find, first, the inverse to (11), that is, the Eulerian-to-Lagrangian mapping

$$\boldsymbol{y} = \boldsymbol{y}(\boldsymbol{x}, t). \tag{13}$$

Then the velocity and the velocity potential can be expressed by the equalities

$$\boldsymbol{v}(\boldsymbol{x}, t) = \boldsymbol{v}_0(\boldsymbol{y}(\boldsymbol{x}, t)), \quad \text{and} \quad s(\boldsymbol{x}, t) = s_0(\boldsymbol{y}(\boldsymbol{x}, t)) + \frac{1}{2} v_0^2(\boldsymbol{y}(\boldsymbol{x}, t))\, t. \tag{14}$$

Another, more convenient, form of these fields expresses these quantities through the Eulerian-to-Lagrangian mapping (13):

$$\boldsymbol{v}(\boldsymbol{x}, t) = \frac{\boldsymbol{x} - \boldsymbol{y}(\boldsymbol{x}, t)}{t}, \tag{15}$$

$$s(\boldsymbol{x}, t) = s_0(\boldsymbol{y}(\boldsymbol{x}, t)) + \frac{(\boldsymbol{y}(\boldsymbol{x}, t) - \boldsymbol{x})^2}{2t}. \tag{16}$$

13.6.3 Jacobian of the Lagrangian-to-Eulerian Mapping

The Jacobian
$$J(\boldsymbol{y}, t) = \left| \frac{\partial \boldsymbol{X}(\boldsymbol{y}, t)}{\partial \boldsymbol{y}} \right| \tag{17}$$

of the Lagrangian-to-Eulerian mapping plays an important role in the analysis of 3-D flows. Substituting the relation (11) into (17) and taking advantage of the assumption that the velocity field is potential, we can write

$$J(\boldsymbol{y}, t) = |\delta_{ij} + s_{ij}\, t|\,, \tag{18}$$

where δ_{ij} is the Kronecker symbol and s_{ij} are components of the symmetric matrix $\hat{s}$, equal to the second-order partial derivatives of the initial velocity field potential

$$\hat{s} = [s_{ij}]\,, \qquad s_{ij}(\boldsymbol{y}) = \frac{\partial^2 s_0(\boldsymbol{y})}{\partial y_i \partial y_j}. \tag{19}$$

As is well known, by a rotation of the coordinate system $\{y_1, y_2, y_3\}$, different for different points $\boldsymbol{y}$, the symmetric matrix $\hat{s}$ can be diagonalized, that is, written in the form

$$\hat{s} = [\lambda_i \delta_{ij}]\,,$$

where $\{\lambda_1, \lambda_2, \lambda_3\}$ are the eigenvalues of the matrix $\hat{s}$. In the local coordinate system specified above, the Jacobian (18) turns out to be

$$J = \prod_{i=1}^{3} (1 + \lambda_i\, t). \tag{20}$$

In what follows, we shall assume that the eigenvalues are numbered in order of increasing magnitude:

$$\lambda_1(\boldsymbol{y}) \leq \lambda_2(\boldsymbol{y}) \leq \lambda_3(\boldsymbol{y})\,.$$

Recall that the Jacobian of the Lagrangian-to-Eulerian mapping $J(\boldsymbol{y}, t)$ has a clearcut geometric interpretation. It is equal to the ratio

$$J(\boldsymbol{y}, t) = \frac{\delta \boldsymbol{v}}{\delta \boldsymbol{v}_0},$$

where $\delta \boldsymbol{v}_0$ is the volume of the infinitesimal domain $\delta \boldsymbol{V}_0$ occupied by the particles at the initial time $t = 0$, and $\delta \boldsymbol{v}$ is the volume of the infinitesimal domain $\delta \boldsymbol{V}$ occupied by the same particles at the current time $t > 0$.

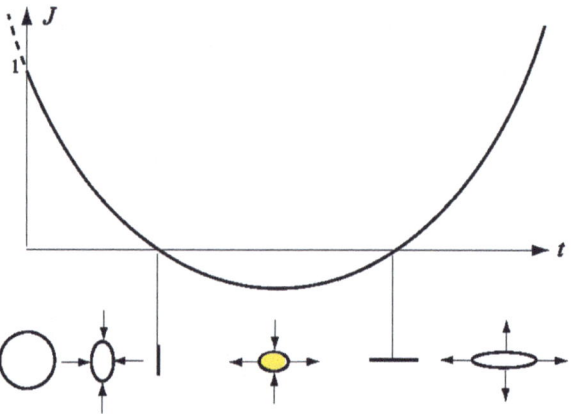

FIGURE 13.6.1
Transformation of an "elementary" 2-D ellipse in the general case $\lambda_1 < \lambda_2 < 0$. Initially, in the interval $0 < t < -1/\lambda_1$, the ellipse is being compressed in both directions. Then in the interval $-1/\lambda_1 < t < -1/\lambda_2$, it expands along the y_1-axis but continues to be compressed along the y_2-axis. Finally, for $t > -1/\lambda_2$, it expands in both directions. The "volume" of the shaded ellipse is negative.

The sign of δv is positive if the corresponding Lagrangian local basis can be transformed into the Eulerian basis using just rotations and dilations; if reflections are also needed, then the sign is negative.

The local Lagrangian basis in which the matrix (19) is diagonal also has a clearcut geometric meaning. Its vectors show the directions of expansion, or compression, of the elementary domain. In other words, if the initial domain δW_0 is a ball, then the domain δW will be an ellipsoid, its principal axes coinciding with the directions of the vectors of the above local basis. A graph illustrating different stages of the temporal evolution of a 2-D ball (disk) is shown in Fig. 13.6.1.

13.6.4 Density of a Multidimensional Flow

Compressions and expansions of the flow lead to changes in its density. In Chap. 2 of Volume 1, we found a solution of the general 3-D continuity equation in the form

$$\rho(\boldsymbol{x}, t) = \int \rho_0(\boldsymbol{y}) \delta\left(\boldsymbol{X}(\boldsymbol{y}, t) - \boldsymbol{x}\right) d^3 y, \tag{21}$$

13.6. Multidimensional Nonlinear PDEs

where $\boldsymbol{x} = \boldsymbol{X}(\boldsymbol{y}, t)$ is the Lagrangian-to-Eulerian mapping generated by the velocity field $\boldsymbol{v}(\boldsymbol{x}, t)$.

In the case of the mapping (11) generated by a flow of uniformly moving particles, one has to distinguish, similarly to the 1-D case, the single-stream and multistream regimes. Within the time interval $t < t_n$, where

$$t_n = -\min_{\boldsymbol{y}} \frac{1}{\lambda_1(\boldsymbol{y})}, \tag{22}$$

the flow is in the single-stream regime, the Jacobian (17) and (20) is strictly positive, and the mapping $\boldsymbol{x} = \boldsymbol{X}(\boldsymbol{y}, t)$ is a one-to-one mapping of $\mathbf{R}^3$ onto $\mathbf{R}^3$. Moreover, in calculating the values of the right-hand side of the equality (21), standard formulas for functionals of the Dirac deltas of composite arguments apply. Thus

$$\rho(\boldsymbol{x}, t) = \frac{\rho_0(\boldsymbol{y}(\boldsymbol{x}, t))}{J(\boldsymbol{y}(\boldsymbol{x}, t), t)}. \tag{23}$$

For $t > t_n$, one sees the appearance of 3-D islets of the multistream motion. In this case, the total density

$$\rho(\boldsymbol{x}, t) = \sum_i^n \frac{\rho_0(\boldsymbol{y}_i(\boldsymbol{x}, t))}{|J(\boldsymbol{y}_i(\boldsymbol{x}, t), t)|} \tag{24}$$

is equal to the sum of the densities of all flows, and it preserves its physical meaning. The above summation is over all n branches of the multivalued, in general, mapping $\boldsymbol{y} = \boldsymbol{y}(\boldsymbol{x}, t)$.

The generalized density field in the 2-D plane and the case of a discontinuous mapping $\boldsymbol{y} = \boldsymbol{y}(\boldsymbol{x}, t)$ will be discussed in the last section of this chapter.

13.6.5 Weak Solutions of the Interface Growth Equation

In some applications, multistream solutions are not physically realizable. In those cases, for $t > t_n$, when a flow of uniformly moving particles becomes multistream and classical solutions of the above-discussed multidimensional first-order partial differential equations no longer exist, one has to replace them with nondifferentiable, or even discontinuous, weak solutions. Let us demonstrate this approach by considering the example of the optical wavefront or the surface of a 3-D combustion region. This brings us back to the problem of propagation of fire fronts.

So, assume that a fire front surrounding a 3-D *combustion region* propagates with unit velocity in a direction perpendicular to the front itself. It is not hard to prove, arguing as in the case of equation (12.2.7), that the fire front

$$z = h(\boldsymbol{x}, t), \tag{25}$$

which is a 2-D surface in a 3-D space, satisfies a nonlinear partial differential equation of the form

$$\frac{\partial h}{\partial t} = \sqrt{1 + (\nabla h)^2}, \tag{26}$$

where $\boldsymbol{x} = \{x_1, x_2\}$ is a 2-D vector in the horizontal plane, and $z = x_3$ is the vertical coordinate.

If the combustion spreads predominantly upward along the z-axis, then the inequality

$$(\nabla h)^2 \ll 1$$

is satisfied, and equation (26) can be simplified. Indeed, if we expand the right-hand side of (26) into a Taylor series with respect to $(\nabla h)^2$, retain only the first two terms of the expansion,

$$\sqrt{1 + (\nabla h)^2} \approx 1 + \frac{1}{2}(\nabla h)^2,$$

and drop the first constant term because it represents a trivial constant-velocity motion of the interface, then we arrive at the approximate equation

$$\frac{\partial h}{\partial t} = \frac{1}{2}(\nabla h)^2, \qquad h(\boldsymbol{x}, t=0) = h_0(\boldsymbol{x}), \tag{27}$$

similar to equation (12.2.9).

Another field closely related to the geometry of the fire front is

$$\boldsymbol{u}(\boldsymbol{x}, t) = -\nabla h(\boldsymbol{x}, t). \tag{28}$$

Its geometric meaning is clear: its magnitude is equal to the tangent of the angle between the normal to the front and the z-axis. Applying the ∇-operator to the terms of (27), we arrive at the following equation for the vector field $\boldsymbol{u}(\boldsymbol{x}, t)$:

$$\frac{\partial \boldsymbol{u}}{\partial t} + (\boldsymbol{u} \cdot \nabla)\boldsymbol{u} = 0, \tag{29}$$

which is a 2-D analogue of equation (1) describing the velocity field of the flow of uniformly moving particles.

13.6. Multidimensional Nonlinear PDEs

As in the case of the flow of uniformly moving particles, equations (27) and (29) describing the interfacial growth can be solved by the method of characteristics. More precisely, they can be transformed into a system of characteristic equations

$$\frac{d\boldsymbol{X}}{dt} = \boldsymbol{U}, \qquad \frac{d\boldsymbol{U}}{dt} = 0, \quad \text{and} \quad \frac{dH}{dt} + \frac{1}{2}U^2 = 0, \tag{30}$$

whose solutions are

$$\boldsymbol{X}(\boldsymbol{y}, t) = \boldsymbol{y} - \nabla h_0(\boldsymbol{y})\, t, \qquad \boldsymbol{U}(\boldsymbol{y}, t) = -\nabla h_0(\boldsymbol{y}), \tag{31}$$

and

$$H(\boldsymbol{y}, t) = h_0(\boldsymbol{y}) - \frac{1}{2}(\nabla h_0(\boldsymbol{y}))^2\, t. \tag{32}$$

Since the vector function

$$\boldsymbol{x} = \boldsymbol{X}(\boldsymbol{y}, t) = \boldsymbol{y} - \nabla h_0(\boldsymbol{y})\, t \tag{33}$$

and its inverse (13) provide a one-to-one mapping of $\mathbf{R}^2$ onto $\mathbf{R}^2$, the expressions (31) and (32) allow us to determine the shape of the desired surface

$$h(\boldsymbol{x}, t) = h_0((\boldsymbol{y}(\boldsymbol{x}, t)) - \frac{(\boldsymbol{y}(\boldsymbol{x}, t) - \boldsymbol{x})^2}{2t} \tag{34}$$

and the field

$$\boldsymbol{u}(\boldsymbol{x}, t) = \frac{\boldsymbol{x} - \boldsymbol{y}(\boldsymbol{x}, t)}{2t} \tag{35}$$

at any point of the $\boldsymbol{x}$-plane. But as soon as the mapping $\boldsymbol{y} = \boldsymbol{y}(\boldsymbol{x}, t)$ becomes multivalued, the real physical surface $h_w(\boldsymbol{x}, t)$ of the combustion region corresponding to the weak solution of equation (27) selects from all branches of the multivalued field $h(\boldsymbol{x}, t)$ the one with the largest magnitude at a given point $\boldsymbol{x}$.

The above-mentioned weak solution of equation (27) can be found by relying on the *global maximum principle*.

At this point, it is helpful to introduce an auxiliary function

$$\mathcal{G}(\boldsymbol{y}; \boldsymbol{x}, t) = h_0(\boldsymbol{y}) - \frac{(\boldsymbol{y} - \boldsymbol{x})^2}{2t} \tag{36}$$

of the vector argument $\boldsymbol{y}$, with $\boldsymbol{x}$ and t considered parameters.

Vanishing of the gradient at $\boldsymbol{x}$, that is, the condition

$$\nabla \mathcal{G}(\boldsymbol{y}; \boldsymbol{x}, t) = 0,$$

is necessary for the smooth function $\mathcal{G}(\boldsymbol{y};\boldsymbol{x},t)$ to have a local extremum at $\boldsymbol{x}$. A direct calculation of the gradient gives

$$\boldsymbol{\nabla} h_0(\boldsymbol{y}) = \frac{1}{t}(\boldsymbol{y}-\boldsymbol{x}).$$

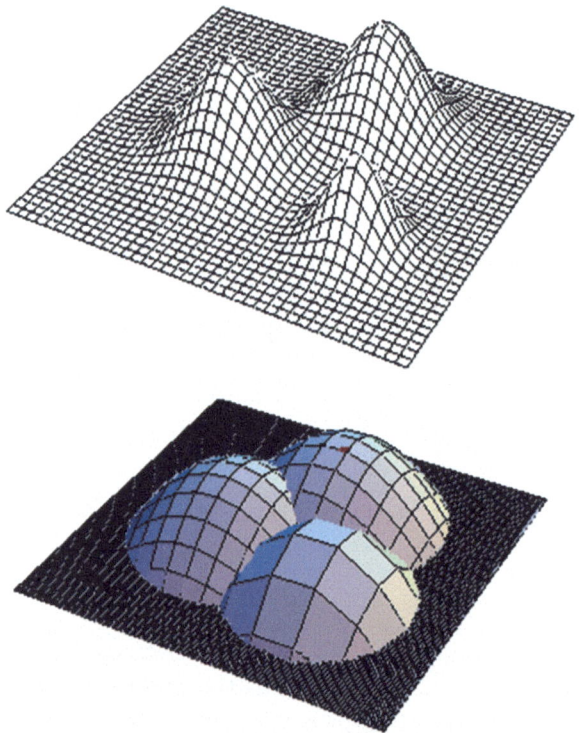

FIGURE 13.6.2
Top: **The initial smooth shape of the growing interface.** *Bottom:* **The same evolving interface after a certain time $t > 0$. The lines of discontinuity of the gradient are clearly visible.**

Comparing this equality with (32)–(34), we observe that all the extreme points of $\boldsymbol{y}(\boldsymbol{x},t)$ turn the mapping (33) into an identity. Moreover, the value of the function $\mathcal{G}(\boldsymbol{y};\boldsymbol{x},t)$ at these points is equal to the value of the Eulerian field $h(\boldsymbol{x},t)$ (34), which represents the elevation of the growing interface. Consequently, substituting the coordinates $\boldsymbol{y}_w(\boldsymbol{x},t)$ of the global maximum of the function (36) into (34) and (35), we obtain the desired weak solutions $h_w(\boldsymbol{x},t)$ and $\boldsymbol{u}_w(\boldsymbol{x},t)$.

Figure 13.6.2 depicts the evolution of a hypothetical initial surface $h_0(x)$ in Fig. 13.6.2 (top), displaying three smooth peaks which, at a certain time $t > 0$, becomes the surface $h(x, t)$ in Fig. 13.6.2 (bottom), which has the form determined by the weak solution of the interface growth equation (27). The graph of the weak solution is constructed as a 3-D parametric plot which uses relations (32), and (33). The *Mathematica* code hides the lower branches of the multivalued functions $h(x, t)$, making only the graph of the weak solution visible. The lines of nondifferentiability are clearly visible; partial derivatives are discontinuous on those lines.

13.7 Exercises

1. Prove that in a single-stream regime, the field $q(x, t)$, see (13.2.9), of the derivative with respect to x of the solution $v(x, t)$ of the Riemann equation (13.1.1) satisfies the inequality

$$q(x,t) < \frac{1}{t}. \tag{1}$$

   Illustrate the above inequality by constructing a graph displaying the dependence on x of the field $tq(x, t)$ in the case of the initial condition

$$v_0(x) = a\sin(kx).$$

2. Find the interval of possible values of the single-stream field $q(x, t)$ in the case in which the initial condition $v_0(x)$ of the Riemann equation satisfies the condition

$$-\nu < v_0'(x) < \mu, \qquad x \in \mathbf{R},$$

   where $\nu > 0$, $\mu > 0$.

3. Suppose that the initial density and the initial velocity of a cloud of uniformly moving particles are equal, respectively, to $\rho_0(x)$ and $v_0(x)$. The initial mass of particles and their center of mass are finite:

$$M = \int \rho_0(y)\,dy < \infty, \qquad x_c = \frac{1}{M}\int y\,\rho_0(y)\,dy < \infty.$$

   Determine the motion

$$\bar{x}(t) = \frac{1}{M}\int x\,\rho(x,t)\,dx$$

of the center of mass of the particle cloud and the time dependence of its dispersion
$$D(t) = \frac{1}{M} \int (x - x_c)^2 \rho(x, t) \, dx \, .$$
Solve the problem by expressing $\bar{x}(t)$ and $D(t)$ via the Fourier images of the density field (13.3.3).

Remark 1. Before we formulate the next problem, the following digression is in order: Our Chap. 13 analysis of the growth of interfaces was based on the approximate equation (13.1.5), which is justified only if the normal to the interface $z = h(x, t)$ deviates little from the z-axis. Our choice was excused by the fact that the approximate equation was easy to study analytically. However, from the geometric point of view it is much easier to construct solutions of the exact interfacial growth equation (12.2.7). Indeed, let us suppose that the interface grows in the direction perpendicular to itself, with constant speed c, and that at $t = 0$, its shape is described by a function $z = h_0(x)$. In this case, to obtain the shape of the growing interface for $t > 0$, it is sufficient to roll a disk of radius ct over the initial shape $z = h_0(x)$. If the radius of the disk is large enough, then the osculating disk will be unable to touch all the points of the initial interface (see Fig. 1), and the osculating disk's center will move along the curve representing the weak solution $h_w(x, t)$ of equation (12.2.7); see Fig. 1.

The above transparent scheme of construction of solutions of the interface growth equation (12.2.7) possesses another valuable feature. It frees us from worrying about whether the initial shape $h_0(x)$ and the obtained solution $h(x, t)$ are represented by single-valued functions. Indeed, the procedure of rolling the circle along any initial shape of the interface preserves the obvious geometric connection of our construction with the physical reality of the interface growing perpendicularly to itself, independently of whether the interface can be represented by a single-valued function in this or other Cartesian coordinate systems.

4. Find a parametric representation of the interface growing perpendicularly to itself in the (x, z)-plane with speed c if at $t = 0$, the interface has the parametric representation
$$x = \zeta(s), \qquad z = \eta(s), \qquad s \in [a, b].$$
Discuss the special case of an explicitly given initial condition $z = h_0(x)$. Compare the solution of the exact equation to the solutions of the equation (13.1.5), valid in the small angle approximation.

13.7. Exercises

5. A contour $\mathcal{L}_0$ is given in the polar coordinate system (r, φ) by the equation $r = \varrho_0(\varphi)$, where $\varrho_0(\varphi)$ is an arbitrary smooth and strictly positive periodic function such that $\varrho_0(0) = \varrho_0(2\pi)$. Track the temporal evolution of this contour in the polar coordinate system assuming that it grows perpendicularly to itself. The solution should rely on a geometric construction.

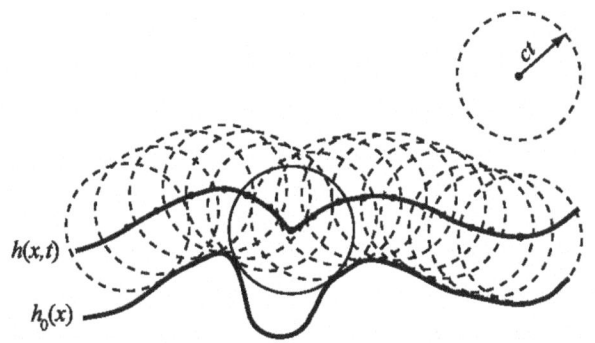

FIGURE 13.7.1
The initial shape $h_0(x)$ of the growing interface and the osculating disk of radius ct. Its center traces the shape of the interface at time $t > 0$. For large enough t, the osculating disk will not able to touch all the points of the initial interface shape, and the center will move along the weak solution $h_w(x,t)$ of the interfacial growth equations (13.1.5) and (12.2.7). A cusp, often encountered in weak solutions, is clearly visible.

6. Using the parametric equations of the contour from Problem 5 (see "Answers and Solutions," Chap. 13, formulas (12)), study its asymptotic shape as $t \to \infty$. Illustrate the results of the asymptotic analysis using the example of the initial contour given by the equation

$$r = 1 + \varepsilon \cos(4\varphi), \qquad \varphi \in [0, 2\pi] \qquad (0 < \varepsilon < 1). \qquad (2)$$

Discuss the dependence of the limit shape on the magnitude of the parameter ε.

7. Visualize a 2-D ice floe bounded by the contour $\mathcal{L}_0$ and floating on the surface of the water. The water surrounding the ice floe can be either cold or warm. We shall assume that in cold water, the floe grows with velocity c perpendicular to the floe's boundary and that in warm water, it melts with the same velocity. Suppose that initially, the floe was frozen by being surrounded by cold water for a period of time, say T. As a result, the floe assumed a new shape with its boundary described by the contour $\mathcal{L}_T$. Subsequently, the water was instantaneously heated, and the floe started melting for another time period T. What conditions on $\mathcal{L}_0$ guarantee that after a cycle of freezing and thawing, the floe returns to its original shape?

8. Suppose that a cone with solid angle Ω is filled with an inert material of 3-D density ϱ, except for the tip of the cone, which we define as the subset of the cone where the distance d from its vertex $\mathcal{O}$ is small, say $d \leq \varepsilon \to 0$. The tip is filled with an explosive of the same density as the inert material. At $t = 0$, the explosive is ignited, acquiring momentum p_0 (momentum density is uniformly distributed throughout the explosive and is equal to p_0/ε). Suppose that the solid angle is small enough ($\Omega \ll 1$) that the material can be assumed to be moving practically along the inner axis r of the cone (see Fig. 13.5.4). Furthermore, assume that in the process of compression, the inert material forms an infinitely thin pancake of sticky particles occupying the whole cross section of the cone. The motion of this pancake can be called a *detonation wave*. Using the ERS principle, find the law of motion of the detonation wave and the rate of growth of the mass of the pancake.

9. Using the global maximum principle, find a weak solution of the Riemann equation (13.1.10) in the case that the initial condition is proportional to the Dirac delta:

$$v_0(x) = s\,\delta(x)\,.$$

Give a physical interpretation of the obtained solution in terms of the flow of inelastically colliding particles.

10. Assume that the initial velocity field of the flow of inelastically colliding particles is

$$v_0(x) = V\chi(-x),$$

and the initial density $\rho_0(x)$ is an absolutely integrable function with total mass

$$\int \rho_0(x)\,dx = M < \infty\,.$$

Derive an equation for the coordinate $x^*(t)$ of the macroparticle created by the flow. Study the asymptotics of this equation as $t \to \infty$.

13.7. Exercises

11. Imagine a 1-D Universe with the very simple universal gravitation law (UGL): The force of gravitational attraction of two bodies is proportional to their masses and independent of their distance. Let us reformulate this UGL in the language of mathematical formulas. Suppose that two bodies, the "left," one of mass M_l, and the "right" one, of mass M_r, are located in the 1-D Universe (x-axis). Denote by F_l the force acting on the "left" body, and by F_r the force acting on the "right" body. According to the UGL, the force of interaction between the two bodies is
$$F_l = -F_r = \gamma M_l M_r,$$
where γ is a "gravitational constant."

Now assume that the matter is continuously distributed in the above universe with initial velocity $v_0(x)$ and density $\rho_0(x)$. Moreover, assume that the total mass of the matter is given by
$$M = \int \rho_0(x)\,dx < \infty.$$

Find the Lagrangian velocity field and the density field of this flow of interacting particles at times before their collisions begin.

12. Suppose that the initial density field in Exercise 11 is
$$\rho_0(x) = \rho_0 \frac{\ell^2}{x^2 + \ell^2}, \tag{27}$$
and that the initial velocity field is identically zero. Construct graphs of the Eulerian velocity field $v(x,t)$ and density field $\rho(x,t)$ for the flow of gravitationally interacting particles at several time instants. Utilizing these graphs, discuss the onset of gravitational instability.

13. Assume that the initial density of the flow of gravitationally interacting particles in Exercise 11 is of the form
$$\rho_0(x) = \rho_0\, g\left(\frac{x}{\ell}\right), \tag{28}$$
where $g(z)$ is a continuous, nonnegative, and even function such that $g(0) = 1$. Find an expression for the Lagrangian velocity and density fields in the limit $\ell \to \infty$ in the case of uniform initial density $\rho_0 = $ const. Assuming that the 1-D universe is initially expanding, that is, $v_0(y) = Hy$, where H is the Hubble constant, find the time of the collapse of the universe.

Chapter 14

Nonlinear Waves and Growing Interfaces: 1-D Burgers–KPZ Models

The present chapter studies behavior of two standard 1-D nonlinear dynamics models described by partial differential equations of order two and higher: the Burgers equation and the related KPZ model. We shall concentrate our attention on the theory of nonlinear fields of hydrodynamic type, where the basic features of the temporal evolution of nonlinear waves can be studied in the context of competition between the strengths of nonlinear and dissipative and/or dispersive effects. Apart from being model equations for specific physical phenomena, Burgers–KPZ equations are generic nonlinear equations that often serve as a testing ground for ideas for analysis of other nonlinear equations. They also produce a striking typically nonlinear phenomenon: shock formation.

14.1 Regularization of First-Order Nonlinear PDEs: Burgers, KPZ, and KdV Equations

The phenomena discussed in Chap. 13 led us to the observation that nonlinear dynamics described in terms of nonlinear first-order partial differential equations may feature a "gradient catastrophe." Of course, the "catastrophe" is just formally mathematical, and signifies a breakdown of the classical solutions of the above-mentioned equations. Nevertheless, regardless of the abstract Armageddon, the physical, economic, and other phenomena we endeavored to model by those equations continue in their happy ways, oblivious to the existential problems of mathematicians.

That disparity between real life and abstract models was overcome in Chap. 13 by the construction of weak solutions. However, physicists are not always pleased with such a formal resolution of the problem. The global principles they employ in the study of the physical world do not lend themselves easily to the exploration of local singularities so essential in the analysis of weak solutions.

Fortunately, there exists another approach to the "gradient catastrophe" problem that relies on the so-called *regularization* of the underlying equations. The regularization method works by adding to the first-order equation higher-order terms that prevent an onset of the "gradient catastrophe" and prolong the "life" of the classical solutions. Moreover, the additional higher-order terms often have a real physical meaning as models of diffusion, dissipation, etc., and their coefficients represent the actual measurable physical parameters essential for the analysis of the phenomena under study.

We shall illustrate the regularization method with examples taken from the theory of interface growth and from nonlinear acoustics.

14.1.1 The Kardar–Parisi–Zhang Equation

Let us consider a curve $z = h(x,t)$ representing (a 1-D analogue of) the interface growing as a result of adhesion of particles deposited on an underlying substrate. The contents of Chap. 13 indicate that if the intensity of the flow of deposited particles is the same in all directions, and the angle between the normal to the interface and the z-axis is small, then in this small angle approximation, the equation describing the temporal evolution of the interface is of the form

$$\frac{\partial h}{\partial t} = \frac{c}{2}\left(\frac{\partial h}{\partial x}\right)^2, \qquad (1)$$

where c is the velocity of interfacial growth in the normal direction. In what follows, to simplify our analysis, we shall assume as a rule that $c = 1$, and will reintroduce this constant only when its magnitude turns out to be essential to understanding the evolution of the interface.

In deriving equation (1), we have assumed that once the deposited particles touch the interface, they stick to it and stop moving. In many physical phenomena, the deposited particles can be subject to further motion such as sliding down the interface's slope under the force of gravity (directed opposite

14.1. Regularization of First-Order Nonlinear PDEs

the z-axis). We shall take this additional motion into account by introducing a new term, $\partial g/\partial x$, into equation (1):

$$\frac{\partial h}{\partial t} + \frac{\partial g}{\partial x} = \frac{1}{2}\left(\frac{\partial h}{\partial x}\right)^2. \tag{2}$$

The new function g introduced above has a transparent physical sense. It represents a *flow* of deposited particles along the $z = h(x,t)$ interface. In the case of particles sliding under the influence of gravity, it is natural to suppose that the rate of sliding increases with the slope of the interface relative to the x-axis. This fact can be modeled by assuming a simple mathematical form of g:

$$g = -\mu\frac{\partial h}{\partial x}. \tag{3}$$

The coefficient μ serves as a measure of the particles' mobility along the interface. Figure 14.1.1 provides a schematic illustration of the process of particles first being deposited on the growing interface and then sliding down the interface.

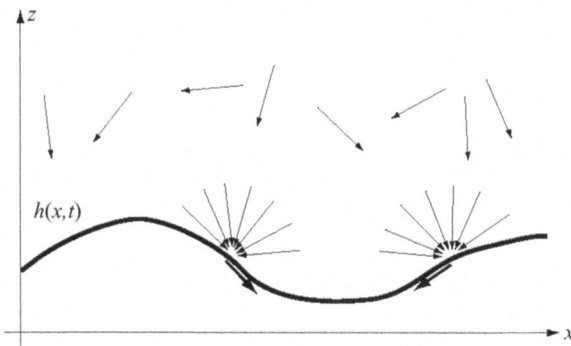

FIGURE 14.1.1
Schematic illustration of the process of isotropic deposition of particles on the interface $z = h(x,t)$. The arrows above the interface represent velocities of particles being deposited on the growing interface. The arrows below the interface indicate the direction of motion of particles sliding down the interface.

In addition to mobility of the particles on the growing interface $z = h(x,t)$, one can also take into consideration the spatial heterogeneity of the incident stream. This extension of the model requires an addition of the term $F(x,t)$ on the right-hand side of equation (2), resulting in the equation

$$\frac{\partial h}{\partial t} = \frac{1}{2}\left(\frac{\partial h}{\partial x}\right)^2 + \mu\frac{\partial^2 h}{\partial x^2} + F(x,t). \tag{4}$$

The above equation is known as the 1-D *Kardar–Parisi–Zhang (KPZ) equation*. It is commonly used as a description of various physical phenomena such as ion-beam deposition of semiconductor films.

Mathematically speaking, the homogeneous version ($F \equiv 0$) of the KPZ equation,

$$\frac{\partial h}{\partial t} = \frac{1}{2}\left(\frac{\partial h}{\partial x}\right)^2 + \mu\frac{\partial^2 h}{\partial x^2}, \tag{5}$$

represents a regularized version of the original equation (1) of the growing interface. We shall see that for $\mu > 0$ and a fairly general initial condition

$$h(x, t=0) = h_0(x), \tag{6}$$

this equation has a unique classical (everywhere twice continuously differentiable) solution valid for every time $t > 0$.

14.1.2 The Burgers Equation

As another example of regularization of a nonlinear first-order partial differential equation, consider the equation of nonlinear acoustics. The discussion in Chap. 13 shows that 1-D nonlinear acoustic waves are described by the Riemann equation

$$\frac{\partial v}{\partial t} + v\frac{\partial v}{\partial x} = 0. \tag{7}$$

In this context, $v(x,t)$ represents the fluctuations of pressure of the medium (for example, the atmosphere) induced by a passing acoustic wave. However, from the physical perspective, equation (7) does not take into account important additional effects such as viscosity. The latter is the principal cause of energy dissipation for an acoustic wave. A first principles derivation of the equation of nonlinear acoustics in a viscous medium leads to the equation

$$\frac{\partial v}{\partial t} + v\frac{\partial v}{\partial x} = \mu\frac{\partial^2 v}{\partial x^2}, \qquad v(x, t=0) = v_0(x), \tag{8}$$

14.1. Regularization of First-Order Nonlinear PDEs

where μ represents the medium's viscosity. From the mathematical point of view, the insertion of viscosity into a physical model leads to a regularization of the nonviscous (inviscid) Riemann equation (7).

Equation (8) is traditionally called the *1-D Burgers equation* in honor of Johannes M. Burgers, who in the 1940s, proposed it as a toy model of strong hydrodynamic turbulence. Indeed, it is a simplified version of the general 3-D *Navier–Stokes equation*

$$\frac{\partial \boldsymbol{v}}{\partial t} + (\boldsymbol{v} \cdot \boldsymbol{\nabla})\boldsymbol{v} = \nu \Delta \boldsymbol{v} - \boldsymbol{\nabla} p + \boldsymbol{F},$$

with the pressure gradient $\boldsymbol{\nabla} p$ and external force $\boldsymbol{F}$ terms omitted. It quickly became apparent that the solutions of equation (8) lack two basic properties of turbulence, namely, transfer of energy between different parts of the spectrum, and coherence at small scales. On the other hand, some other properties of strong hydrodynamic turbulence such as the inertial quadratic nonlinearity and viscosity are present in the Burgers model.

Meanwhile, many other applications of the Burgers equation have been found in areas such as polymer theory and astrophysics. They are mainly due to the fact that the Burgers equation describes a nonlinear wave propagation in a nondispersive medium with weak dissipation, and that the shock fronts behave like inelastic (sticky) particles. This is very different from the behavior of the soliton solutions of other nonlinear equations such as the Korteweg–de Vries equation, which will be discussed in a later section.

Remark 1. The Burgers equation describes the evolution of a gradient in the KPZ model. A termwise differentiation of the KPZ equation (5) with respect to x yields the Burgers equation (8) for the field

$$v(x,t) = -\frac{\partial h(x,t)}{\partial x}.$$

Thus the field $v(x,t)$ can be interpreted not only as fluctuations in the acoustic pressure, but also as the gradient of the growing interface.

Remark 2. The Burgers equation as a conservation law. The Burgers equation also often arises in the following generic situation: Consider a flow of $u(t,x)$ (say, describing the density per unit length of a certain quantity) along the real line with the flux of this quantity through the section at x described by another function $\phi(t,x)$. Assume that the flow is subject to a *conservation law*

$$\frac{\partial}{\partial t}\int_{x_0}^{x_1} u(t,x)\,dx + \phi(t,x_1) - \phi(t,x_0) = 0,$$

when $x_0 < x_1$. If we assume that the flux

$$\phi(t,x) = \Phi(u(t,x))$$

depends on the local density only, then as $x_0 \to x_1$, the above conservation law leads to an equation of Riemann type:

$$\frac{\partial u}{\partial t} + \Phi'(u)\frac{\partial u}{\partial x} = 0.$$

If the flux function is permitted to depend additionally on the gradient of the density u, say

$$\phi(t,x) = \Phi(u(t,x)) - \nu\frac{\partial}{\partial x}u(t,x),$$

then the above conservation law leads to the equation

$$\frac{\partial u}{\partial t} + \Phi'(u)\frac{\partial u}{\partial x} = \nu\frac{\partial^2 u}{\partial x^2},$$

of which the Burgers equation is a special case.

Remark 3. Inviscid limit. In what follows, we shall study the behavior of solutions of the Burgers equation in the *inviscid, or zero-viscosity, limit*, that is, for $\mu \to 0_+$, and discover the so-called *Cheshire cat* effect: the viscosity disappears, but its influence remains. More precisely, as $\mu \to 0_+$, the solution of the Burgers equation converges to a weak solution of the Riemann equation obtained via the global minimum principle. The Cheshire cat effect demonstrates a phenomenon that is characteristic of nonlinear equations: the introduction of a term containing higher derivatives qualitatively changes the behavior of the solutions, even in the case of vanishingly small coefficients.

14.1.3 The Korteweg–de Vries Equation

Regularization of the Riemann equation by addition of a viscosity term is not always acceptable from a physical perspective. The point is that the viscosity *dissipates* energy, but in the study of propagation of nonlinear waves, this dissipation phenomenon is not what is most important. What is more essential is the *dispersion* effect, which makes different harmonic components of a nonlinear wave propagate at different group velocities.

14.1. Regularization of First-Order Nonlinear PDEs

We shall model the interaction of nonlinear and dispersive effects by adding to the nonlinear equation (7) a term characterizing dispersion phenomena for linear waves. Recall that in Chap. 11, we studied propagation of linear waves in dispersive media, which was mathematically described by the dispersion equation (11.1.8). The behavior of a linear 1-D wave $u(x,t)$ in such a medium was described by the integral expression (11.1.10),

$$u(x,t) = \int \tilde{u}_0(\kappa) \exp\left(i[\kappa x - W(\kappa)t]\right) d\kappa, \tag{9}$$

where $\tilde{u}_0(\kappa)$ is the Fourier image of the initial field $u_0(x)$. Observe that the right-hand side of the equality (9) can be treated as a solution of an integrodifferential equation that can be written in the following symbolic form:

$$\frac{\partial u}{\partial t} + iW\left(\frac{\partial}{i\partial x}\right)u = 0. \tag{10}$$

For the wave to propagate without energy dissipation, it is necessary and sufficient that $W(\kappa)$ be a real function of a real argument κ. We also require that a real-valued initial condition $u_0(x)$ result in a real-valued field $u(x,t)$. This forces $W(\kappa)$ to be an odd function of κ.

Suppose that $u(x,t)$ is a sufficiently smooth function of x with Fourier image that essentially vanishes outside a small neighborhood of $\kappa = 0$, and let $W(k)$ be a smooth function of its argument. Then we can approximate $W(\kappa)$ in (9) and (10) by a few terms of its Taylor expansion around $\kappa = 0$:

$$W(\kappa) = c\kappa + \mu\kappa^3 + \cdots.$$

The first term is responsible for a straightforward propagation of the wave without any change of its shape. It can be easily eliminated by changing the variables to a moving coordinate system. The remaining second term permits replacement of the integrodifferential equation (10) by the purely differential equation

$$\frac{\partial u}{\partial t} = \mu\frac{\partial^3 u}{\partial x^3}.$$

Combining it with the Riemann equation gives the equation

$$\frac{\partial u}{\partial t} + u\frac{\partial u}{\partial x} = \mu\frac{\partial^3 u}{\partial x^3}, \tag{11}$$

which is called the *Korteweg–de Vries (KdV)* equation. The KdV equation takes into account the combined effects of nonlinearity and dispersion. The coefficient μ reflects the magnitude of the dispersion effects and plays a completely different role from that of the dissipation coefficient μ in the Burgers equation.

14.2 Basic Symmetries of the Burgers and KPZ Equations and Related Solutions

We shall begin with an analysis of the basic properties of the Burgers equation related to its fundamental symmetries. Later on, we shall produce a general analytic formula for solutions of the Burgers equation, but we will not hurry to present it. Instead, we shall start with a systematic investigation of intrinsic structural properties of the equation.

Note that only rarely can one count on finding explicit general solutions of a nonlinear equation. In their absence, the investigator tries to get some idea about what they are like, studies typical special cases, and using this information tries to assemble a coherent general picture applying qualitative and asymptotic methods. We shall implement this approach in the case of the KPZ and Burgers equations, where the qualitative and asymptotic methods can be compared with explicit analytic solutions, thus reassuring us about the effectiveness of the former.

14.2.1 Invariance Under Translations, Reflections, and Galilean Transformations

The approaches of mathematicians and physicists to what often seems to be the same problem can be strikingly different. However, there is always enough in common to permit a successful collaboration, since both sides are united in search of *symmetries*, or *invariances under transformations*, enjoyed by the phenomena under investigation and the equations representing them. An explicit acknowledgment of these symmetries permits a deeper understanding of the nature of the physical phenomena and can be useful in the process of selection of the appropriate mathematical models. Moreover, knowledge of symmetries helps in the search for solutions of nonlinear equations. With this in mind, we shall now embark on a study of symmetries of the KPZ and Burgers equations.

Our first observation is that if the field $v(x,t)$ satisfies the Burgers equation, then the translated fields $v(x+a,t)$ and $v(x,t+\tau)$, where a and τ are arbitrary space and time shifts, also satisfy the same equation. In what

14.2. Symmetries of Burgers and KPZ Equations

follows, we shall symbolically denote the presence of a symmetry by the symbol $\iff$. Thus the invariances indicated above can be written as the relations

$$\begin{aligned} v(x,t) &\iff v(x+a, t+\tau), \\ h(x,t) &\iff h(x+a, t+\tau). \end{aligned} \quad (1)$$

The second symmetry relation refers to the translational invariance enjoyed by the homogeneous KPZ equation.

In the context of a growing interface, the above symmetries can be augmented by translational symmetry along the z-axis: if the field $h(x,t)$ is a solution of the homogeneous KPZ equation, then for an arbitrary constant H, the field $h(x,t) + H$ is also a solution of the same equation. In other words,

$$h(x,t) \iff h(x,t) + H. \quad (1a)$$

The Burgers equation also enjoys the following *anti-invariance under reflections*, which is sometimes called *antisymmetry under reflections*: if $v(x,t)$ satisfies the Burgers equation, then so does $-v(-x,t)$,

$$v(x,t) \iff -v(-x,t). \quad (2)$$

On the other hand, the KPZ equation enjoys the standard *invariance under reflection* property, that is,

$$h(x,t) \iff h(-x,t). \quad (2a)$$

This symmetry represents the fact that the interfacial growth is the same in both directions of the base line.

In addition to translation and reflection symmetries, the Burgers equation (14.1.8) enjoys *Galilean invariance*, which is of fundamental physical importance. We shall explain this concept assuming that $u(x,t)$ represents the velocity in a 1-D flow of a medium. Let $\tilde{u}(x',t)$ be the velocity field of a medium in the coordinate system x' moving with constant velocity V. Then the same velocity field in the stationary coordinate system x such that $x' = x - Vt$ will have the form

$$u(x,t) = V + \tilde{u}(x - Vt, t).$$

Galilean invariance for the Burgers equation means that

$$v(x,t) \iff V + v(x - Vt, t). \quad (3)$$

It is easy to verify that the KdV equation is also invariant under Galilean transformations.

Let us show how knowledge of symmetries of the Burgers equation helps in finding its solutions. The Galilean transformations (3) permit us to automatically produce from a single solution of the Burgers equation an infinite family of solutions. Moreover, knowledge of the solution of the Burgers equation satisfying the initial condition $\tilde{v}_0(x)$ permits an automatic determination of the solution satisfying the initial condition

$$v_0(x) = \tilde{v}_0(x) + V,$$

where V is an arbitrary constant.

It is also clear that the Galilean invariance of the Burgers equation implies the related symmetry of the homogeneous KPZ equation:

$$h(x,t) \iff \theta x + h(x - c\theta t, t). \tag{3a}$$

To facilitate understanding of the essence of this relationship, we have explicitly included in it the velocity c of the interfacial growth. Since the physical origins of the KPZ equation are very different from those of the Burgers equation, the above relationship has a geometric interpretation not directly related to the Galilean principle: the homogeneous KPZ equation (14.1.5) is invariant under rotation of the Cartesian coordinate system (x, z) by an (infinitesimally small) angle θ.

14.2.2 Scale Invariance and the Reynolds Number

In addition to the symmetry properties discussed in the preceding subsection, it is useful to know how the evolution of the solutions of the Burgers equation is affected by the changing scales of the initial field. To investigate this issue, assume that the initial field $v_0(x)$ has a characteristic spatial scale ℓ and a characteristic magnitude U, that is,

$$v_0(x) = U\, u_0\left(\frac{x}{\ell}\right), \tag{4}$$

where $u_0(s)$ is a dimensionless function of a dimensionless argument.

Passing in the Burgers equation (14.1.8) to the new dimensionless coordinate s and the dimensionless field $u(s,t)$ related to the original quantities via the formulas

$$s = \frac{x}{\ell}, \quad u(s,t) = \frac{v(s\ell, t)}{U}, \tag{5}$$

14.2. Symmetries of Burgers and KPZ Equations

we arrive at a new form of the Burgers equation,

$$\frac{\partial u}{\partial t} + \frac{U}{\ell} u \frac{\partial u}{\partial s} = \frac{\mu}{\ell^2} \frac{\partial^2 u}{\partial s^2}, \qquad u(s, t=0) = u_0(s). \tag{6}$$

First, observe that in the special case of $\mu = 0$, the above equation takes the form

$$\frac{\partial u}{\partial t} + \frac{U}{\ell} u \frac{\partial u}{\partial s} = 0, \qquad u(s, t=0) = u_0(s).$$

The solution of the latter will not depend on the scales of the initial field if the unit of time measurement is taken to be $T = \ell/U$, or in other words, if we pass to the dimensionless time

$$\tau = \frac{t}{T} = \frac{\ell}{U} t. \tag{7}$$

In this case, the above equation is transformed into the canonical Riemann equation

$$\frac{\partial u}{\partial \tau} + u \frac{\partial u}{\partial s} = 0, \qquad u(s, t=0) = u_0(s).$$

Therefore, in the absence of viscosity ($\mu = 0$), the identical in shape, but different in spatial scale and magnitude initial condition $v_0(x)$ generate solutions $v(x, t)$ that vary in time in a similar fashion. The only feature differentiating them is their rate of evolution. And it is not surprising that this *scale invariance* of the Riemann equation extends to its weak, discontinuous solutions as well.

In presence of viscosity ($\mu \neq 0$), equation (6) is transformed into the equation

$$\frac{\partial u}{\partial \tau} + u \frac{\partial u}{\partial s} = \frac{\mu}{\ell U} \frac{\partial^2 u}{\partial s^2}, \qquad u(s, t=0) = u_0(s), \tag{8}$$

so that the behavior of the solutions of the Burgers equation becomes much more complicated and depends qualitatively on the magnitude of the dimensionless parameter

$$\mathrm{R} = \frac{U\ell}{\mu}, \tag{9}$$

which appears on the right-hand side of equation (8). Traditionally, this parameter is called the *acoustic Reynolds number*, or simply the *Reynolds number*. A small Reynolds number ($\mathrm{R} \ll 1$) indicates that the influence of the viscosity is large, the diffusive term on the right-hand side of the Burgers equation (14.1.8) dominates the nonlinear term on the left-hand side, and the equation can be well approximated by the linear diffusion equation

$$\frac{\partial v}{\partial t} = \mu \frac{\partial^2 v}{\partial x^2}. \tag{10}$$

For this reason, the Burgers equation is sometimes called the *nonlinear diffusion equation*. For large Reynolds numbers, the evolution of the field $v(x,t)$ is dominated by propagating nonlinear effects. In other words, one can think of the Reynolds number as a measure of the influence of the nonlinearity on the behavior of the field $v(x,t)$.

Let us take a look at the Reynolds number from yet another point of view. The number can be represented as the ratio of two spatial scales,

$$R = \frac{\ell}{\delta},$$

where ℓ is an external, in relation to the Burgers equation, spatial scale dictated by the initial conditions, while

$$\delta = \frac{\mu}{U} \tag{11}$$

is the internal characteristic spatial scale, intrinsic to the structure of the Burgers equation itself. In what follows, we shall call ℓ and δ the *internal* and *external* scales, respectively.

The concepts of the internal and external scales are difficult to express in a rigorous mathematical language. But they are valuable in heuristic studies of nonlinear fields and in particular, are widely used in the theory of strong atmospheric turbulence. Also, if we select a time-dependent exterior scale $\ell(t)$ and an interior scale $\delta(t)$ of the evolving nonlinear field, we can then define the evolving, time-dependent Reynolds number

$$R(t) = \frac{\ell(t)}{\delta(t)}. \tag{12}$$

14.2.3 Special Solutions: The Hubble Expansion

Until now, we have studied only general properties of the Burgers equation. To prepare us for an in-depth investigation of its solutions, we shall now try to complete the general picture, and to improve our intuitive understanding of the equation, by finding a supply of special solutions. We will not neglect even the simplest of them. As in the tale of Aladdin's lamp, paying attention to simple things can produce handsome payoffs.

14.2. Symmetries of Burgers and KPZ Equations

First, let us investigate whether a function

$$v(x,t) = \beta(t)\, x \qquad (\beta(t=0) = \alpha)$$

linear in the spatial variable x can be a solution of the Burgers equation. Substituting it in (14.1.8), we immediately see that it satisfies the Burgers equation (and the Riemann equation as well) whenever the coefficient $\beta(t)$ is a solution of the ordinary differential equation

$$\frac{d\beta}{dt} + \beta^2 = 0, \qquad \beta(t=0) = \alpha,$$

which has the obvious solution

$$\beta(t) = \frac{\alpha}{1+\alpha t}.$$

Thus we arrive at the following special solution of the Burgers equation:

$$v(x,t) = \frac{\alpha\, x}{1+\alpha\, t}. \tag{13}$$

This seemingly trivial expression deserves a detailed analysis, which will help us understand the structure of solutions of the Burgers equation satisfying arbitrary initial conditions.

In this context, imagine a 1-D expanding (for $\alpha > 0$) Universe. Formula (13) describes the motion of particles moving away from the origin with uniform velocity under the assumption that their initial velocity is given by $v_0(y) = \alpha y$, where y is the initial (Lagrangian) coordinate of a particle. Indeed, the motion of such particles is described by the expressions

$$v = \alpha\, y, \qquad x = y + \alpha\, y t.$$

Eliminating y from these two equations, we obtain the relation (13), which is often called the *Hubble expansion solution*.[1]

The characteristic property of the Hubble expansion is that over time, the velocity field $v(x,t)$ in (13) loses information about the initial velocity profile, which in our case means a loss of information about the coefficient α. Indeed, as $t \to \infty$, the field v approaches the field

$$v(x,t) \sim \frac{x}{t}, \tag{14}$$

which is independent of α.

---

[1] Edwin Hubble was an astronomer who in the 1920s, discovered that the universe is continuously expanding, with galaxies moving away from each other.

For $\alpha < 0$, the solution (13) describes not an expansion but a compression of the universe. This situation leads to the so-called *gradient catastrophe*, in which the velocity field becomes infinite everywhere in finite time. As we will see later on, similar, but local, gradient catastrophes are typical of solutions of the Burgers equation with small values of the coefficient μ.

As a next step, consider a more general model of the 1-D Universe in which the simple Hubble expansion considered above is augmented by other perturbations. In other words, we shall try to solve the Burgers equation with the initial condition
$$v(x, t = 0) = \alpha x + \hat{v}_0(x),$$
where the first term on the right-hand side takes into account the Hubble expansion and the second term is a perturbation of the Hubble velocity field. Not to belabor the point, we shall seek a solution of the Burgers equation in the form of a sum
$$v(x,t) = \frac{\alpha x}{1 + \alpha t} + \hat{v}(x,t), \tag{15}$$
where the first summand corresponds to the Hubble expansion. Astrophysicists traditionally call the second summand *peculiar velocity*. It reflects the evolution of perturbations corresponding to the dilation of the initial velocity fluctuations. Let us explore the evolution of the peculiar velocity in a coordinate system expanding with the fluctuations. Thus the new coordinate system z is related to the old coordinate system x via the obvious equation
$$z = \frac{x}{1 + \alpha t}.$$
In the new coordinate system, the solution velocity field takes the form
$$v(x,t) = \alpha z + u(z,t), \tag{16}$$
where
$$u(z,t) = \hat{v}\left(z(1 + \alpha t), t\right).$$
Substituting (16) into (14.1.8), we arrive at the following equation for the peculiar velocity:
$$\frac{\partial u}{\partial t} + \frac{1}{1 + \alpha t} u \frac{\partial u}{\partial z} = -\frac{\alpha u}{1 + \alpha t} + \frac{\mu}{(1 + \alpha t)^2} \frac{\partial^2 u}{\partial z^2}.$$
The above equation makes it clear that the expanding fluctuations diminish the influence of the nonlinearity, weaken the field's dissipation, but also

14.2. Symmetries of Burgers and KPZ Equations

attenuate (for $\alpha > 0$) the field itself. The last effect can be intuitively explained by the fact that "particles" moving with high velocity overtake the Hubble expansion, so that the difference between the peculiar velocity and Hubble expansion becomes smaller.

We will take attenuation of the peculiar velocity into account by introducing the substitution

$$u(z,t) = \frac{w(z,t)}{1+\alpha t},$$

thus arriving, finally, at the equation

$$\frac{\partial w}{\partial t} + \frac{1}{(1+\alpha t)^2} w \frac{\partial w}{\partial z} = \mu \frac{1}{(1+\alpha t)^2} \frac{\partial^2 w}{\partial z^2}.$$

Note the similarity of coefficients for the nonlinear and dissipative terms. It reflects the fact that after transition to the new time

$$\tau = \frac{t}{1+\alpha t}, \tag{17}$$

which takes into account the simultaneous attenuation of the nonlinear effects and viscosity in the expanding background, we "miraculously" return to the original Burgers equation

$$\frac{\partial w}{\partial \tau} + w \frac{\partial w}{\partial z} = \mu \frac{\partial^2 w}{\partial z^2}, \qquad w(z, \tau=0) = \hat{v}_0(z). \tag{18}$$

Implementation of all of the above transformations yields the relation

$$\hat{v}(x,t) = \frac{1}{1+\alpha t} w\left(\frac{x}{1+\alpha t}, \frac{t}{1+\alpha t}\right). \tag{19}$$

Now it is clear that the interaction of the perturbation field $\hat{v}(x,t)$ with the dilating (for $\alpha > 0$) background leads to attenuation of the field's magnitude and growth of its spatial scale. For this reason, Hubble expansion slows down the evolution of the peculiar field, which, in the absence of expansion, is equal to $w(x,t)$. As a result, during the infinite time interval $t \in (0, \infty)$, one realizes only a portion of the evolution of the field $w(x,t)$, namely that corresponding to the finite time interval $(0, t^*)$, where $t^* = 1/\alpha$.

In the opposite case, $\alpha < 0$, a global gradient catastrophe takes place: an avalanche-type growth of the magnitude of the peculiar field occurs, squeezing the scales of the initial perturbation. Thus the course of events is accelerated dramatically. More precisely, all stages of the evolution of the perturbation $w(x,t)$ on an infinite time interval are played out in the finite time interval $t \in (0, t^*)$, where $t^* = 1/|\alpha|$.

14.2.4 Special Solutions: Stationary Waves

Stationary, or *traveling, waves* are important special solutions of nonlinear partial differential equations. In the 1-D case, they are of the form

$$v(x,t) = v(x - Vt),$$

where $v = v(z)$ depends on only one variable, $z = x - Vt$. Stationary waves move with constant velocity V without changing their shape.

To find stationary solutions of the Burgers equation, it suffices, in view of its Galilean invariance, to find a static, time-independent solution and then "move it" using the relationship (3).

A static solution $v(x)$ must satisfy the equation

$$\frac{d}{dx}\left[\frac{1}{2}v^2(x) - \mu\frac{dv(x)}{dx}\right] = 0, \qquad (20)$$

which has been written in a form immediately providing us with the first integral,

$$\mu\frac{dv(x)}{dx} + \frac{1}{2}\left[U^2 - v^2(x)\right] \equiv 0, \qquad (21)$$

where U is an arbitrary constant of integration, which coincides, obviously, with the extremal value of $v(x)$. Because of the translational invariance of the Burgers equation, it is easy to see that as the second constant of integration we can use an arbitrary shift a of the solution along the x-axis: $v = v(x-a)$. Thus we arrive at a final expression for the static field:

$$v(x) = -U\tanh\left(\frac{Ux}{2\mu}\right). \qquad (22)$$

Finally, an arbitrary stationary wave solution of the Burgers equation can be constructed by applying to (22) the Galilean transformation (3) and an arbitrary shift:

$$v(x,t) = V - U\tanh\left(\frac{U}{2\mu}(x - Vt - a)\right). \qquad (23)$$

Dimensionless graphs of $u(x) = v(x)/U$ are shown in Fig 14.2.1 for several values of the viscosity coefficient μ.

It is evident that the stationary wave solution describes a transition from the maximum value v_- (on the left) of the field to its minimum value v_+ (on the right), where

$$v^- = v(x = -\infty, t) = V + U, \qquad v^+ = v(x = \infty, t) = V - U.$$

14.2. Symmetries of Burgers and KPZ Equations

It follows from (23) that the effective width of the transition zone is equal to the internal scale δ, see (11), and that the velocity V of the stationary wave and its amplitude U are related to the above maximal and minimal values by the familiar relationships

$$V = \frac{v^+ + v^-}{2}, \qquad U = \frac{v^+ - v^-}{2}. \tag{24}$$

Recall that we already encountered these relationships in the context of weak solutions of the Riemann equation obtained via the global minimum principle. There, similar equalities expressed the velocity and amplitude of the discontinuity of the weak solution $v_w(x,t)$ in terms of the field's values immediately to the left, and right, of the jump.

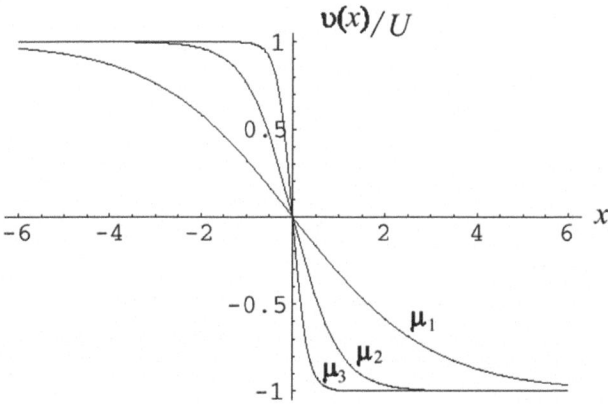

FIGURE 14.2.1
Static dimensionless solutions $v(x)/U$ of the Burgers equation are shown here in a fixed scale on the x-axis for three values of the viscosity coefficient: $\mu_1, \mu_2 = \mu_1/3, \mu_3 = \mu_2/3$. The spatial scale on the x-axis was selected so that the transition from the maximum $v^- = 1$ to the minimum $v^+ = -1$ was best illustrated for the middle value μ_2 of the viscosity coefficient.

A direct relationship between the stationary wave solution and the local behavior of the weak solutions at a jump point is completely understandable if one observes that for $\mu \to 0_+$, the stationary solution (23) weakly converges to the shock wave

$$v_w(x,t) = V - U \operatorname{sign}(x - Vt - a). \tag{25}$$

For this reason, one sometimes says that the introduction of the viscosity ($\mu > 0$) leads to a *smoothing out* of jumps in a weak solution $v_w(x,t)$ of the Riemann equation. If the internal scale of a jump of the weak solution $v_w(x,t)$ is much smaller than the external scale ($\delta(t) \ll \ell(t)$), then the fine structure of the jump can be reconstructed by means of the stationary solution (23) and relations (24), substituting into them the current values $v^{\pm}(t)$ to the right and to the left of the jump of the weak solution $v_w(x,t)$.

14.2.5 Special Solutions: Khokhlov's Formula

The stationary solutions discussed above have a fixed shape. In this section we shall obtain a richer, variable-shaped, family of solutions of the Burgers equation by replacing w in the formulas (19) and (15) by the static solution $v(x)$ from (22):

$$v(x,t) = \frac{1}{1+\alpha t}\left[\alpha x - U\tanh\left(\frac{Ux}{2\mu(1+\alpha t)}\right)\right]. \qquad (26)$$

This rather simple formula, first studied by Rem Khokhlov, a Russian physicist and one of the founders of the nonlinear theory of acoustic waves, illuminates the important interaction of nonlinearity and viscosity in solutions of the Burgers equation.

To better understand the essential features of Khokhlov's solution, we shall first look at its "skeleton," as represented by the weak limit, as $\mu \to 0_+$, of the exact solution (26):

$$v_w(x,t) = U(t)\left[\frac{x}{\ell} - \text{sign}(x)\right], \qquad (27)$$

where

$$U(t) = \frac{U}{1+Ut/\ell}, \qquad \ell = \frac{U}{\alpha}. \qquad (28)$$

Here $U(t)$ is the time-dependent amplitude of the jump, and ℓ is the distance from the origin to the zeros of the weak solution (27). It is natural to consider ℓ an external scale of the fields (26) and (27). Substituting the current jump amplitude $U(t)$, see (28), into (11), we obtain the relevant internal scale:

$$\nu(t) = \frac{\mu}{U(t)} = \frac{\mu}{U}\left(1+\frac{Ut}{\ell}\right). \qquad (29)$$

The ratio of the external and internal scales,

$$\text{R}(t) = \frac{\ell}{\nu(t)} = \text{R}\left(1+\frac{U}{\ell}t\right)^{-1}, \qquad \text{R} = \text{R}(t=0) = \frac{U\ell}{\mu} \qquad (30)$$

14.2. Symmetries of Burgers and KPZ Equations

is called the *time-dependent Reynolds number*. Its value decreases monotonically in time. In the theory of acoustic waves one often speaks about the dissipation of a viscous medium dampening the influence of nonlinear effects on shaping the profile of the acoustic field.

Finally, let us take a look at an important particular case of Khokhlov's solution obtained formally from (26) by letting $U \to \infty$ and $\alpha \to \infty$ in such a way that their ratio remains constant and equal to $\ell < \infty$. Clearly, the resulting limit,

$$v(x,t) = \frac{1}{t}\left[x - \ell \tanh\left(\frac{\ell x}{2\mu t}\right)\right], \tag{31}$$

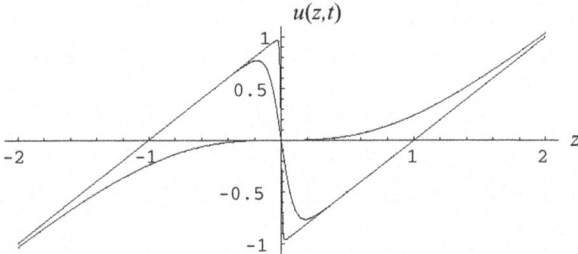

FIGURE 14.2.2
The dimensionless special Khokhlov's solution (33) of the Burgers equations shown for different values R = 100, 10, 1 of the Reynolds number (which is inversely proportional to time). The larger the Reynolds number, the closer the graph becomes to the limiting shock wave solution $u(z) = z - \text{sign}(z)$.

is also an exact solution of the Burgers equation. Its time-dependent internal scale and Reynolds number are

$$\delta = \frac{\mu t}{\ell}, \qquad R = \frac{\ell^2}{\mu t}. \tag{32}$$

From the physicist's viewpoint, the solution (31) describes a universal profile that for arbitrary α and U is the limit of Khokhlov's solution as $t \to \infty$. A graph of the dimensionless field (31),

$$u(z,t) = \frac{t}{\ell} v(z\ell, t) = z - \tanh(Rz), \tag{33}$$

is shown in Fig. 14.2.2 for different values of t. The spatial variable $z = x/\ell$ has been nondimensionalized as well.

14.2.6 Self-Similar Solutions

In this section we shall find *self-similar solutions* of the KPZ and Burgers equations. Their theoretical importance is due to the fact that they determine the large-time asymptotic behavior of many other solutions.

Let us begin by finding self-similar solutions of the homogeneous KPZ equation (14.1.5). Following an earlier developed approach for the self-similar solutions of the linear diffusion equation (10.2.1), we shall look for a solution of equation (14.1.5) of the form

$$h(x,t) = f(\rho) = f(t^n x), \qquad (34)$$

where $f(\rho)$ is a smooth and bounded function depending on one variable $\rho = t^n x$.

Substituting (34) into (14.1.5), we get

$$nt^{n-1} x f' = \frac{1}{2} t^{2n} (f')^2 + \mu t^{2n} f'',$$

where the prime denotes differentiation with respect to ρ. Dividing both sides by t^{2n}, we obtain the equation

$$n \frac{x}{t^{n+1}} f' = \frac{1}{2} (f')^2 + \mu f'',$$

which is consistent with the self-similarity requirement only if $xt^{-n-1} = \rho = xt^n$, which, in turn, is possible only if $n = -1/2$, in which case $\rho = x/\sqrt{t}$. Consequently, $f(\rho)$ must satisfy the equation

$$2\mu g' = g^2 - \rho g, \qquad g = -f'. \qquad (35)$$

Now, (35) can be reduced to a linear equation by considering an auxiliary function $p = 1/g$. Indeed, dividing both sides of equation (35) by $1/p^2$, we obtain the linear first-order equation

$$2\mu p' = \rho p - 1$$

for which the integrating factor is $\exp(-\int \rho\, d\rho/2\mu) = \exp(-\rho^2/4\mu)$. Thus, its solution is

$$p(\rho) = \sqrt{\frac{\pi}{\mu}} \exp\left(\frac{\rho^2}{4\mu}\right) \left[C - \Phi\left(\frac{\rho}{2\sqrt{\mu}}\right) \right],$$

where the auxiliary function is

$$\Phi(z) = \frac{1}{\sqrt{\pi}} \int_{-\infty}^{z} e^{-w^2}\, dw = \frac{1}{2} \left[1 + \mathrm{erf}(z)\right]. \qquad (36)$$

14.2. Symmetries of Burgers and KPZ Equations

As a result,

$$g(\rho) = \frac{1}{p(\rho)} = \sqrt{\frac{\mu}{\pi}} \frac{\exp\left(-\frac{\rho^2}{4\mu}\right)}{C - \Phi\left(\frac{\rho}{2\sqrt{\mu}}\right)}. \tag{37}$$

To find the desired function $f(\rho)$, we need to integrate the right-hand side of the expression (37) with respect to ρ. Let us select the constant of integration so that $f(-\infty) = 0$ and express the constant C in terms of the geometrically more significant constant $S = -f(\infty)$. A simple calculation shows that under the above constraints,

$$f(\rho) = 2\mu \ln\left[1 - \left(1 - e^{-R}\right)\Phi\left(\frac{\rho}{2\sqrt{\mu}}\right)\right], \tag{38}$$

where

$$R = \frac{S}{2\mu}. \tag{39}$$

Hence, the self-similar solutions of the homogeneous KPZ equation are of the form

$$h_R(x,t) = f\left(\frac{x}{2\sqrt{\mu t}}\right). \tag{40}$$

Differentiating (40) with respect to x, we obtain the self-similar solutions of the Burgers equation:

$$v = -\frac{\partial h}{\partial x} \quad \Rightarrow \quad v(x,t) = \sqrt{\frac{\mu}{\pi t}}\, \mathcal{V}_R\left(\frac{x}{2\sqrt{\mu t}}\right), \tag{41}$$

where the "shape" functions are

$$\mathcal{V}_R(z) = \frac{(1 - e^{-R})\, e^{-z^2}}{1 - (1 - e^{-R})\, \Phi(z)}. \tag{42}$$

The graphs of $\mathcal{V}_R(z)$, as functions of the variable z, are shown in Fig. 14.2.3 for several values of the parameter R.

To understand the deeper nature of the above self-similar solutions, we shall now consider mechanisms that lead to their formation. First of all, let us check what kind of initial conditions are associated with such solutions. Note that the weak limit of (40), as $t \to 0+$, is proportional to the Heaviside unit step function

$$h(x, t = 0+) = -S\chi(x),$$

and that the self-similar solution (41) of the Burgers equation weakly converges, as $t \to 0+$, to the Dirac delta,

$$v(x, t = 0+) = S\,\delta(x). \qquad (43)$$

Next, let us elucidate the role of the parameter R. First, observe that by applying the scaling (5) and (7) to the Burgers equation, we get the equation (8). With new scaling, the Dirac delta initial condition is transformed into the initial condition

$$u_0(s) = \frac{S}{U\ell}\,\delta(s).$$

This solution is independent of the scales of the initial condition only if

$$U\ell \sim S.$$

Comparing this equality with (39) and (9), we arrive at the conclusion that R in (39) can be treated as a *Reynolds number of the singular initial conditions* (43).

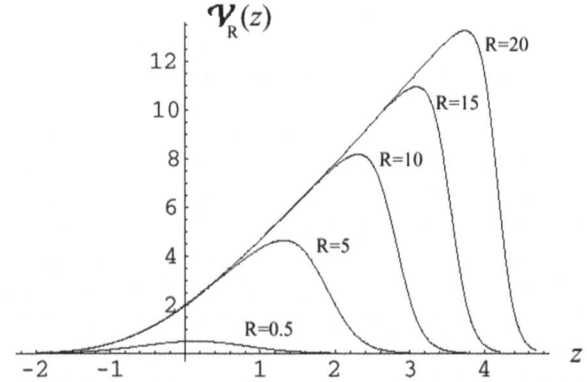

FIGURE 14.2.3
Graphs of the function $\mathcal{V}_R(z)$ for $R = 0.5, 5, 10, 15, 20$. It is clear that the larger R is, the more asymmetric (triangular) the shape of the self-similar solution of the Burgers equation becomes.

To reinforce the above statement, we note that the main asymptotics of $v(x,t)$, see (41) and (42),

$$v(x,t) \sim \frac{S}{2\sqrt{\pi\mu t}} \exp\left(-\frac{x^2}{4\mu t}\right) \qquad (R \to 0), \qquad (44)$$

14.2. Symmetries of Burgers and KPZ Equations

coincides with the main asymptotics of the self-similar solution of the linear diffusion equation (10). This should come as no surprise, since at small Reynolds numbers, the nonlinear effects are negligible. Let us also note that a similar "linearized" asymptotics,

$$v(x,t) \sim \sqrt{\frac{\mu}{\pi t}}(1 - e^{-R}) \exp\left(-\frac{x^2}{4\mu t}\right) \quad (z \to -\infty), \tag{45}$$

is encountered for every R, as long as the values of the function $\Phi(z)$ are small enough. Indeed, for all $z < 0$, they are.

Next, let us discuss the behavior of the self-similar solution (41-42) for large values of the Reynolds number ($R \gg 1$) and for $z \gg 1$. For this purpose, we need to know the asymptotics of $\Phi(z)$ as $z \to \infty$. It is not hard to show that

$$\Phi(z) \sim 1 - \frac{1}{2z\sqrt{\pi}} e^{-z^2} \quad (z \to \infty). \tag{46}$$

Substituting the above expression into (41)–(42), we obtain

$$v(x,t) \approx \frac{x}{t} \frac{1}{2z\sqrt{\pi}\exp(z^2 - R) + 1}, \quad z = \frac{x}{2\sqrt{\mu t}}. \tag{47}$$

It follows from (47) that as long as the first term in the denominator remains small, that is,

$$2z\sqrt{\pi}\exp(z^2 - R) \ll 1, \tag{48}$$

the asymptotics are the familiar, linear in x, asymptotics of the solutions of the Burgers equation,

$$v(x,t) \approx \frac{x}{t}. \tag{49}$$

The critical value z_* at which the linear growth (49) is replaced by an exponential decay is determined by the transcendental equation

$$2z_*\sqrt{\pi}e^{z_*^2} = e^R.$$

A rough estimate of its solution gives

$$z_* \approx \sqrt{R} = \sqrt{\frac{S}{2\mu}}. \tag{50}$$

With this information in hand, it is possible to estimate the dependence of the characteristic scale of the self-similar solution on the variable t:

$$z_* \approx \frac{\ell}{2\sqrt{\mu t}} \quad \Rightarrow \quad \ell(t) \approx \sqrt{2St}. \tag{51}$$

In turn, replacing x in (49) by the characteristic scale $\ell(t)$ of the self-similar solution, see (51), we discover its characteristic amplitude

$$U(t) \approx \frac{\ell(t)}{t}. \tag{52}$$

The above information about $\ell(t)$ and $U(t)$ now can be used to evaluate the time dependence of the Reynolds number. Indeed, taking into account (9), (51), and (52), we have[2]

$$\mathrm{R} = \frac{\ell^2(t)}{2\mu t} \approx \frac{2S}{\mu} = \mathrm{const}. \tag{53}$$

This bring us to an important conclusion: *The Reynolds number of a self-similar solution of the Burgers equation is independent of time.*

For large Reynolds numbers ($\mathrm{R} \gg 1$), we can also clarify the shape of the self-similar solution of the Burgers equation in a transition zone. There, expression (47) can be rewritten in the form

$$v(x,t) \approx \frac{\ell(t)}{t} \frac{1}{e^{z^2 - z_*^2} + 1}. \tag{55}$$

Since in this case, we have $z_* \gg 1$, expression (55) can be simplified even more. Indeed, let us represent z as $z = z_* + s$ and replace the difference $z^2 - z_*^2$ in (55) by $2z_* s$. As a result, we obtain

$$v(x,t) \approx U(t) \frac{1}{e^{2z_* s} + 1}.$$

Since

$$\frac{1}{e^{2a} + 1} = \frac{1}{2} - \frac{1}{2}\tanh(a),$$

we have

$$v(x,t) \approx U(t) - U(t)\tanh\left(\frac{U(t)}{2\mu}[x - \ell(t)]\right), \qquad x \gg \ell(t). \tag{56}$$

Observe that up to the replacement of the second term by V and of the uniform motion $X(t) = Vt + a$ of the jump by $X(t) = \ell(t)$ (the real law of

---

[2]Note that the Reynolds number (53) is four times the size of the R previously introduced in (39). There is no inconsistency here, since the Reynolds number has a semiqualitative character and can be enlarged or reduced a little without changing the essence of its meaning. The definition of R in (39) was selected in a way that was convenient for the subsequent formulas. In what follows, we will stick to the latter definition.

14.2. Symmetries of Burgers and KPZ Equations

motion for a stationary wave solution), the above expression coincides with expression (23).

In summary, one can think of the self-similar solutions of the Burgers equation as being assembled, like a Lego blocks toy, from known particular solutions (45), (49), and (56):

$$v(x,t) \approx \begin{cases} \sqrt{\dfrac{\mu}{\pi t}} \exp\left(-\dfrac{x^2}{4\mu t}\right), & x \ll 0, \\ \dfrac{x}{t}, & 0 < x < \ell, \\ U - U \tanh\left(\dfrac{U}{2\mu}(x-\ell)\right), & x \gg \ell. \end{cases} \quad (57)$$

The first piece is a solution of the linear diffusion equation (10) with initial condition

$$v_0(x) = 2\mu\delta(x) = \frac{2}{R} S \delta(x).$$

The second piece coincides with the characteristic incline (49) and reflects the influence of the nonlinearity. Finally, the third piece describes a sharp decay of a self-similar solution, mimicking the shape of a stationary wave solution.

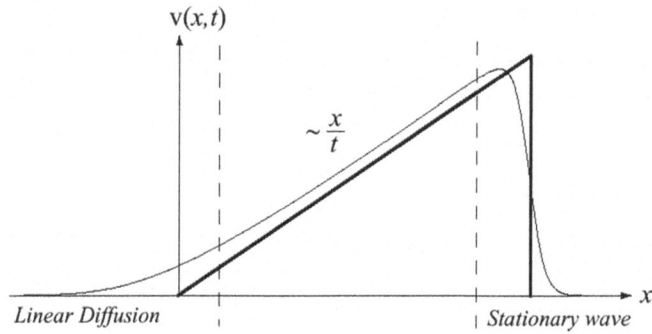

FIGURE 14.2.4
A self-similar solution (41) of the Burgers equation, at a fixed time $t > 0$, is shown here for $R = 50$. Three characteristic pieces of the solution reflecting, respectively, the linear diffusion, the shock wave structure, and the stationary wave behavior are indicated. Also, the triangular "skeleton" of the self-similar solution, reflecting the case of $R = \infty$, is shown.

It is also worth noticing that since

$$\int v(x,t)\, dx = \text{const} \qquad (58)$$

is an obvious invariant of solutions of the Burgers equation, the area under the graph of the self-similar solution remains constant in time and is equal to S. Also, for $R \to \infty$, the self-similar solution (41) and (57) converges weakly to the triangular weak solution

$$v_w(x,t) = \frac{x}{t}\left[\chi(x) - \chi(x-\ell)\right] \qquad (59)$$

of the Riemann equation obtained via the global minimum principle. Following this terminology, for $R \gg 1$, the self-similar solution (41) of the Burgers equation will also be referred to as the *triangular solution*. A graph of the triangular solution $v(x,t)$, see (41), and (57) of the Burgers equation, with all three characteristic pieces indicated, is shown in Fig. 14.2.4.

14.3 General Solutions of the Burgers Equation

Having discussed various special solutions and their heuristics, we are now ready to study the general solutions of the Burgers equation. To kill two birds with one stone and solve the Burgers equation (14.1.8) and the equation (14.1.5) of the growing interface at the same time, we shall initially take a look at the KPZ equation (14.1.4).

14.3.1 The Hopf–Cole Formula

The inhomogeneous KPZ equation

$$\frac{\partial h}{\partial t} = \frac{1}{2}\left(\frac{\partial h}{\partial x}\right)^2 + \mu\frac{\partial^2 h}{\partial x^2} + F(x,t), \qquad h(x,t=0) = h_0(x), \qquad (1)$$

is closely related to the Burgers equation. Although the two equations describe completely different physical phenomena, they are mathematical cousins. Indeed, the substitution

$$v(x,t) = -\frac{\partial h(x,t)}{\partial x} \qquad (2)$$

14.3. General Solutions of Burgers Equation

and a termwise differentiation of the KPZ equation (1) with respect to x leads directly to the inhomogeneous Burgers equation

$$\frac{\partial v}{\partial t} + v\frac{\partial v}{\partial x} = \mu\frac{\partial^2 v}{\partial x^2} + f(x,t), \qquad v(x, t=0) = v_0(x), \qquad (3)$$

where

$$f(x,t) = -\frac{\partial F(x,t)}{\partial x}, \qquad v_0(x) = -\frac{\partial h_0(x)}{\partial x}. \qquad (4)$$

Moreover, both equations are very close relatives of the linear diffusion equation

$$\frac{\partial \varphi}{\partial t} = \mu\frac{\partial^2 \varphi}{\partial x^2} + \frac{1}{2\mu} F(x,t)\,\varphi. \qquad (5)$$

Indeed, write $\varphi(x,t)$ in the form

$$\varphi(x,t) = \exp\left(\frac{h(x,t)}{2\mu}\right). \qquad (6)$$

Then differentiating via the chain rule gives

$$\frac{\partial \varphi}{\partial t} = \frac{1}{2\mu}\frac{\partial h}{\partial t} e^{h/2\mu}, \qquad \frac{\partial^2 \varphi}{\partial x^2} = \frac{1}{2\mu}\left[\frac{1}{2\mu}\left(\frac{\partial h}{\partial x}\right)^2 + \frac{\partial^2 h}{\partial x^2}\right] e^{h/2\mu}.$$

Substituting (6) and the above equalities into (5) and canceling the common factors, we discover that if $\varphi(x,t)$ is a solution of the diffusion equation (5), then $h(x,t)$ appearing in the exponent in (6) obeys the KPZ equation (1).

Thus, every solution of the KPZ equation (1) can be expressed in terms of a solution of the linear diffusion equation (5) via the logarithmic transformation

$$h(x,t) = 2\mu \ln \varphi(x,t). \qquad (7)$$

To ensure that this solution satisfies the initial condition indicated in (1), it is necessary that equation (5) be solved taking into account the initial condition

$$\varphi_0(x) = \exp\left(\frac{h_0(x)}{2\mu}\right). \qquad (8)$$

Formula (7), reducing the nonlinear KPZ equation (1) to a linear diffusion equation (5), is called the *Hopf–Cole substitution*.

Differentiating (7) with respect to x and taking (2) into account, we arrive at the conclusion that the general solution of the inhomogeneous Burgers equation (3) is expressed in terms of the solution of the initial value problem (5) and (8) as follows:

$$v(x,t) = -2\mu \frac{\partial}{\partial x} \ln \varphi(x,t), \tag{9}$$

or equivalently,

$$v(x,t) = -\frac{2\mu}{\varphi} \frac{\partial \varphi}{\partial x}. \tag{9a}$$

In particular, the Hopf–Cole substitution (9) reduces the homogeneous Burgers equation

$$\frac{\partial v}{\partial t} + v \frac{\partial v}{\partial x} = \mu \frac{\partial^2 v}{\partial x^2}, \qquad v(x, t=0) = v_0(x), \tag{10}$$

to the linear diffusion equation

$$\frac{\partial \varphi}{\partial t} = \mu \frac{\partial^2 \varphi}{\partial x^2}, \qquad \varphi_0(x) = \exp\left(-\frac{s_0(x)}{2\mu}\right), \tag{11}$$

where

$$s_0(x) = \int^x v_0(x')\, dx' \tag{12}$$

is the initial potential of the solution of the Burgers equation. We have encountered it in the previous chapters. The solution of the initial value problem for the linear diffusion equation (11) is a convolution of the initial data with the Gaussian kernel,

$$\varphi(x,t) = \frac{1}{2\sqrt{\pi \mu t}} \int \varphi_0(y) \exp\left(-\frac{(y-x)^2}{4\mu t}\right) dy, \tag{13}$$

and has been discussed in Chap. 10. In our case, it takes the form

$$\varphi(x,t) = \frac{1}{2\sqrt{\pi \mu t}} \int \exp\left(-\frac{1}{2\mu t}\left[s_0(y)t + \frac{(y-x)^2}{2}\right]\right) dy. \tag{13a}$$

Note that the expression in the exponent contains the function (13.4.10),

$$\mathcal{G}(y;x,t) = s_0(y)t + \frac{1}{2}(y-x)^2, \tag{14}$$

which we have already encountered in the global minimum principle. Utilizing this notation, the solution $\varphi(x,t)$ can be written in a more compact form,

$$\varphi(x,t) = \frac{1}{2\sqrt{\pi \mu t}} \int \exp\left(-\frac{\mathcal{G}(y;x,t)}{2\mu t}\right) dy, \tag{15}$$

which is known as the *Hopf–Cole formula*.

14.3.2 Averaged Lagrangian Coordinate

In cases in which the integral (9a) can be evaluated analytically, solution of the Burgers equation can be found via a direct substitution of the known function $\varphi(x,t)$ into the right-hand side of the equality (9a). However, an analysis of general properties of solutions of the Burgers equation is easier if they are written in a more convenient form that is obtained by differentiation of the integral in (13) with respect to x, and substitution of the resulting expression into (9a):

$$v(x,t) = \frac{x - \{y\}(x,t)}{t}. \tag{16}$$

The above brace notation $\{\cdots\}$ stands for the operation of spatial averaging with the weight function provided by the nonnegative normalized kernel,

$$f(y;x,t) = \frac{\exp\left[-\frac{1}{2\mu}G(y;x,t)\right]}{\int \exp\left[-\frac{1}{2\mu}G(y;x,t)\right]dy}. \tag{17}$$

In other words, the brace operation applied to the function $g(y)$ is defined by the equality

$$\{g(y)\}(x,t) := \int g(y)f(y;x,t)\,dy. \tag{18}$$

Formula (16) demonstrates a close connection between solutions of the Burgers equation and the classical and weak solutions of the Riemann equation; see, respectively, (12.1.12) and (13.4.6). Indeed, the function $\{y\}(x,t)$ is a natural generalization of the Lagrangian coordinate in the case $\mu > 0$. For this reason, we shall call $\{y\}(x,t)$ the *average Lagrangian coordinate*. Two of its fundamental properties, analogous to the properties of the Lagrangian coordinates, are listed below.

(1) *The average Lagrangian coordinate $\{y\}(x,t)$ is a nondecreasing function of x.*

Indeed, observe first that by direct differentiation of the equalities (17) and (18), we obtain the following relation:

$$\frac{\partial}{\partial x}\{g(y)\} = \frac{1}{2\mu t}\left[\{g(y)y\} - \{g(y)\}\{y\}\right].$$

Substituting $g(y) = y$, we arrive at the inequality

$$\frac{\partial \{y\}}{\partial x} = \frac{1}{2\mu t}\left[\{y^2\} - \{y\}^2\right] = \frac{1}{2\mu t}\{(y - \{y\})^2\} \geq 0 \tag{19}$$

which proves our assertion. The monotonicity property of the average Lagrangian coordinate, together with (16), implies that every solution of the Burgers equation satisfies the inequality

$$\frac{\partial v(x,t)}{\partial x} \leq \frac{1}{t}. \tag{20}$$

(2) Let $\{y\}(x,t)$ and $\{y\}(x,t|V)$ be the average Lagrangian coordinates corresponding to the initial fields $v_0(x)$ and $v_0(x) + V$, respectively. Then

$$\{y\}(x,t|V) = \{y\}(x - Vt, t). \tag{21}$$

The above property follows from the Galilean invariance of solutions of the Burgers equation (16) and from the fact that the function x/t is also invariant under Galilean transformations:

$$V + \frac{x - Vt}{t} \equiv \frac{x}{t}.$$

Finally, let us observe that the weak limit $\lim_{\mu \to 0_+} f(y; x, t)$ is given by

$$\lim_{\mu \to 0_+} f(y; x, t) = \delta(y - y_w(x, t)), \tag{22}$$

where $y_w(x,t)$ is the coordinate of the global minimum of the function (14). Consequently, as $\mu \to 0_+$, the solution of the Burgers equation converges to the weak solution (13.4.12) of the Riemann equation found via the Oleinik–Lax global minimum principle.

14.4 Evolution and Characteristic Regimes of Solutions of the Burgers Equation

In this section we shall review typical scenarios of evolution of solutions of the Burgers equation under various initial conditions. We shall restrict our attention to the strongly nonlinear regimes corresponding to large Reynolds numbers and to transition of solutions of the Burgers equation to the linear regime.

14.4. Evolution and Characteristic Regimes for Burgers Equation

14.4.1 Self-Similar Solutions Revisited

Let us return to the solution (14.3.16) of the Burgers equation. It is not difficult to prove that it exists for all $t \in (0, \infty)$ and that it is a unique solution of the initial value problem (14.3.10), as long as for $x \to \pm\infty$, the growth of $v_0(x)$ is sublinear, that is,

$$|v_0(x)| < A + B|x|^\gamma \qquad (x \in \mathrm{R},\ A, B < \infty,\ 0 \le \gamma < 1). \tag{1}$$

However, the above condition of the global existence and uniqueness of solutions of the Burgers equation does not preclude us from seeking other solutions corresponding to initial conditions that do not satisfy (1). Often, "exotic" solutions of such type provide information that is valuable both theoretically and in applications.

Here, a good example is Khokhlov's solution (14.2.26), its special case (14.2.31), or even the elementary solution (14.2.10) satisfying the linear initial condition $v_0(x) = \alpha x$. An "exotic" character of the latter is demonstrated by the fact that for $\alpha < 0$, the field (14.2.10) explodes in finite time, becoming infinite everywhere at $t^* = 1/|\alpha|$. But even this "end-of-the-world" scenario can serve as a convenient model for the gradient catastrophe familiar from the previous chapters.

Below, we shall provide an example of a solution of the Burgers equation with a singular initial condition. It will provide us with useful hints about the behavior of a wide class of solutions of the Burgers equation.

Example 3. Source solution. A solution $v(x,t)$ of the Burgers equation with initial condition

$$v_0(x) = S\,\delta(x) \tag{2}$$

is often called a *point source solution*. We already know its form, because we discovered earlier that for $t \to 0$, the self-similar solution (14.2.41) converges weakly to the Dirac delta.

The initial potential corresponding to the initial condition (2) is

$$s_0(x) = S\,\chi(x),$$

where $\chi(x)$ stands for the usual Heaviside unit step function. Thus the initial condition for the auxiliary linear diffusion equation (14.3.11) has the form

$$\varphi_0(x) = 1 - (1 - e^{-\mathrm{R}})\chi(x).$$

Substituting it into the solution (14.3.13) of the diffusion equation, we obtain

$$\varphi(x,t) = 1 - (1 - e^{-\mathrm{R}})\Phi\left(\frac{x}{2\sqrt{\mu t}}\right), \tag{3}$$

where R is the initial Reynolds number, familiar from the self-similar solution (14.2.39), and $\Phi(z)$ is the auxiliary function (14.2.36).

Now we can find the desired solution of the Burgers equation with the singular initial condition by substituting (3) into (14.3.9a). Naturally, it coincides with the self-similar solution (14.2.41), but the time spent on finding it via a different route has not been wasted. Indeed, the present approach will now permit us to apply our knowledge of properties of solutions of the linear diffusion equation to obtain a better understanding of the universal importance of self-similar solutions of the Burgers equation; up to this point, they were considered to be just another type of special solution. More specifically, relying on known properties of solutions of the linear diffusion equation, we arrive at the following result: *If the initial field $v_0(x)$ is integrable, with*

$$\int v_0(x)\,dx = S,$$

then for $t \to \infty$, the corresponding solution $v(x,t)$ of the Burgers equation converges to the self-similar solution (14.2.41), with Reynolds number $R = S/2\mu$.

The above fact illustrates how self-similar solutions can serve as a powerful instrument in the analysis of asymptotic properties of arbitrary solutions of nonlinear equations.[3]

Remark 1. Commonsense intuition can be wrong. Note that the above assertion seems to contradict common sense, which suggests that because of dissipation effects, an acoustic wave in a viscous medium ($\mu \neq 0$) should have the current Reynolds number converging to zero ($R \to 0$, $t \to \infty$), and that sooner or later, the field should approach the linear evolution stage. However, we can see that this intuition is wrong: The characteristic amplitude of the field $U(t)$ (14.2.52) does indeed converge to zero at $t \to \infty$, but at the same time, and at the same rate, the exterior scale $\ell(t)$ (14.2.51) is growing. As a result, the Reynolds number remains constant, and the nonlinearity continues to determine the field's profile.

---

[3]See also E. Zuazua, Weakly nonlinear large time behavior in scalar convection–diffusion equation, *Differential and Integral Equations* 6 (1993), 1481–1491.

14.4.2 Approach to the Linear Regime

Let us now rehabilitate the reputation of common sense by observing that if the initial condition is wavelet-like, that is,

$$\int v_0(x)\,dx = 0, \tag{4}$$

then for every arbitrarily large initial Reynolds number R, the solution of the Burgers equation is asymptotically described by the linear diffusion equation. In other words, the nonlinear stage of the evolution of the field $v(x,t)$ is replaced by the linear stage.

Let us begin by finding the asymptotics of solutions of the Burgers equation in the linear stage. Observe that inequality (4), applied to the potential $s_0(x)$ of the initial field $v_0(x)$, implies that the limiting values of $s_0(x)$ as $x \to \pm\infty$ are identical:

$$\lim_{x \to -\infty} s_0(x) = \lim_{x \to \infty} s_0(x).$$

Since the potential is defined only up to an arbitrary constant, we can assume that these limits are equal to zero. A physicist would say that the potential is *localized* in a certain area of the x-axis. To make subsequent considerations more rigorous, let us also assume that the initial potential $s_0(x)$ is compactly supported, which means that it vanishes outside a certain finite interval $|x| < \ell$. In this case, the initial condition of the corresponding linear diffusion equation (14.3.11) will have the form

$$\varphi_0(x) = 1 + q_0(x), \tag{5}$$

where

$$q_0(x) = \exp\left(-\frac{s_0(x)}{2\mu}\right) - 1 \tag{6}$$

is a compactly supported function as well, vanishing outside the interval $|x| \geq \ell$.

The theory of linear diffusion equations implies that for

$$t \gg t_d, \qquad t_d = \frac{\ell^2}{4\mu}, \tag{7}$$

the solution of (14.3.11) with initial condition (5) is described by the asymptotic formula

$$\varphi(x,t) \sim 1 + Q(t) \exp\left(-\frac{x^2}{4\mu t}\right), \tag{8}$$

where
$$Q(t) = \frac{Q_0}{2\sqrt{\pi \mu t}}, \qquad Q_0 = \int q_0(y)\, dy. \tag{9}$$

Substituting (8) into the right-hand side of (14.3.9a), we obtain the following asymptotics of the solution of the Burgers equation:
$$v(x,t) \sim \frac{x}{t} \frac{Q(t)}{\exp\left(\frac{x^2}{4\mu t}\right) + Q(t)}. \tag{10}$$

The function $Q(t)$ appearing in the above formulas decreases monotonically to zero as $t \to \infty$. Therefore, for $t \gg t_l$, where t_l is the characteristic time of entry into the linear stage defined by the condition
$$Q(t_l) \simeq 1 \quad \Rightarrow \quad t_l = \frac{Q_0^2}{4\pi \mu}, \tag{11}$$

the asymptotics of the solution of the Burgers equation (14.3.10) will practically coincide with those of the solution of the linear diffusion equation
$$v(x,t) \sim \frac{x}{t} \frac{Q_0}{2\sqrt{\pi \mu t}} \exp\left(-\frac{x^2}{4\mu t}\right). \tag{12}$$

14.4.3 N Waves and U Waves

The above, somewhat formal, considerations permitted us to find the characteristic time t_l (11) of entry of the solution of the Burgers equation into the linear stage. However, it turns out that the time itself, as well as the mechanism of transition to the linear regime, strongly depend on the shape of the initial field $v_0(x)$. In this context it is fruitful to introduce two types of nonlinear waves: *N-waves* and *U-waves*. They will permit us to illustrate qualitatively different evolution scenarios and transitions to the linear stage.

Example 1. N-wave solution. Suppose that the initial field is the difference of two Dirac deltas:
$$v_0(x) = S\left[\delta(x+\ell) - \delta(x-\ell)\right]. \tag{13}$$

Its potential $s_0(x)$, a rectangular function, and the function $q_0(x)$ defined in (6) are as follows:
$$s_0(x) = \left[\chi(x+\ell) - \chi(x-\ell)\right], \tag{14}$$
$$q(x) = \left(e^{-R} - 1\right)\left[\chi(x+\ell) - \chi(x-\ell)\right].$$

Plots of the initial condition (13) and the functions (14) are shown in Fig. 14.4.1.

14.4. Evolution and Characteristic Regimes for Burgers Equation

The expression (14) explicitly contains the initial Reynolds number $R = S/2\mu$; see (14.2.39). In the most interesting case for us, $(R \gg 1)$, one can obtain from (14), (9), and (11) that

$$q(x) \simeq -[\chi(x+\ell) - \chi(x-\ell)] \quad \Rightarrow \quad Q_0 \simeq -2\ell \quad \Rightarrow \quad t_l \simeq \frac{4}{\pi} t_d.$$

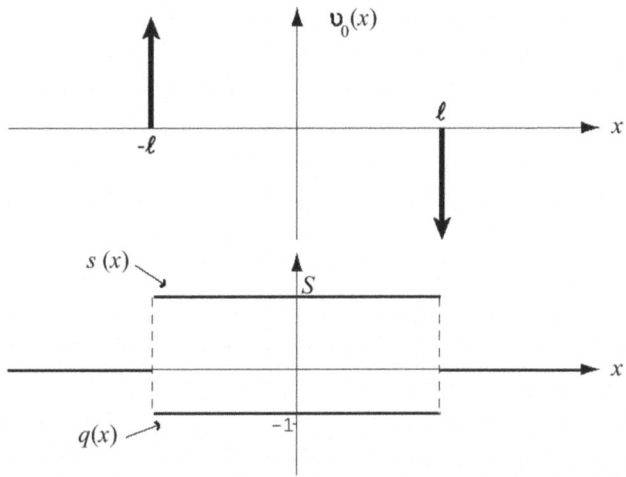

FIGURE 14.4.1
A singular initial field $v_0(x)$, see (13), (*top*), and the corresponding fields $s_0(x)$ and $q_0(x)$ (*bottom*). Bold vertical arrows in the top graph symbolize the Dirac delta components of the initial condition (13).

So in this case, the transition time to the linear regime is close to the characteristic time t_d of the linear diffusion; see (7).

Now let us trace the evolution of the field $v(x,t)$ corresponding to the initial condition (13) employing the exact solution of the Burgers equation

$$v(x,t) = -\frac{2\mu}{\ell} \sqrt{\frac{R}{\pi\tau}} \times \qquad (15)$$

$$\frac{(1-e^{-R}) \exp\left(-R\frac{z^2+1}{\tau}\right) \sinh\left(\frac{2Rz}{\tau}\right)}{1 - (1-e^{-R})\left[\Phi\left(\sqrt{\frac{R}{\tau}}(1-z)\right) - \Phi\left(-\sqrt{\frac{R}{\tau}}(1+z)\right)\right]},$$

where the dimensionless time and space coordinates are

$$z = \frac{x}{\ell}, \qquad \tau = \frac{2St}{\ell^2},$$

and the function Φ is determined by the equality (14.2.36). Graphs of the field $v(x,t)$ as a function of z for different values of τ and for the initial Reynolds number R = 25 are given in Fig. 14.4.2.

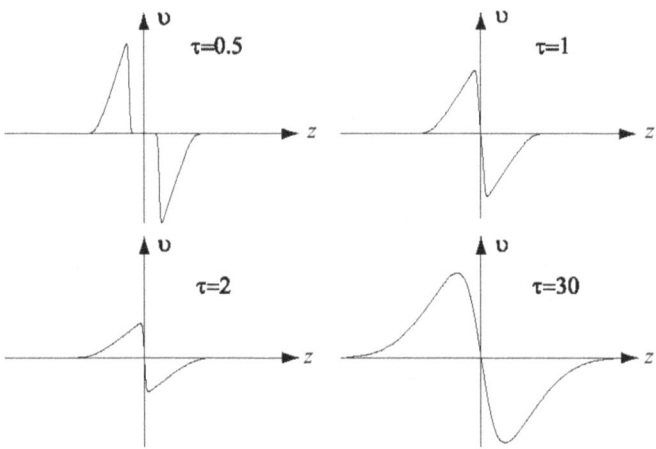

FIGURE 14.4.2
The N-wave solution of the Burgers equation at various times τ. The initial condition is that of (13), and the initial Reynolds number is R=25. The scales of the horizontal axes in all four pictures are identical, but the scales on the vertical axes are different. The vertical scale in the last picture corresponding to $\tau = 30$ is much greater than that on the first three to make it possible to see the shape of the solution, drastically attenuated by dissipation, of the Burgers equation at the linear stage.

At time $\tau = 0.5$, the field consists of two triangular waves moving toward each other and practically noninteracting. At $\tau = 1$, the fronts of the triangular waves meet and begin to annihilate each other. By $\tau = 2$, their mutual absorption has substantially reduced their amplitudes. Nevertheless, the absolute value of the parameter (9),

$$Q(\tau) \simeq -\sqrt{\frac{4\text{R}}{\pi\tau}},$$

is still large ($|Q(\tau = 1)| \simeq 4$), and the nonlinear stage of the evolution continues. For $\tau = 1$ and $\tau = 2$, the characteristic profiles of the field $v(x,t)$ resemble the letter N. Hence the name N-waves. Finally, for $\tau = 30$, when $|Q(\tau = 30)| \simeq 1$, the process of mutual destruction of colliding triangular

14.4. Evolution and Characteristic Regimes for Burgers Equation

waves is completed by a transition to the linear regime, and the graph of $v(x, t)$ has a shape close to the profile of the solution (12) of the linear diffusion equation.

Example 2. U-wave solution. An example of the U-wave solution of the Burgers equation will be obtained by considering an initial condition of the form
$$v_0(x) = S\left[\delta(x - \ell) - \delta(x + \ell)\right]. \tag{16}$$

The corresponding exact solution is given by

$$v(x, t) = -\frac{2\mu}{\ell} \sqrt{\frac{R}{\pi \tau}} \times$$
$$\frac{\left(1 - e^R\right) \exp\left(-R\frac{z^2+1}{\tau}\right) \sinh\left(\frac{2Rz}{\tau}\right)}{1 - (1 - e^R)\left[\Phi\left(\sqrt{\frac{R}{\tau}}(1 - z)\right) - \Phi\left(-\sqrt{\frac{R}{\tau}}(1 + z)\right)\right]}.$$

At first glance, this solution looks similar to the N-wave solution (15). However, on closer inspection, it turns out that its behavior is drastically different. In the present case, the triangular waves generated by the Dirac-delta-shaped components of the initial conditions (16) run away from each other (see Fig. 14.4.3). Thus, unlike the N-waves, the triangular components of the U-waves interact only through their tails. For large initial Reynolds numbers, this interaction is significantly weaker than the mutually destructive collision of the fronts of N-waves. As a result, the nonlinear stage of U-waves lasts much longer than that of N-waves.

Let us confirm the above conclusions by a quantitative evaluation of the transition time (11) to the linear regime. Indeed, in this case,

$$t_l = \frac{4}{\pi}\left(e^R - 1\right) t_d.$$

Thus, for the same (large) initial Reynolds number, the nonlinear stage of a U-wave lasts incomparably longer—approximately 10^{10} times as long—than the similar stage of an N-wave.

Remark 1. Universality of N-waves and U-waves. The importance of the concepts of N-waves and U-waves is worth reemphasizing. These two wave types describe elements of the universal nonlinear structures that arise in the process of nonlinear evolution of fields satisfying different initial conditions. Mathematically, the universality of N- and U-waves can be expressed by the fact that the asymptotic formula (10) does not depend on subtle details

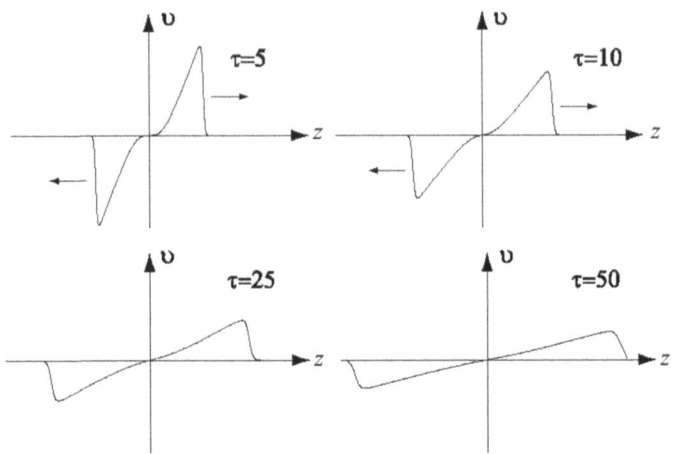

FIGURE 14.4.3
U-wave solution of the Burgers equation at various times τ. The initial condition is that of (16), and the initial Reynolds number is R = 25.

of the initial condition $v_0(x)$, but only on the waves' integral characteristic Q_0; see (9). If the latter is positive, then the field eventually assumes the N-wave shape. If $Q_0 < 0$, then the field converges to a U-wave. Relying on the universal asymptotic formula (10), Fig. 14.4.4 shows a graph of an exact solution of the Burgers equation with initial condition (16) for R = 25 and $\tau = 100$ and compares it with a graph of the corresponding U-wave. Their similarity is obvious.

14.4.4 Approximation of Sawtooth Waves

Until now, we have been tracking the evolution of a solution of the Burgers equation for a given initial condition $v_0(x)$ all by itself. But it is also useful to study the above evolution in parallel with the evolution of solutions of the linear diffusion equation (11), for the same initial condition $\varphi_0(x)$. We shall employ this approach to discuss the concept of *sawtooth waves*, which play a key role in the analysis of the asymptotic behavior of nonlinear fields described by the Burgers equation. We shall begin with the following illuminating example:

Example 1. Sum of two Dirac deltas as initial data. Consider the initial condition

$$\varphi_0(x) = \delta(x+\ell) + Q^2 \delta(x-\ell). \tag{17}$$

14.4. Evolution and Characteristic Regimes for Burgers Equation

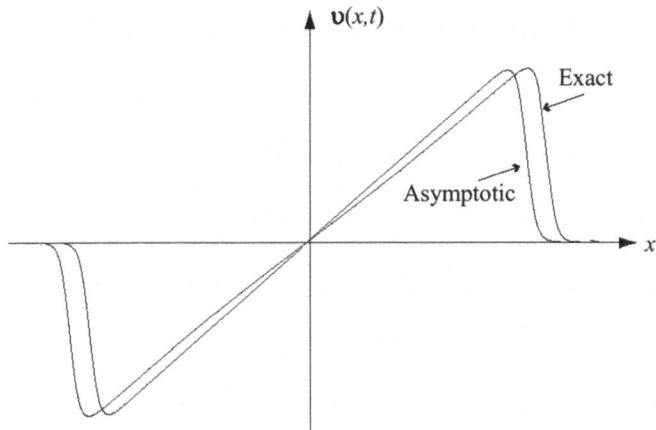

FIGURE 14.4.4
An exact solution of the Burgers equation with initial condition (16) for R = 25 and $\tau = 100$, compared with the graph of the U-wave (10) for the corresponding value $Q \simeq 8 \cdot 10^9$.

In this case, up to a factor independent of x, the solution of the linear diffusion equation (14.3.11) is of the form

$$\varphi(x,t) = \exp\left(-\frac{(x+\ell)^2}{4\mu t}\right) + Q^2 \exp\left(-\frac{(x-\ell)^2}{4\mu t}\right).$$

Substituting it into (14.3.9a), we obtain

$$v(x,t) = \frac{x}{t} - \frac{\ell}{t} \frac{Q^2 \exp\left(\frac{x\ell}{2\mu t}\right) - \exp\left(-\frac{x\ell}{2\mu t}\right)}{Q^2 \exp\left(\frac{x\ell}{2\mu t}\right) + \exp\left(-\frac{x\ell}{2\mu t}\right)}.$$

The physical meaning of the parameter Q can be explained by writing it in the form

$$Q = \exp\left(-\frac{s}{2\mu}\right). \tag{18}$$

As a result, we finally obtain

$$v(x,t) = \frac{1}{t}\left[x - \ell \tanh\left(\frac{\ell}{2\mu t}(x - Vt)\right)\right], \qquad V = \frac{s}{\ell}.$$

This is the familiar solution (14.2.31) of the Burgers equation moving with velocity V.

So, starting with a somewhat artificial initial condition (17), we have arrived at a physically meaningful solution of the Burgers equation. Relying on this modest success, let us now consider an initial condition $\varphi_0(x)$ in the form of a superposition of several Dirac deltas,

$$\varphi_0(x) = \sum_k Q_k \, \delta(x - y_k), \tag{19}$$

and, by analogy with (18), set the coefficients

$$Q_k = \exp\left(-\frac{s_k}{2\mu}\right). \tag{20}$$

The resulting solution field $v(x,t)$ will be called a *sawtooth wave*, and we shall write it in the form (14.3.16),

$$v(x,t) = \frac{x - \{y\}(x,t)}{t}, \tag{21}$$

where $\{y\}(x,t)$ is the averaged Lagrangian coordinate, which in this particular case is

$$\{y\}(x,t) = \frac{\sum_k y_k \exp\left(-\frac{1}{2\mu t}\left[s_k t + \frac{(x-y_k)^2}{2}\right]\right)}{\sum_k \exp\left(-\frac{1}{2\mu t}\left[s_k t + \frac{(x-y_k)^2}{2}\right]\right)}. \tag{22}$$

To better understand the geometric meaning of the above solution, let us extract the skeleton of the averaged Lagrangian coordinate $\{y\}(x,t)$ and the corresponding skeleton of the sawtooth wave $v(x,t)$. Obviously, as $\mu \to 0_+$, for a given point x it is sufficient to be concerned only with the largest terms in the numerator and the denominator in (22). These dominating terms can be identified by construction of the so-called *critical parabolas*

$$\Pi_k(x) = s_k t + \frac{(x - y_k)^2}{2} \tag{23}$$

and selection of the one that has the minimal value for x. Accordingly, $\{y\}(x,t)$ turns out to be a step function,

$$\{y\}(x,t) = y_k,$$

where k is the number of the selected parabola.

14.4. Evolution and Characteristic Regimes for Burgers Equation

At the points where the minimal critical parabolas intersect, the skeleton of the function $\{y\}(x,t)$ has a jump. The coordinates of the points $x_{k,m}(t)$ of the intersection of the kth and mth parabolas can be easily found. Indeed,

$$\Pi_k(x) = \Pi_m(x) \quad \Rightarrow \quad x_{k,m}(t) = x^0_{k,m} + V_{k,m} t,$$

$$x^0_{k,m} = \frac{y_k + y_m}{2}, \quad V_{k,m} = \frac{s_m - s_k}{M_{km}}, \quad M_{k,m} = y_m - y_k \quad (k < m). \tag{24}$$

If the parabolas Π_k and Π_m immediately to the left and to the right of the point $x_{k,m}(t)$ are below other parabolas, then the function $\{y\}(x,t)$ has a jump at $x_{k,m}(t)$, and

$$\{y\}(x_{k,m}(t) - 0, t) = y_k, \qquad \{y\}(x_{k,m}(t) + 0, t) = y_m \qquad (k < m).$$

Graphs of the critical parabolas in (23) illustrating construction of the step function $\{y\}(x,t)$ are shown in Fig. 14.4.5.

If the velocities $V_{k,m}$ of jumps (shocks) are not equal, then some of them merge, creating a jump that moves with yet another velocity. Let us take a look at the rules of jump merging.

Suppose that in a small neighborhood $t \in (t_{k,m,n} - \varepsilon, t_{k,m,n})$ of the time $t_{k,m,n}$ of the intersection of straight lines $x_{k,m}(t)$ and $x_{m,n}(t)$ and in a small vicinity of the point of intersection $x_{k,m,n}$, the minimal parabolas are respectively Π_k, Π_m, and Π_n ($k < m < n$). This means that at the points $x_{k,m}(t)$ and $x_{m,n}(t)$, the function $\{y\}(x,t)$ has a discontinuity. At time $t_{k,m,n}$, the middle parabola Π_m is everywhere above the other parabolas, and instead of two jumps, there arises a new jump with coordinate $x_{k,n}(t)$ and velocity $V_{k,n}$. It follows from (24) that the velocity of the jump created by the merger of the two previous jumps is determined by the velocities of the latter by the formula

$$V_{k,n} = \frac{s_n - s_k}{y_n - y_k} = \frac{M_{k,m} V_{k,m} + M_{m,n} V_{m,n}}{M_{k,m} + M_{m,n}}. \tag{25}$$

This formula coincides with the law governing the velocities of the particles of masses $M_{k,m}$ and $M_{m,n}$, respectively, in the case of perfectly inelastic collisions.

The fact that the velocities of jumps are constant between collisions and that they change at the point of impact according to the law of inelastically colliding particles allows us to describe the behavior of the sawtooth wave in the language of a 1-D stream of inelastically colliding particles. The skeletons of the sawtooth field $v(x,t)$, constructed via the formula (14.3.16), are shown

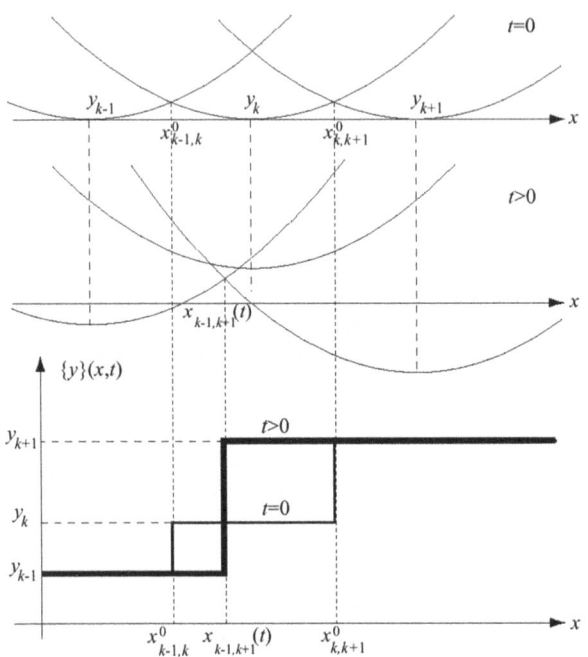

FIGURE 14.4.5
Construction of osculating parabolas of the skeleton of the function $\{y\}(x,t)$. The parabolas in the upper part of the picture are drawn for $t = 0$, and those in the middle for $t > 0$. One can see that the latter emerged from a competitive struggle for the right to achieve the least value. As a result, instead of two jumps of the function $\{y\}(x,0)$ at the points $x^0_{k-1,k}$ and $x^0_{k,k+1}$, there is only one jump of $\{y\}(x,t)$ at the point $x_{k-1,k+1}(t)$.

in Fig. 14.4.6. A typical pattern of trajectories of merging jumps can be seen in Fig. 14.4.7.

So far, we have explored only the skeleton of the sawtooth wave corresponding to the limit $\mu \to 0_+$. However, the expression (22) for the average coordinate $\{y\}(x,t)$ describes additionally the viscosity-driven ($\mu > 0$) phenomenon of jumps being smoothed out and a transition to the linear stage of evolution. The latter occurs when for every x, the adjacent terms in the sums (22) have comparable magnitudes.

Sawtooth waves can be used as an approximation tool. To see how this can be done, let us recall the initial condition of the simple diffusion equation (14.3.11). Let s_k be the minimum value of the initial potential $s_0(x)$, attained

14.4. Evolution and Characteristic Regimes for Burgers Equation

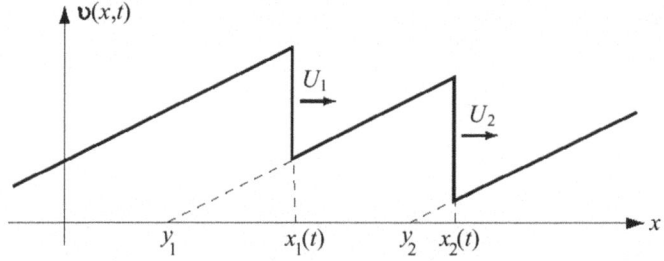

FIGURE 14.4.6
The skeleton of a sawtooth wave created as a result of a nonlinear transformation of the initial field. The zeros of the linear portions of the graph, as well as the jump coordinates, are indicated.

at $x = y_k$. Then for small μ and in a neighborhood of the minimum, one can approximate this function by the Gaussian function

$$\varphi_0^k(x) \sim \exp\left[-\frac{s_k}{2\mu} - \frac{\sigma_k}{4\mu}(x - y_k)^2\right], \qquad (26)$$

where

$$\sigma_k = \sigma(y_k), \qquad \sigma(x) = \frac{\partial^2 s_0(x)}{\partial x^2} = \frac{\partial v_0(x)}{\partial x}. \qquad (27)$$

In what follows, we shall assume that there exists an ε such that for all k,

$$\sigma_k > \varepsilon > 0.$$

Substituting (26) into (14.3.13), we obtain the corresponding solution of the linear diffusion equation

$$\varphi^k(x, t) = \frac{1}{\sqrt{1 + \sigma_k t}} \exp\left[-\frac{\sigma_k t}{1 + \sigma_k t} \frac{(x - y_k)^2}{4\mu t}\right]. \qquad (28)$$

At time t satisfying the inequality

$$\varepsilon t \gg 1, \qquad (29)$$

the function $\varphi^k(x, t)$ (28) will not change much if the right-hand side of (26) is replaced by the Dirac delta:

$$\varphi_0^k(x) \sim \sqrt{\frac{4\pi\mu}{\sigma_k}} \exp\left(-\frac{s_k}{2\mu}\right) \delta(x - y_k).$$

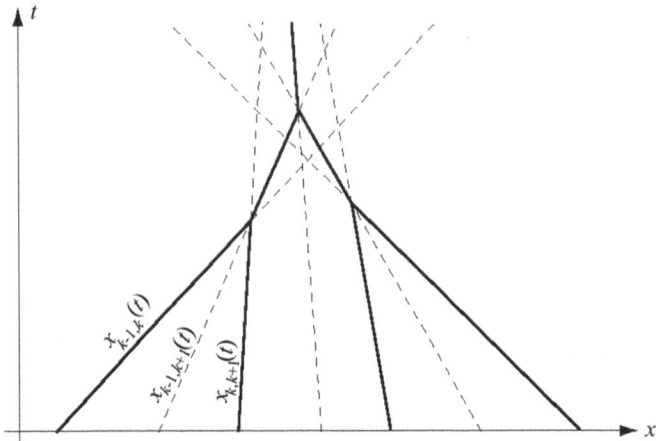

FIGURE 14.4.7
Pattern of trajectories of merging jumps. The solid lines mark trajectories of jump points before and after they underwent collisions with other jump points.

Combining the contributions of all the minima of the initial potential and taking into account the fact that in the case of a solution of the Burgers equation, the initial potential is defined up to an arbitrary constant factor, we arrive at the following assertion: *If the condition (29) is satisfied, then the solution of the Burgers equation can be expressed via the formula (14.3.9a), where $\varphi(x,t)$ is the solution of the linear diffusion equation with initial condition (19), and*

$$Q_k = \frac{1}{\sqrt{\sigma_k}} \exp\left(-\frac{s_k}{2\mu}\right).$$

The factor appearing in front of the exponential function can be treated as a small correction to s_k:

$$\frac{1}{\sqrt{\sigma_k}} \exp\left(-\frac{s_k}{2\mu}\right) = \exp\left(-\frac{s'_k}{2\mu}\right), \qquad s'_k = S_k + \frac{\mu}{2} \ln \sigma_k,$$

which can be neglected for small μ. As a result, we arrive at the initial conditions (19) and (20).

14.4.5 Quasiperiodic Waves

In this subsection we shall consider solutions of the Burgers equation with—obviously physically important—periodic initial data, and study them in the sawtooth wave approximation.

14.4. Evolution and Characteristic Regimes for Burgers Equation

To begin with, let us take a look at the case in which the initial field is a simple harmonic function,

$$v_0(x) = a\sin(kx) \quad \Rightarrow \quad s_0(x) = -\frac{a}{k}\cos(\kappa x). \tag{30}$$

The question of applicability of the sawtooth approximation needs to be clarified first. In this case, the characteristic curvature is the same at all the minimum points of the initial potential and equals

$$\sigma_k = \varepsilon = a\kappa = \frac{1}{t_n}, \quad t_n = \frac{1}{a\kappa},$$

where t_n is the time of the jump creation. Accordingly, one can rewrite the condition (29) in the form

$$t \gg t_n. \tag{31}$$

Let us find a sawtooth wave corresponding to the initial harmonic field (30). Since the minima $s_0(x)$ of the potential are identical, the values s_k in (22) can be assumed to be equal to zero. Moreover, $y_k = 2\pi k/\kappa$. As a result, the expression for the sawtooth wave approximating the original harmonic wave is of the form

$$v(x,y) = \frac{a}{\tau}\left(z - 2\pi\frac{\sum_k k \exp\left[-\frac{\mathrm{R}}{8\tau}(z - 2\pi k)^2\right]}{\sum_k \exp\left[-\frac{\mathrm{R}}{8\tau}(z - 2\pi k)^2\right]}\right), \tag{32}$$

where

$$z = \kappa x, \quad \tau = a\kappa t, \quad \mathrm{R} = \frac{2a}{\mu\kappa}. \tag{33}$$

Plots of the field $v(x,t)$ in (32), for different values of τ, are shown in Fig. 14.4.8.

The Reynolds number of the initial harmonic wave varies in time. If $\mathrm{R}\gg 1$, then there are three, clearly separated, stages of evolution of the field $v(x,t)$. For times

$$t < t_n,$$

there are no jumps, and the current Reynolds number is close to the initial one:

$$\mathrm{R}(t) \simeq \frac{a\ell}{\mu} = \frac{\pi a}{\mu\kappa} \simeq \mathrm{R} \quad \left(\ell = \frac{\pi}{\kappa}\right).$$

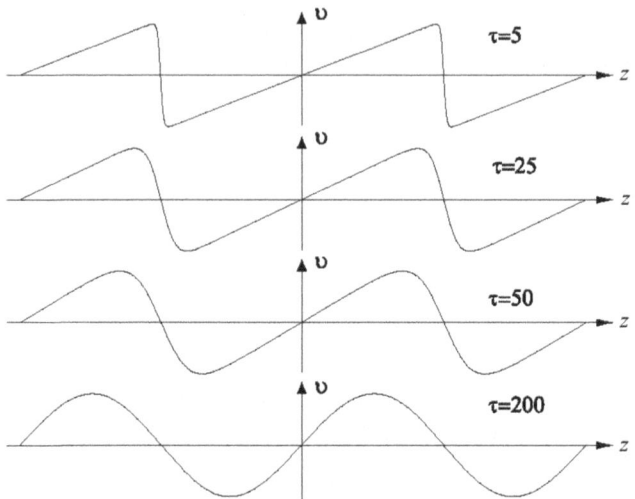

FIGURE 14.4.8
The sawtooth wave in (32) at R = 100. The vertical scales of the plots are selected so that the shape of the field $v(x,t)$ is visible. As τ becomes larger, the jumps are smoothed out, and the field assumes a sine-like shape, characteristic of the linear stage of the evolution.

For $t \gg t_n$, the size of the jump $U(t)$ and its width $\Delta(t)$ change according to the formulas

$$U(t) \simeq \frac{\ell}{t}, \quad \text{and} \quad \Delta(t) \simeq \frac{2\mu}{U(t)},$$

so that the Reynolds number decays like $1/t$,

$$\mathrm{R}(t) \simeq \frac{U(t)\ell}{\mu} \simeq \frac{\ell^2}{\mu t}.$$

Finally, for $t \sim t_l$, where t_l is the time when the jump's width becomes comparable to the half-period of the field $v(x,t)$,

$$\Delta(t_l) \simeq \ell \quad \Rightarrow \quad t_l \simeq \frac{\pi^2}{2\mu\kappa^2} \simeq \mathrm{R}\, t_n,$$

we see the onset of the linear stage, and the Reynolds number tends to zero quickly. A semiqualitative plot of the current Reynolds number $\mathrm{R}(t)$ of the initial harmonic field $v(x,t)$ is shown in Fig. 14.4.9.

14.4. Evolution and Characteristic Regimes for Burgers Equation

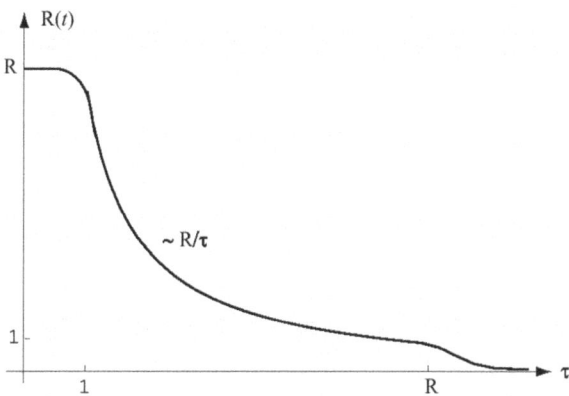

FIGURE 14.4.9
Semiqualitative plot of the dependence of the current Reynolds number of a periodic field on the dimensionless time $\tau = a\kappa t$. It is assumed that the initial Reynolds number is large, $R \gg 1$.

Remark 1. N-wave scenario for sawtooth waves. Observe that for $R \gg 1$, the sawtooth wave that evolved from the initial harmonic field carries features of both N- and U-waves. Since the mechanism of energy dissipation is, in the first case, much more effective, evolution of the initial harmonic wave and its entry into the linear stage follow the N-wave scenario.

To conclude this section, we would like to turn reader's attention to the fact that the expression (32) does not contain any information on the shape of the initial field. This means that if the initial periodic field were a superposition of several harmonic fields, then its asymptotics would be dictated by the wave with the longest period. Moreover, the eventual "victory" of the large-scale component would be accelerated by a merger of jumps corresponding to the waves with shorter periods. This effect is illustrated in Fig. 14.4.10, which shows plots of solutions of the Burgers equation with the initial conditions

$$v_0(x) = a \left[\sin(\kappa x) + \frac{1}{4} \sin\left(\frac{1}{4}\kappa x\right) \right], \tag{34}$$

for $R = 50$. The definitions of the initial Reynolds number and other dimensionless parameters were borrowed from (33).

The figure shows that the initial discontinuities correspond to the small-scale components of the initial condition (34). Then, as a result of modulation of the jumps' velocity by the large-scale component, the jumps merge, eventually forming a single jump corresponding to the period of the large-scale component of the initial condition.

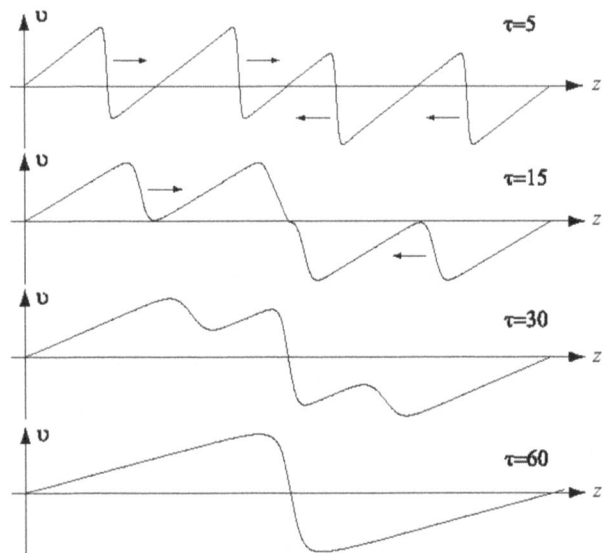

FIGURE 14.4.10
Evolution of the solution of the Burgers equation with the initial condition (34) and R = 50. Merging of jumps leads to creation of a sawtooth wave with period equal to the period of the large-scale component of the initial field.

Finally, our analysis of evolution of the initially periodic and quasiperiodic fields can be augmented by the observation that depending on the structure of the initial field, the current Reynolds number not only can decrease or remain constant in time, but can also increase. We shall see these different behaviors in the following example.

Example 1. Increasing current Reynolds number for periodic data. Assume that the initial field is a sum of two harmonic waves,

$$v_0(x) = a\left[\sin(\kappa x) + \beta^{\gamma-1}\sin(\beta\kappa x)\right] \qquad (\beta \ll 1,\ \gamma > 0). \qquad (35)$$

Introduce, by analogy with the procedures discussed before, the characteristic times of nonlinearity and of entry in the linear stage for the second harmonic component. They are, respectively,

$$t_n(\beta) \simeq \beta^{-\gamma} t_n, \qquad t_l(\beta) \simeq \beta^{-2} t_l \simeq \beta^{-2} \mathrm{R}\, t_n.$$

14.4. Evolution and Characteristic Regimes for Burgers Equation

In the case that
$$t_l \ll t_n(\beta) \quad \Rightarrow \quad \mathrm{R}\,\beta^\gamma \ll 1,$$
for $t < t_l$, the second summand does not significantly influence the evolution of the shape of the first periodic wave, and until time $t \sim t_l$, when this wave is almost fully dissipated, the resulting field is approximately equal to
$$v(x, t \sim t_l) \simeq a\,\beta^{\gamma-1} \sin(\beta\kappa x).$$
Its Reynolds number is of the form
$$\mathrm{R}(t_n(\beta)) \simeq \frac{t_l(\beta)}{t_n(\beta)} \simeq \mathrm{R}\,\beta^{\gamma-2}.$$
So, under the condition
$$0 < \gamma < 2, \tag{36}$$
the current Reynolds number at the beginning of the nonlinear stage of evolution of the large-scale component in (35) becomes larger than the initial Reynolds number.

Example 2. Increasing current Reynolds number for quasiperiodic data. Now consider a more complex, Weierstrass-function-like, quasiperiodic initial field,
$$v_0(x) = a \sum_{k=1}^{\infty} \beta^{(\gamma-1)k} \sin(\beta^k \kappa x + \theta_k), \tag{37}$$
where θ_k are arbitrary phase shifts. If the restrictions on β listed above are satisfied and γ satisfies the inequalities (36), then by time
$$t^k \simeq \beta^{-k\gamma} t_n,$$
we see the onset of the nonlinear stage for the kth components of the sum (37). Because of dissipation, the preceding summands have stopped contributing to the evolution, and the subsequent summands form a quasi-constant background. The corresponding current Reynolds number is
$$\mathrm{R}(t^k) \simeq \mathrm{R}\,\beta^{k(\gamma-2)}.$$
Eliminating β from the last two relations, we find that under the condition (36), the Reynolds number grows in time according to the power law
$$\mathrm{R}(t) \simeq \mathrm{R}\left(\frac{t}{t_n}\right)^{\frac{2-\gamma}{\gamma}}.$$

14.5 Exercises

1. The functional
$$\int u(x,t)\,dx = \text{const}$$
is an invariant of every solution of the KdV equation (14.1.11). The same is true for solutions of the Burgers equation. Show that
$$\int u^2(x,t)\,dx = \text{const}$$
is also an invariant of solutions of the KdV equation. This fact can be interpreted as an energy conservation law for KdV waves. In the proof, assume that for all $t > 0$ and as $x \to \pm\infty$, the field $u(x,t)$ and its first derivative with respect to x vanish, and that the second derivative is bounded everywhere.

2. Although the stationary wave (14.2.22) does not propagate, its lack of motion is caused by a dynamic equilibrium of the inertial and dissipative processes. Therefore, it is illuminating to calculate the rate of decay of the (infinite) energy
$$\Gamma = \mu \int \left(\frac{\partial v}{\partial x}\right)^2 dx \tag{1}$$
of the stationary wave. Analyze the dependence of Γ on the amplitude of U-waves and on the coefficient of viscosity μ.

3. Obviously, the initial condition of a stationary solution of the Burgers equation coincides with the stationary solution (14.2.22) itself. Find and discuss the solution of the Burgers equation for the initial data equal to the sum of two stationary waves,
$$v_0(x) = -U_1 \tanh\left(\frac{U_1}{2\mu}(x-\ell_1)\right) - U_2 \tanh\left(\frac{U_2}{2\mu}(x-\ell_2)\right). \tag{2}$$

4. Use the concept of the skeleton of the solution
$$v(x,t) = -\frac{U_+ \sinh\left(\frac{U_+}{2\mu}(x-\ell_+)\right) + \exp\left(-\frac{U_1 U_2}{\mu}t\right) U_- \sinh\left(\frac{U_-}{2\mu}(x-\ell_-)\right)}{\cosh\left(\frac{U_+}{2\mu}(x-\ell_+)\right) + \exp\left(-\frac{U_1 U_2}{\mu}t\right) \cosh\left(\frac{U_-}{2\mu}(x-\ell_-)\right)}.$$
of the Burgers equation to explain the origin of the formula
$$\ell_+ = \frac{U_2 \ell_2 + U_1 \ell_1}{U_2 + U_1}.$$

14.5. Exercises

Take advantage of the solution to the above Exercise 3 in the chapter "Answers and Solutions." For the sake of concreteness, assume
$$U_1 > 0, \qquad U_2 > 0, \qquad \ell_2 > \ell_1.$$

5. Find an exact solution of the Burgers equation with the initial condition
$$v_0(x) = S\left[\delta(x) + \delta(x - \ell)\right] \qquad (\ell > 0, \quad S > 0). \qquad (3)$$
Construct plots of $v(x,t)$ for $R = S/2\mu \gg 1$ and several values of the dimensionless time $\tau = 2St/\ell^2$.

6. Find the skeleton of the solution of the previous problem. Using the fact that a jump's velocity is the average of the values of the field $v(x,t)$ to the left and to the right of the jump, write equations of motion of the jumps and estimate the time of merger of two triangular waves into a single triangular wave.

7. Find the form of jumps of the averaged Lagrangian coordinate $\{y\}(x,t)$, see (14.4.19), taking into account the competition of two dominating summands in the vicinity of the jump.

8. Using the well-known formula
$$e^{R\cos(\kappa x)} = I_0(R) + 2\sum_{n=1}^{\infty} I_n(R)\cos(n\kappa x), \qquad (4)$$
where $I_n(z)$ is the modified Bessel function of order n, find the asymptotic behavior (at the linear stage) of the solution of the Burgers equation with a harmonic initial condition
$$v_0(x) = a\sin(\kappa x). \qquad (5)$$
Explore the dependence of the solution field's amplitude on the initial amplitude a at the linear stage.

9. Solve the Burgers equation in the case of
$$\varphi_0(x) = x^2.$$

10. The equality (14.3.21) is a consequence of a similar property of the function $f(y;x,t)$, see (14.3.17):
$$f(y;x,t|V) = f(y;x - Vt, t), \qquad (6)$$
where $f(y;x,t)$ corresponds to the initial condition $v_0(x)$, and $f(y;x,t|V)$ to the initial condition $v_0(x) + V$. Prove (6).

Chapter 15
Other Standard Nonlinear Models of Higher Order

This chapter builds on the material of Chap. 14 and reviews other standard nonlinear models that can be described by partial differential equations. We begin with the model equations of gas dynamics, expand the Burgers–KPZ model to the multidimensional case and the related concentration fields, study in detail the Korteweg–de Vries (KdV) equations, in which one can observe the creation of solitary waves (solitons), and finally, discuss nonlinear flows in porous media.

There are new ingredients here. The KdV equation includes partial derivatives of order three, and the porous medium equation contains nonlinearity intertwined with the highest (second) derivative. The latter complicates the game considerably. All the other nonlinear equations we have considered thus far were linear in the highest derivative; in the mathematical literature, such equations are called *quasilinear*. The porous medium equation contains the Laplacian superimposed on top of a nonlinearity and thus is classified as a *strongly nonlinear* equation. Nevertheless, the reader will notice that the nonlinearities we consider are always quadratic or of power type. The quadratic ones are the simplest possible, obtained by retaining the first nonlinear term in the Taylor expansions of more complex, and perhaps more realistic, nonlinearities that describe various physical phenomena.

15.1 Model Equations of Gas Dynamics

In this section we shall discuss modifications and generalizations of the Burgers equations that admit solutions in closed form. They also happen to play a role as model equations of gas dynamics. For now, we will stay in the one-dimensional universe.

15.1.1 Model of 1-D Polytropic Gas

Recall that the equation

$$\frac{\partial v}{\partial t} + v\frac{\partial v}{\partial x} = \mu\frac{\partial^2 v}{\partial x^2} \tag{1}$$

was proposed by Johannes Burgers as a model of strong hydrodynamic turbulence that takes into account competing influences of inertial nonlinearity and viscosity. However, from the physics perspective, the Burgers equation does not provide an accurate description of the gas dynamics phenomena. Most importantly, it does not take into account pressure forces that prevent excessive compression of the particles. To model these forces, let us modify equation (1) by an additional term to obtain the equation

$$\frac{\partial v}{\partial t} + v\frac{\partial v}{\partial x} + \frac{1}{\rho}\frac{\partial P}{\partial x} = \mu\frac{\partial^2 v}{\partial x^2}. \tag{2}$$

The quantity $P(x,t)$ introduced above represents the pressure of the model gas, and $\rho(x,t)$ its density.

In the remainder of this section, we shall restrict our attention to the case of so-called *polytropic gas*, whereby gas pressure and gas density are tied together by the relation

$$P(x,t) = \frac{\kappa^2}{\gamma}\rho^\gamma(x,t), \qquad \kappa > 0, \quad \gamma > 0. \tag{3}$$

Equations (2)–(3) describe the behavior of just the velocity field and are not closed, since they contain three unknown fields, $v(x,t), P(x,t)$ and $\rho(x,t)$. Fortunately, we also know that the velocity and density fields are tied together by the continuity equation

$$\frac{\partial \rho}{\partial t} + \frac{\partial}{\partial x}(\rho v) = \nu\frac{\partial^2 \rho}{\partial x^2}, \tag{4}$$

where ν is the coefficient of molecular diffusion, which takes into account the influence of the Brownian motion of gas molecules on the evolution of the gas density.

15.1. Model Equations of Gas Dynamics

The set of equations (2)–(4) gives a much more adequate mathematical model of nonlinear dynamics of 1-D compressible gas than the Burgers equation. The value of this model in the study of nonlinear gas dynamics is enhanced by the fact that for special values

$$\gamma = 3, \quad \text{and} \quad \nu = \mu, \tag{5}$$

equations (2)–(4) have an explicit analytic solution for a broad class of initial conditions

$$v(x, t = 0) = v_0(x), \qquad \rho(x, t = 0) = \rho_0(x). \tag{6}$$

This fact becomes clear once we rewrite (2) and (4), taking (5) into account, as a set of equations

$$\frac{\partial v}{\partial t} + v\frac{\partial v}{\partial x} + c\frac{\partial c}{\partial x} = \mu\frac{\partial^2 v}{\partial x^2}, \qquad \frac{\partial c}{\partial t} + v\frac{\partial c}{\partial x} + c\frac{\partial v}{\partial x} = \mu\frac{\partial^2 c}{\partial x^2}. \tag{7}$$

The quantity

$$c(x,t) = \kappa\rho(x,t) \tag{8}$$

introduced above is called the *local sound velocity*. For convenience, we shall also introduce two auxiliary fields

$$u_\pm(x,t) = v(x,t) \pm c(x,t), \tag{9}$$

so that the original velocity and density fields can be expressed by the formulas

$$v(x,t) = \frac{u_+(x,t) + u_-(x,t)}{2}, \qquad \rho(x,t) = \frac{u_+(x,t) - u_-(x,t)}{2\kappa}. \tag{10}$$

Adding and subtracting equations (7), we discover that each of the auxiliary fields (9) obeys its own Burgers equation

$$\frac{\partial u_\pm}{\partial t} + u_\pm \frac{\partial u_\pm}{\partial x} = \mu \frac{\partial^2 u_\pm}{\partial x^2}, \tag{11}$$

with the initial conditions

$$u_\pm(x, t = 0) = v_0(x) \pm \kappa\rho_0(x). \tag{12}$$

From the perspective of physics, the most fundamental case is that of the constant initial density

$$\rho(x, t = 0) = \rho_0 = \text{const}. \tag{13}$$

The crux of the matter is that, in this particular case, the solutions obey the *momentum conservation law*

$$\int \rho(x,t) v(x,t) \, dx = \text{const.} \tag{14}$$

For this reason, we shall analyze this case in greater detail and observe first that the initial conditions (12) of the Burgers equations (11) assume the form

$$u_\pm(x, t=0) = v_0(x) \pm c_0 \qquad (c_0 = \kappa \rho_0 = \text{const}),$$

and due to the Galilean invariance (14.2.3), the desired solutions of equations (11) have the form

$$u_+(x,t) = u(x - c_0 t, t) + c_0, \qquad u_-(x,t) = u(x + c_0 t, t) - c_0. \tag{15}$$

The remaining auxiliary field $u(x,t)$ satisfies the Burgers equation with the standard initial condition

$$\frac{\partial u}{\partial t} + u \frac{\partial u}{\partial x} = \mu \frac{\partial^2 u}{\partial x^2}, \qquad u(x, t=0) = v_0(x). \tag{16}$$

Substituting (15) into (10), we can now express solutions of the model gas dynamics equations (7) through a solution of the Burgers equation (16):

$$v(x,t) = \frac{1}{2} [u(x - c_0 t, t) + u(x + c_0 t, t)] \tag{17}$$

and

$$\rho(x,t) = \rho_0 \left[1 - \frac{u(x + c_0 t, t) - u(x - c_0 t, t)}{2 c_0} \right]. \tag{18}$$

Let us verify that the above solutions obey the momentum conservation law (14). Indeed, multiplying the right-hand sides of (17) and (18), we obtain

$$\rho(x,t) \, v(x,t) = \frac{\rho_0}{2} [u(x - c_0 t, t) + u(x + c_0 t, t)]$$

$$+ \frac{1}{2\kappa} [u^2(x - c_0 t, t) - u^2(x + c_0 t, t)]. \tag{19}$$

Integrating both sides of the above equality with respect to x in infinite limits, we find that the integral of the last term on the right-hand side vanishes in view of its symmetry. The integral of the first term is independent of time, because the quantity

$$\int u(x,t) \, dx = \int v_0(x) \, dx = \text{const}$$

15.1. Model Equations of Gas Dynamics

is an invariant of the solution of the Burgers equation (16). Thus, the total momentum of the model gas does not depend on time and equals

$$\int \rho(x,t)v(x,t)\,dx = \rho_0 \int v_0(x)\,dx.$$

15.1.2 Physical Properties of a Model Gas

It is useful to begin a study of any physical phenomenon with an analysis of its intrinsic times and scales. In this spirit, let us introduce the characteristic magnitude U and the characteristic scale ℓ of the initial field $v_0(x)$. The evolution of a solution of the Burgers equation depends qualitatively on the relationship between the characteristic times of evolution of the nonlinear and dissipative effects:

$$t_n = \frac{\ell}{U}, \qquad t_d = \frac{\ell^2}{\mu}. \tag{20}$$

The ratio of these times is the initial Reynolds number

$$\mathrm{R} = \frac{t_d}{t_n} = \frac{U\ell}{\mu}.$$

The motion of gas particles is also determined by another characteristic time,

$$t_a = \frac{\ell}{c_0} \tag{21}$$

which we shall call the *acoustic time* of propagation of the initial perturbations. Its ratio to the characteristic time t_n of the nonlinear evolution is called the *Mach number*:

$$\mathrm{M} = \frac{t_a}{t_n} = \frac{U}{c_0}. \tag{22}$$

If the Mach number is small, then the density fluctuations around the initial gas density ρ_0 are also small:

$$\delta\rho(x,t) = \rho(x,t) - \rho_0 = \rho_0 \frac{u(x-c_0 t, t) - u(x+c_0 t, t)}{2c_0}. \tag{23}$$

In other words, in this case, the pressure forces prevent creation of large gas density fluctuations. Moreover, for times much smaller than the characteristic nonlinearity and dissipation times ($t \ll \min\{t_n, t_d\}$), the field $u(x,t)$ in (17) can be replaced by the initial field $v_0(x)$. The corresponding solution (17) will describe, characteristically for linear acoustics, the propagation of waves of fixed shape with sound velocity c_0,

$$v(x,t) = \frac{1}{2}[v_0(x - c_0 t) + v_0(x + c_0 t)].$$

In the limit $\mu \to 0_+$, the field $u(x,t)$ on the right-hand sides of (17) and (18) has to be replaced by the weak solutions $u_w(x,t)$ of the Riemann equations obtained in the previous chapter via the global minimum principle. The relevant expressions

$$v_w(x,t) = \frac{1}{2}[u_w(x - c_0 t, t) + u_w(x + c_0 t, t)],$$

$$\rho(x,t) = \rho_0 \left[1 - \frac{u_w(x + c_0 t, t) - u_w(x - c_0 t, t)}{2c_0}\right], \tag{24}$$

can be viewed as a weak solution of the system of 1-D polytropic gas equations

$$\frac{\partial v}{\partial t} + v\frac{\partial v}{\partial x} + \kappa^2 \rho \frac{\partial \rho}{\partial x} = 0, \qquad \frac{\partial \rho}{\partial t} + \frac{\partial}{\partial x}(\rho v) = 0, \tag{25}$$

with the polytropic exponent $\gamma = 3$. Observe that they satisfy the momentum conservation law. In this case, for large times ($t \sim t_n$), the fields $v(x,t)$ and $\rho(x,t)$ become discontinuous, but they remain bounded as a result of pressure forces that prevent creation of domains of high density.

Example 1. Khokhlov's solution. For large Mach numbers, a better understanding of the behavior of the generalized velocity and density fields (24) in the vicinity of shocks (jumps) can be acquired if we trace their evolution in the case in which the auxiliary field $u(x,t)$ is described by Khokhlov's solution. In the inviscid limit ($\mu = 0_+$), the field $u(x,t)$ coincides with the skeleton of Khokhlov's solution (14.2.27). To analyze that skeleton, it is convenient to pass to the dimensionless space and time coordinates

$$z = \frac{x}{\ell}, \qquad \tau = \frac{Ut}{\ell},$$

and consider the dimensionless velocity and density fields

$$V(z,\tau) = \frac{v}{U}, \qquad R(z,\tau) = \frac{\rho}{\rho_0}.$$

15.1. Model Equations of Gas Dynamics

From (24) and (14.2.27) we see that the above fields are as follows:

$$V(z,\tau) = \frac{1}{2(1+\tau)}\left[g\left(z+\frac{\tau}{M}\right) + g\left(z-\frac{\tau}{M}\right)\right],$$
$$R(z,\tau) = 1 - \frac{M}{2(1+\tau)}\left[g\left(z+\frac{\tau}{M}\right) - g\left(z-\frac{\tau}{M}\right)\right], \quad (26)$$
$$g(z) = z - \text{sign}(z).$$

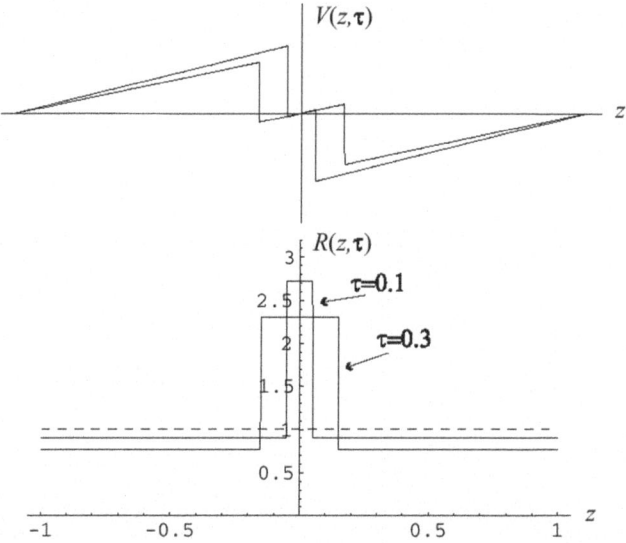

FIGURE 15.1.1
Graphs of the generalized velocity and density fields (26) of a polytropic gas for Mach number $M = 2$. They are shown here for times $\tau = 0.1$ and $\tau = 0.3$.

Their plots are shown in Fig. 15.1.1. It is clear that in the vicinity of the point $z = 0$, where the jump of the initial velocity field was located, an area of increased density is created. For larger times, the pressure forces lead to an expansion of this area and to a reduction of the density within it.

Finally, let us discuss the behavior of a model gas in the limiting case of Mach number $M \to \infty$. Physically, this limit corresponds to a pressureless gas, whose evolution is determined solely by a competition between the inertial nonlinearity and viscosity. In this case, equations of the model gas take the form

$$\frac{\partial v}{\partial t} + v\frac{\partial v}{\partial x} = \mu\frac{\partial^2 v}{\partial x^2}, \qquad \frac{\partial \rho}{\partial t} + \frac{\partial}{\partial x}(\rho v) = \mu\frac{\partial^2 \rho}{\partial x^2}, \qquad (27)$$

and the density field is described by the formula

$$\rho(x,t) = \rho_0 \left[1 - t\frac{\partial v(x,t)}{\partial x}\right], \qquad (28)$$

which follows from the formulas (17) and (18) as $c_0 \to 0$. For $\mu > 0$, the density field (28) is uniformly bounded. However, in contrast to the previous case, in which an increase of the density was restricted by pressure forces, in the present case, the boundedness of the density is caused by viscosity and molecular diffusion.

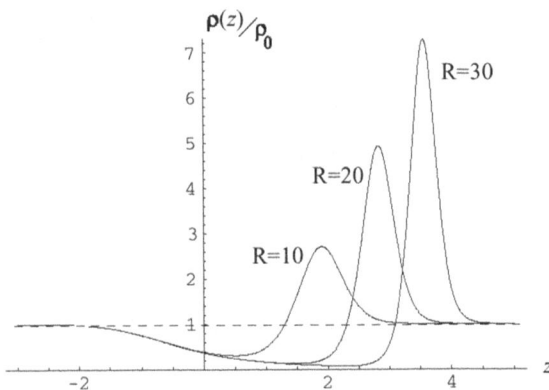

FIGURE 15.1.2
Density plots for a pressureless gas with initial momentum applied at the origin. The resulting motion leads to creation of a high-density area trailed to the left by a low density tail. As the Reynolds number increases (i.e., μ decreases), the maximum of the density increases as well.

Example 2. Pressureless gas with a given initial momentum. Suppose that at the initial time, at the point $x = 0$, a gas has momentum $\rho_0 S$. Then the velocity of a pressureless gas is described by the expressions (14.2.41) and (14.2.42). The density field (28) of such a gas is self-similar and given by the formula

$$\rho(x,t) = \rho_0 \left[1 - \frac{1}{1\sqrt{\pi}}\frac{d\mathcal{V}_R(z)}{dz}\right], \qquad z = \frac{x}{2\sqrt{\mu t}}.$$

Sample plots of this density, for different values of the Reynolds number $R = S/\mu$, are shown in Fig. 15.1.2.

15.2 Multidimensional Nonlinear Equations

Until now, we have considered nonlinear partial differential equations in 1-D space. However, in reality, many nonlinear fields of physical interest evolve in 3-D space, and in this section we turn our attention to a study of multidimensional problems. Despite their enormous importance and long history, many fundamental issues in this area are not completely understood, and a full exposition of what is known would require a separate multivolume treatise. For this reason, we shall restrict our attention to a brief review of the most natural generalizations of the 1-D problems discussed above, mainly in the context of the KPZ and Burgers equations.[1]

15.2.1 The KPZ Equation Revisited

Suppose that a surface grows with velocity c along a specified axis z in 3-D space. The plane perpendicular to the z-axis will be called the $\boldsymbol{x}$-plane. Assuming that there is a one-to-one correspondence between points of the $\boldsymbol{x}$-plane and the evolving surface, the latter can be described by the function

$$z = ct + h(\boldsymbol{x}, t). \tag{1}$$

The first term on the right-hand side represents the growth of the flat surface perpendicular to the z-axis, and the second takes into account the surface's fluctuations and other effects influencing the surface growth rate. For simplicity's sake let us assume that the constant growth rate is $c = 1$. Assuming that the angles between the surface gradient vectors and the z-axis are small, the elevation function $h(\mathbf{x}, t)$ obeys, in this small angle approximation, a multidimensional analogue of the equation (14.1.4):

$$\frac{\partial h}{\partial t} = \frac{1}{2}(\nabla h)^2 + \mu \Delta h + F(\boldsymbol{x}, t). \tag{2}$$

---

[1] For more and in-depth information, see, S. Gurbatov, A. Malakhov and A. Saichev, *Nonlinear Random Waves and Turbulence in Nondispersive Media: Waves, Rays and Particles*, Manchester University Press, Manchester–New York, 1991, and W.A. Woyczyński, *Burgers–KPZ Turbulence: Göttingen Lectures*, Springer-Verlag, Berlin–Heidelberg, 1998.

Here $F(\boldsymbol{x}, t)$ takes into account the nonuniformity of the stream of particles deposited on the surface, and the term containing μ reflects the diffusive mobility of the particles along the surface.

The logarithmic substitution

$$h(\boldsymbol{x}, t) = 2\mu \ln \varphi(\boldsymbol{x}, t), \tag{3}$$

similar to that for equation (14.1.7), reduces equation (2) to the linear diffusion equation

$$\frac{\partial \varphi}{\partial t} = \mu \Delta \varphi + \frac{1}{2\mu} F(\boldsymbol{x}, t) \varphi, \tag{4}$$

which needs to be solved with the initial conditions

$$\varphi(\boldsymbol{x}, t = 0) = \exp\left(\frac{h_0(\boldsymbol{x})}{2\mu}\right). \tag{5}$$

A complete analysis of the properties of solutions of equation (4) and relevant solutions of the KPZ equation is outside the scope of this book. In this section, we shall provide an explicit solution of equations (2) and (4) only in the homogeneous case, that is, when $F(\boldsymbol{x}, t) \equiv 0$. In this case,

$$\varphi(\boldsymbol{x}, t) = \left(\frac{1}{2\sqrt{\pi \mu t}}\right)^n \int \cdots \int \exp\left(\frac{1}{2\mu t}\left[h_0(\boldsymbol{y})t - \frac{(\boldsymbol{y} - \boldsymbol{x})^2}{2}\right]\right) d^n y. \tag{6}$$

Here n is the dimension of the $\boldsymbol{x}$-space; in our context, $n = 2$.

15.2.2 Multidimensional Burgers Equation

Let us introduce the gradient

$$\boldsymbol{v}(\boldsymbol{x}, t) = -\nabla h(\boldsymbol{x}, t) \tag{7}$$

of the surface defined by the equation (1). It is easy to check that if $h(\boldsymbol{x}, t)$ satisfies the KPZ equation (2), then the vector field $\boldsymbol{v}(\boldsymbol{x}, t)$ satisfies the multidimensional Burgers equation

$$\frac{\partial \boldsymbol{v}}{\partial t} + (\boldsymbol{v} \cdot \nabla) \boldsymbol{v} = \mu \Delta \boldsymbol{v} + \boldsymbol{f}(\boldsymbol{x}, t), \qquad \boldsymbol{f}(\boldsymbol{x}, t) = -\nabla F(\boldsymbol{x}, t).$$

An analytic solution of the homogeneous KPZ equation, in combination with the equality (7), gives us an exact analytic solution of the homogeneous Burgers equation,

$$\frac{\partial \boldsymbol{v}}{\partial t} + (\boldsymbol{v} \cdot \nabla) \boldsymbol{v} = \mu \Delta \boldsymbol{v}, \tag{8}$$

15.2. Multidimensional Nonlinear Equations

in the case of a *potential* initial field

$$v(x, t = 0) = v_0(x). \quad (9)$$

The potentiality of this field means that there exists a scalar potential $s_0(x)$ such that

$$v_0(x) = \nabla s_0(x). \quad (10)$$

In this case, the solution $v(x,t)$ of the Burgers equation remains potential for every $t > 0$, and is described by the expression

$$v(x,t) = \frac{x - \{y\}(x,t)}{t}, \quad (11)$$

which follows from (6) and (7). By analogy with (14.3.16), the braces $\{\ldots\}$ denote here the spatial average with respect to the weight function

$$f(y; x, t) = \frac{\exp\left[-\frac{1}{2\mu t} G(y; x, t)\right]}{\int \ldots \int \exp\left[-\frac{1}{2\mu t} G(y; x, t)\right] d^n y}, \quad (12)$$

where

$$G(y; x, t) = s_0(y)\, t + \frac{(y-x)^2}{2}. \quad (13)$$

Recall more generally that in our present context, the space average $\{g(y)\}$ is defined as the n-tuple integral

$$\{g(y)\}(x,t) := \int \ldots \int g(y)\, f(y; x, t)\, d^n y. \quad (14)$$

15.2.3 Concentration Field

As noted above, the nonlinear KPZ equation (2) can be reduced, with the help of the substitution (3), to a linear parabolic equation (4). This is a lucky coincidence that does not occur often in the realm of nonlinear partial differential equations, although the obvious way to produce such examples is to start with a linear equation and see what happens to it after a nonlinear substitution. However, the latter approach seldom produces nonlinear equations that have a useful physical interpretation. Nevertheless, as will be shown below, this path leads occasionally to interesting models.

In the hunt to be conducted below we shall be informed by the fact that the logarithmic substitution (3) reduces the physically important KPZ equation (2) to the linear parabolic equation (4), and we shall hope that similar substitutions applied to a *system* of linear parabolic equations will produce nonlinear equations with a promising interpretation in the theory of interfacial growth or gas dynamics.

Let us begin by considering a pair of interdependent linear parabolic equations

$$\frac{\partial \varphi}{\partial t} = \mu \Delta \varphi + a\,\varphi + b\,\psi, \qquad \frac{\partial \psi}{\partial t} = \mu \Delta \psi + d\,\varphi + e\,\psi. \tag{15}$$

We shall apply the substitution (3) to the first equation. For this purpose, multiply the first equation (15) by $2\mu/\varphi$. As a result, we arrive at the equation

$$\frac{\partial h}{\partial t} = \frac{1}{2}(\nabla h)^2 + \mu \Delta h + 2\mu\,a + 2\mu\,b\,C, \tag{16}$$

where

$$C(\mathbf{x}, t) = \frac{\psi(\mathbf{x}, t)}{\varphi(\mathbf{x}, t)}. \tag{17}$$

Now let us find an equation satisfied by the field $C(\mathbf{x}, t)$ (23). First, write out the time derivative of this field:

$$\frac{\partial C}{\partial t} = \frac{1}{\varphi}\frac{\partial \psi}{\partial t} - \frac{\psi}{\varphi^2}\frac{\partial \varphi}{\partial t}. \tag{18}$$

Substituting for φ and ψ the right-hand sides of the equations (15), one obtains

$$\frac{\partial C}{\partial t} = \mu\left(\frac{\Delta \psi}{\varphi} - \frac{\psi \Delta \varphi}{\varphi^2}\right) + d + (e-a)\,C - b\,C^2. \tag{19}$$

To complete the operation, it is necessary to express, through C and h, the first term of the right-hand side. This can be accomplished by noticing that a formula analogous to (18) holds for the gradient of the field C:

$$\nabla C = \frac{\nabla \psi}{\varphi} - \frac{\psi \nabla \varphi}{\varphi^2}. \tag{20}$$

In turn, taking the scalar product of both sides of this equality with the vector ∇, we find that

$$\Delta C = \left(\frac{\Delta \psi}{\varphi} - \frac{\psi \Delta \varphi}{\varphi^2}\right) - \frac{2}{\varphi}\left[\frac{(\nabla \varphi \cdot \nabla \psi)}{\varphi} - \frac{\psi(\nabla \varphi \cdot \nabla \varphi)}{\varphi^2}\right]. \tag{21}$$

15.2. Multidimensional Nonlinear Equations

Taking into account the relation (20) and the expression for the gradient of field h,
$$\nabla h = 2\mu \frac{\nabla \varphi}{\varphi},$$
we can rewrite formula (21) in the convenient form
$$\frac{\Delta \psi}{\varphi} - \frac{\psi \Delta \varphi}{\varphi^2} = \Delta C + \frac{1}{\mu}(\nabla h \cdot \nabla) C. \tag{22}$$

In view of this relation, equation (19) now assumes the form
$$\frac{\partial C}{\partial t} = \mu \Delta C + (\nabla h \cdot \nabla) C + d + (e-a) C - b C^2, \tag{23}$$
containing only the fields h and C.

In this fashion, we have arrived at a closed system of nonlinear equations (16) and (23), whose solutions can be obtained explicitly from the solution of the linear parabolic equations (15). The arguments in favor of the physical relevance of the above system, for arbitrary parameters a, b, d, e, are outside the scope of this book. However, we will provide an example of such an interpretation of equations (16) and (23) in the special case that

$$a = e = \frac{1}{2\mu} F(\boldsymbol{x}, t), \qquad b \equiv 0, \qquad d = d(\boldsymbol{x}, t). \tag{24}$$

In addition, as we have done before, we shall pass from the field h to the potential field $\boldsymbol{v}(\boldsymbol{x}, t)$ (7), interpreting it as a gas velocity field. In view of (16) and (23), the relevant equations for the fields $\boldsymbol{v}(\boldsymbol{x}, t)$ and $C(\boldsymbol{x}, t)$ have the form

$$\frac{\partial \boldsymbol{v}}{\partial t} + (\boldsymbol{v} \cdot \nabla) \boldsymbol{v} = \mu \Delta \boldsymbol{v} + \boldsymbol{f}(\boldsymbol{x}, t), \qquad \frac{\partial C}{\partial t} + (\boldsymbol{v} \cdot \nabla) C = \mu \Delta C + d(\boldsymbol{x}, t). \tag{25}$$

We are already familiar with the first one, so at this point, we shall show only that the second equation also has a physical meaning. Indeed, it describes the evolution of the *concentration* $C(\boldsymbol{x}, t)$ of a passive tracer suspended in a gas moving with the velocity field $\boldsymbol{v}(\boldsymbol{x}, t)$. Recall that the passive tracer concentration is equal to the ratio of the passive tracer density and the medium density and that it does not vary with compression or rarefaction of the medium. Therefore, disregarding the influence of the molecular diffusion, the passive tracer concentration satisfies the following equation:

$$\frac{DC}{Dt} = 0 \quad \Rightarrow \quad \frac{\partial C}{\partial t} + (\boldsymbol{v} \cdot \nabla) C = 0.$$

Equation (25) for the field C takes into account, in addition to the hydrodynamic transport of passive tracer, molecular diffusion with diffusivity μ and the presence of the passive tracer's sources described by the function $d(\boldsymbol{x}, t)$. Also, note that the first equation in (25) includes a potential force field

$$\boldsymbol{f}(\boldsymbol{x}, t) = -\nabla F(\boldsymbol{x}, t),$$

which influences the motion of the medium.

The above calculations show that the solution of the coupled equations (26) with the initial conditions

$$\boldsymbol{v}(\boldsymbol{x}, t = 0) = \boldsymbol{v}_0(\boldsymbol{x}), \qquad C(\boldsymbol{x}, t = 0) = C_0(\boldsymbol{x}), \qquad (26)$$

can be reduced by substitutions

$$\boldsymbol{v} = -2\mu \frac{\nabla \varphi}{\varphi}, \qquad C = \frac{\psi}{\varphi}, \qquad (27)$$

to a solution of the linear equations of parabolic type

$$\frac{\partial \varphi}{\partial t} = \mu \Delta \varphi + \frac{1}{2\mu} F(\boldsymbol{x}, t) \varphi, \qquad \frac{\partial \psi}{\partial t} = \mu \Delta \psi + \frac{1}{2\mu} F(\boldsymbol{x}, t) \psi + d(\boldsymbol{x}, t) \varphi, \quad (28)$$

with the initial conditions

$$\varphi_0(\boldsymbol{x}, t = 0) = \exp\left(-\frac{s_0(\boldsymbol{x})}{2\mu}\right), \quad \psi_0(\boldsymbol{x}) = C_0(\boldsymbol{x}) \exp\left(-\frac{s_0(\boldsymbol{x})}{2\mu}\right). \quad (29)$$

Here $s_0(\mathbf{x})$ is the potential of the initial velocity field, connected with the latter via the equality (10).

In particular, in the absence of external forces ($\boldsymbol{f} \equiv 0$) and sources of passive tracer ($d \equiv 0$), solutions of equations (26) take the form

$$\boldsymbol{v}(\boldsymbol{x}, t) = \frac{\boldsymbol{x} - \{\boldsymbol{y}\}(\boldsymbol{x}, t)}{t}, \qquad C(\boldsymbol{x}, t) = \{C_0(\boldsymbol{y})\}(\boldsymbol{x}, t), \qquad (30)$$

where the braces signify, as before, spatial averaging defined by equalities (12) and (14).

15.3 KdV Equation and Solitons

15.3.1 From Riemann to the KdV Equation

Let us begin by returning to the canonical Riemann equation

$$\frac{\partial v}{\partial t} + v \frac{\partial v}{\partial x} = 0, \qquad v(x, t = 0) = v_0(x) \qquad (1)$$

15.3. KdV Equation and Solitons

for the field $v(x,t)$, which we interpret as the velocity of a hydrodynamic flow of particles, each of them moving uniformly along the x-axis with constant velocity. Accordingly, $v_0(x)$ describes the dependence of the particles' velocities on their spatial coordinate x at the initial time $t = 0$. As an illustration, let us consider first the Gaussian initial velocity

$$v_0(x) = V_0 \exp\left(-\frac{x^2}{2\ell^2}\right). \tag{2}$$

In Fig. 12.1.1, we have already seen that due to different velocities of different particles, the right-hand side of the velocity field profile $v(x,t)$ becomes, in the course of time $t > 0$, progressively steeper. Since the Riemann equation (1) describes the motion of noninteracting particles, the steepening fronts lead, in finite time, to the so-called gradient catastrophe described in Chap. 13 and subsequently to appearance of the multistream regime shown in Fig. 13.2.3 for the same Gaussian initial profile (2). In reality, in most applications of the Riemann equation, the arrival of the gradient catastrophe and the multistream regime means that the Riemann equation no longer adequately describes the particular physical phenomena under consideration; gradient catastrophes and multistream regimes are obviously prohibited for them. In these cases one has to replace the Riemann equation by an equation that can take into account more subtle physical effects that prevent the infinite steepening of the field $v(x,t)$ and its multistream behavior.

Recall that in Chap. 14, we already discussed a regularization of the Riemann equation by taking into account the viscosity effects and replacing equation (1) by the Burgers equation

$$\frac{\partial v}{\partial t} + v\frac{\partial v}{\partial x} = \mu\frac{\partial^2 v}{\partial x^2}, \quad v(x,0) = v_0(x). \tag{3}$$

The main characteristic peculiarity of the field $v(x,t)$ satisfying the Burgers equation is dissipation of energy of the field $v(x,t)$ due to the viscous term appearing at the end of equation (3). Nevertheless (see Sect. 14.1.2), many nonlinear fields satisfy the energy conservation law while being subject to dispersive effects. The simplest equation taking into account both the nonlinearity and the dispersion effects that at the same time is suitable for diverse physical applications is the Korteweg–de Vries (KdV) equation (14.1.11), which we rewrite here in the form

$$\frac{\partial v}{\partial t} + v\frac{\partial v}{\partial x} + \gamma\frac{\partial^3 v}{\partial x^3} = 0, \quad v(x,0) = v_0(x). \tag{4}$$

It is worthwhile to stress again the following qualitative difference between the Burgers and KdV equations: The Burgers equation describes dissipative processes. Mathematically, this means that the viscosity constant μ must be positive ($\mu > 0$). Consequently, the initial value problem (3) is mathematically well posed only if one solves it for increasing time, i.e., for $t > 0$. On the other hand, due to the nature of dispersion laws governing solutions of the KdV equation, both directions of time are mathematically equivalent in (4), while the dispersive parameter γ might be either positive or negative ($\gamma \gtrless 0$). As a result, solutions of the KdV equation (4) enjoy the symmetry (invariance) property

$$v(x,t;\gamma) \quad \Longleftrightarrow \quad -v(-x,t;-\gamma), \quad -\infty < t + \infty. \tag{5}$$

The relation $\Longleftrightarrow$ means that the left-hand side in (5) satisfies the KdV equation if and only if its right-hand side does. Note that the dependence of the solution $v(x,t,\gamma)$ on the parameter γ was explicitly indicated because its role in the symmetry property (5) is essential.

One can say that relation (5) (excluding the changing sign of the parameter γ) resembles the symmetry property (14.2.2) of the solutions to the Burgers equation. On the other hand, reflection in time produces the following symmetry property for the KdV equation:

$$v(x,t) \quad \Longleftrightarrow \quad v(-x,-t). \tag{6}$$

But it has no counterpart for the Burgers equation. An additional useful corollary follows from (6): if the initial field $v_0(x)$ is even, then the backward-in-time ($t < 0$) solution $v_-(x,t)$ of the initial value problem (4) is given by the formula

$$v_-(x,t) = v_+(-x,-t), \quad t < 0,$$

where $v_+(x,t)$ is the corresponding forward-in-time ($t > 0$) solution.

Finally, we emphasize that Galilean invariance (14.2.3) is enjoyed by the solutions of both the Burgers and KdV equations. Indeed,

$$v(x,t) \quad \Longleftrightarrow \quad V + v(x - Vt, t). \tag{7}$$

Another obvious symmetry property of both the Burgers and KdV equations is their translational invariance (14.2.1). In particular,

$$v(x,t) \quad \Longleftrightarrow \quad v(x+d,t), \tag{8}$$

where d is an arbitrary shift in the space variable.

15.3.2 Canonical Forms of the KdV Equation

Until now, equations (3) and (4) have been considered descriptions of physical phenomena. For the Burgers equation (3), it meant that the argument t had the dimension of time (in the language of dimensionality relations, this assertion is written $[t] = T$), the argument x had the dimension of the spatial coordinate ($[x] = L$), while the solution $v(x,t)$ was a velocity field satisfying the following dimensionality relation:

$$[v(x,t)] = \frac{[x]}{[t]} = \frac{L}{T}.$$

Consequently, since all terms of the Burgers equation must have the same dimensionality, the dimension of the viscosity coefficient is

$$[\mu] = \frac{[x^2]}{[t]} = \frac{L^2}{T}.$$

Although the dimensionality of the solutions of KdV-like equations has in different physical applications different interpretations, we suppose, for the sake of concreteness, that the solution of the KdV equation (4) has the dimensionality of velocity as well,[2] and we obtain that the dimensionality of the coefficient γ is

$$[\gamma] = \frac{L^3}{T}.$$

However, from a mathematical perspective, it is inconvenient to study competition between linear and nonlinear effects (viscosity or dispersion and inertial nonlinearity in the cases considered above) in the presence of dimensional coefficients. Thus it makes sense to introduce dimensionless counterparts of the above-mentioned coefficients. To implement this strategy, we will represent the initial field $v_0(x)$ in the form (14.2.4):

$$v_0(x) = U u_0\left(\frac{x}{\ell}\right), \qquad (9)$$

where U and ℓ are respectively the characteristic velocity and the characteristic scale of the initial field $v_0(x)$, while $u_0(s)$ is a dimensionless function of a dimensionless argument. In the case of the Gaussian initial condition (2),

$$u_0(s) = \exp\left(-\frac{s^2}{2}\right). \qquad (10)$$

---
[2] Note that the form of the first two terms in (4) forces $v(x,t)$ to have the dimensionality of velocity.

The dimensionless versions of the time t, spatial coordinate x, and "velocity" $v(x,t)$ can then be written in the form of the following relations:

$$\tau = \frac{\ell}{U} t, \qquad s = \frac{x}{\ell}, \qquad u(s,\tau) = \frac{v(x,t)}{U}.$$

Passing in the KdV equation (4) to the dimensionless field $u(s,\tau)$, we can now rewrite (4) in the nondimensional form

$$\frac{\partial u}{\partial \tau} + u \frac{\partial u}{\partial s} + \gamma \frac{\partial^3 u}{\partial s^3} = 0, \qquad u(s,0) = u_0(s). \tag{11}$$

Here, the nondimensional factor

$$\gamma = \frac{1}{D}, \qquad \text{where} \qquad D = \frac{U\ell^2}{\gamma}, \tag{12}$$

is an analogue of the Reynolds number R introduced in (14.2.9) (assuming that $\gamma > 0$).[3] If D is small, then the linear dispersive effects are dominant as compared to nonlinearity. On the other hand, if D is large, then the field $u(s,\tau)$ (and $v(x,t)$) becomes strongly nonlinear.

One can interpret equation (11) as the canonical form of the KdV equation in which the coefficient D in (12) explicitly takes into account competition between linear dispersion and nonlinearity, while the initial field $u_0(s)$ is "frozen" in the sense that its characteristic amplitude and spatial scale are both equal to one.

There is another, even more convenient from a mathematical perspective, canonical form of the KdV equation, which "freezes" not the initial condition, but the equation itself. To derive this canonical form, we introduce the following change of variables:

$$u = \alpha w, \qquad \tau = \beta \theta, \qquad \alpha, \beta \neq 0.$$

As a result, the initial value problem for the new renormalized field $w(s,\theta)$ takes the form

$$\frac{\partial w}{\partial \theta} + \alpha\beta w \frac{\partial w}{\partial s} + \beta\gamma \frac{\partial^3 w}{\partial s^3} = 0, \qquad w(s,0) = \frac{1}{\alpha} u_0(s).$$

---

[3]For clarity's sake, note that the first equality in (12) defines a dimensionless dispersive factor, while the right-hand side of the second equation contains its dimensional counterpart.

15.3. KdV Equation and Solitons

To eliminate the coefficient D from the KdV equation, we take

$$\beta\gamma = 1 \quad \Rightarrow \quad \beta = \frac{1}{\gamma} = D.$$

Now let us choose the coefficient in front of the nonlinear term to be equal to σ:

$$\alpha\beta = \sigma \quad \Rightarrow \quad \alpha = \sigma\gamma = \frac{\sigma}{D}.$$

Consequently, the initial value problem for the new field $w(s,\theta)$ can be written in the form

$$\frac{\partial w}{\partial \theta} + \sigma w \frac{\partial w}{\partial s} + \frac{\partial^3 w}{\partial s^3} = 0, \quad w(s,0) = \frac{1}{\sigma\gamma}u_0(s). \tag{13}$$

In the mathematical literature, for certain historical reasons, one often sets $\sigma = -6$, so that the above equation takes the form

$$\frac{\partial w}{\partial \theta} + \frac{\partial^3 w}{\partial s^3} = 6w\frac{\partial w}{\partial s}, \quad w(s,0) = -C\, u_0(s), \quad C = \frac{D}{6} > 0. \tag{14}$$

Obviously, once we find a solution $w(s,\theta)$ of the initial value problem (13), then the solution $u(s,\tau)$ of the initial value problem (11) can be obtained from the relation

$$u(s,\tau) = \sigma\gamma w(s, \gamma\tau). \tag{15}$$

In the commonly considered case $\sigma = -6$, we have $u(s,\tau) = -6\gamma w(s, \gamma\tau)$.

15.3.3 Numerical Explorations: Soliton-Like Solutions

The solutions to the Burgers have been studied in detail in the previous chapters, and by now, we have a pretty good understanding of their behavior. Now, to gain similar intuition about solutions of the KdV equation, we are going to solve the initial value problem (14) numerically and then discuss the characteristic features of these solutions.

We begin by studying the solution of the KdV equation (11) for the standardized Gaussian initial condition (10) and for $\gamma = 0.1$ (D = 10); Fig. 15.3.1 shows plots of this solution for different time instants.

A quick look at Fig. 15.3.1 shows that $u(s,\tau)$ consists of a small-amplitude but long oscillating tail and a relatively strong pulse moving with constant velocity without changing its shape. The pulse is usually called the *soliton* part of the solution of the KdV equation. Its presence is due to a balance

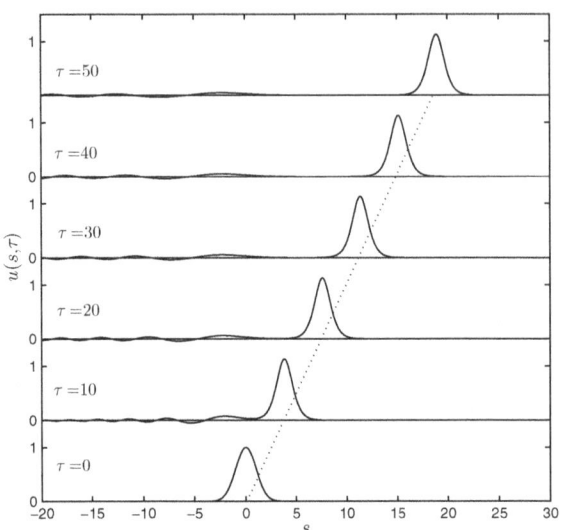

FIGURE 15.3.1
Plots of the solution $u(s,\tau)$ of the initial value problem (11), (10), for $\tau = 0, 10, 20, 30, 40, 50$, and for $D = 10$ ($\gamma = 0.1$). The main part of the solution is a pulse (soliton) moving with uniform velocity without changing its shape.

between the inertial nonlinearity and dispersive effects that is inherent for waves obeying the KdV equation.

The existence of solitons is a remarkable feature of diverse nonlinear waves in dispersive media. Below, we shall discuss in some detail properties of solitons in the context of KdV equations. In later subsections we shall find the exact analytic shape of the KdV solitons, but for now, let us continue the numerical investigation of the field $u(s,\tau)$ and its characteristic properties. First, let us elucidate the influence of the nonlinearity on the behavior of solutions of the KdV equation. We already observed that the larger the value of the number D (12) (the smaller the factor γ in equation (11)), the stronger becomes the influence of nonlinearity on the field $u(s,\tau)$. We shall study this phenomenon quantitatively.

Figure 15.3.2 shows plots of the field $u(s,\tau)$ for different values of τ and for $D = 25$ ($\gamma = 0.04$). In this, more nonlinear, case, not one but two solitons arise. They have different amplitudes, and they move uniformly with different velocities. Additionally, notice that the soliton with the higher amplitude moves faster. Numerical calculations for even larger values of the constant D show that the more nonlinear field $u(s,\tau)$ is (in our case, the larger the

15.3. KdV Equation and Solitons

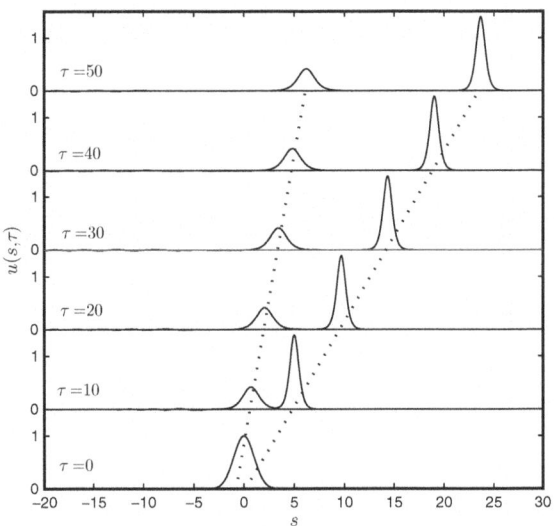

FIGURE 15.3.2
Plots of the solution $u(s, \tau)$ of the initial value problem (11), (10), in the case $D = 25$, and for $\tau = 0, 10, 20, 30, 40, 50$. The initial unimodal Gaussian field splits into two solitons, both moving uniformly but with different velocities.

number D), the larger is the number of solitons generated by the initial unimodal field $w(s, 0)$ (in our case, the unimodal Gaussian field (10)).

So, what about interactions of different solitons? The above exploration indicates that as long as the larger (and faster) soliton is located to the right of the smaller (slower) soliton, the two move happily, independently of each other, preserving their shapes; the distance between them increases. However, if the larger soliton is located to the left of the smaller soliton, the former will start approaching the latter, and eventually the two will collide. To see what happens at the collision time and thereafter, we have calculated numerically a solution of the KdV equation (11), taking as the initial field the sum of two separated Gaussian peaks,

$$u(s, 0) = 2 \exp\left(-\frac{s^2}{2}\right) + \exp\left(-\frac{(s - \ell)^2}{2}\right), \qquad \ell = 10. \qquad (16)$$

As it turns out, in this case there are three qualitatively different stages of evolution of the solution of the KdV equation. We shall review them in the case $D = 10$.

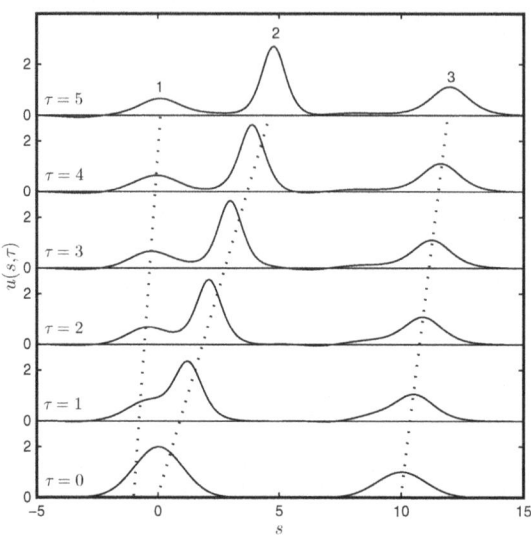

FIGURE 15.3.3a
Plots of a numerically calculated solution $u(s,\tau)$ of the initial value problem (11), (16), for **D** $= 10$, and $\tau = 0, 1, 2, 3, 4, 5$. **The first Gaussian peak (*on the left*) splits into two solitons, while the second peak (*on the right*) generates a single soliton.**

Figure 15.3.3a shows the plots of the field $u(s,\tau)$ in the first stage $\tau \in (0, 5)$ when the original Gaussian peak on the left splits into two solitons, while the smaller Gaussian peak on the right generates a single soliton. As a result, at $\tau \gtrsim 5$, we have three solitons present, which are marked 1, 2, and 3 (from left to right).

In the second stage, for $\tau \in (5, 30)$ (depicted in Fig. 15.3.3b), the largest soliton (number 2), which moves with velocity higher than that of soliton number 3, catches up with the third soliton and then overtakes it. Due to nonlinearity, the resulting shape of the colliding solitons is not equal to the shape of the solitons' sum. After the collision, at times $\tau \gtrsim 25$, the original shapes and velocities of the solitons return. However, a more accurate tracking of the colliding solitons' positions reveals a subtle shift disturbing their uniform motion as if they had become tangled up during their collision. We shall discuss this phenomenon in more detail in a later subsection.

Figure 15.3.3c shows the field $u(s,\tau)$ in the last, third, stage, when solitons are arranged, from left to right, by their increasing amplitudes (and velocities) and move uniformly without collisions.

15.3. KdV Equation and Solitons

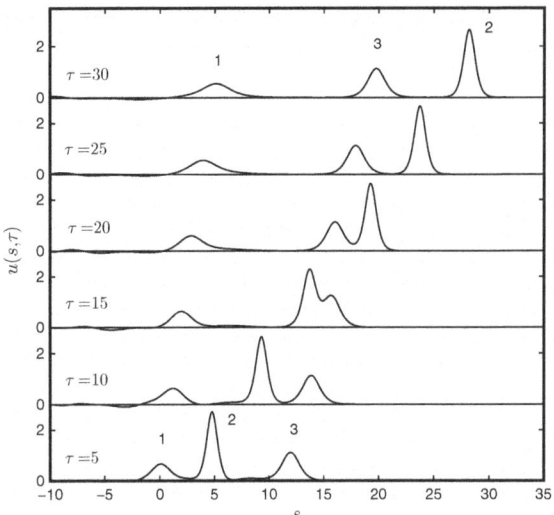

FIGURE 15.3.3b
Plots of a numerically calculated solution $u(s,\tau)$ of the initial value problem (11), (16), for $D = 10$, and $\tau \in (5, 30)$. The second and third solitons are colliding nonlinearly, but after the collision, they return to their original shapes.

Remark 1. Normalization of the initial data. The appearance of the three solitons triggered by the initial field (16) in the case of a moderate value $D = 10$ of the coefficient D was the result of the fact that the initial condition (16) was not normalized in the sense of relation (9), where $u_0(s)$ had unit amplitude, and unit scale. So, one can argue that instead of increasing D, an appropriate normalization of the initial condition (16) can produce similar effects.

15.3.4 Numerical Explorations: The Nonsoliton Case

Until now we have discussed solutions of the KdV equation (11) assuming that the initial field $u_0(s)$ was positive everywhere. It turns out that in the opposite case of a negative or alternating-sign initial condition, the behavior of the solution $u(s,\tau)$ of the KdV equation changes dramatically. This can be observed in Figs. 15.3.4a and 15.3.4b, which show a solution of (11) in the case of a negative initial condition.

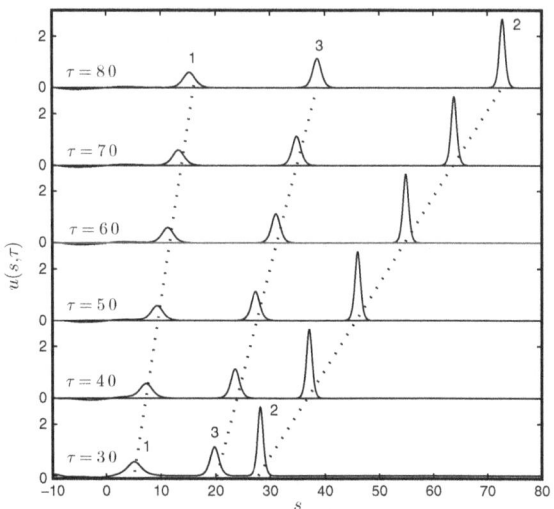

FIGURE 15.3.3c
Plots of a numerically calculated solution $u(s,\tau)$ of the initial value problem (11), (16), for $D=10$, and $\tau > 30$. After their nonlinear collision, solitons 2 and 3 reverse their order but regain their original shapes.

Figure 15.3.4a shows the solution $u(s,\tau)$ corresponding to the negative initial condition (17) at time instants $\tau = 0,1,2,3,4,5$. Instead of solitons, we see only an emerging oscillating tail, which becomes longer and longer in the course of time. In other words, the negative initial field $u_0(s)$ becomes the source of oscillating waves moving to the left of the initial Gaussian peak. The above-mentioned characteristic behavior of solutions of the KdV equation is seen even better in Fig. 15.3.4b, which shows the evolution of the field $u(s,\tau)$ for larger times.

$$u_0(s) = -\exp\left(-\frac{s^2}{2}\right). \tag{17}$$

At a nonrigorous physical level, an explanation of the qualitatively different behaviors of the solution of the KdV equation $u(s,\tau)$ in the cases of positive and negative initial field $u_0(s)$ is not hard to obtain. Indeed, consider the linearized KdV equation

$$\frac{\partial u}{\partial \tau} + \gamma \frac{\partial^3 u}{\partial s^3} = 0.$$

15.3. KdV Equation and Solitons

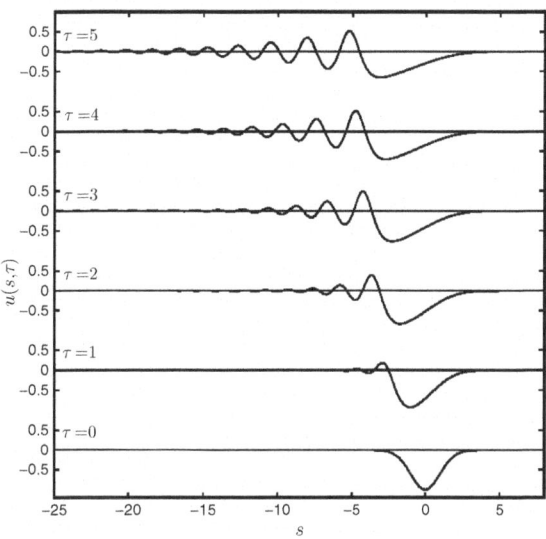

FIGURE 15.3.4a
Plots of the KdV field $u(s,\tau)$ at time instants $\tau = 0, 1, 2, 3, 4, 5$ in the case $D = 10$ ($\gamma = 0.1$) and the negative initial condition (17).

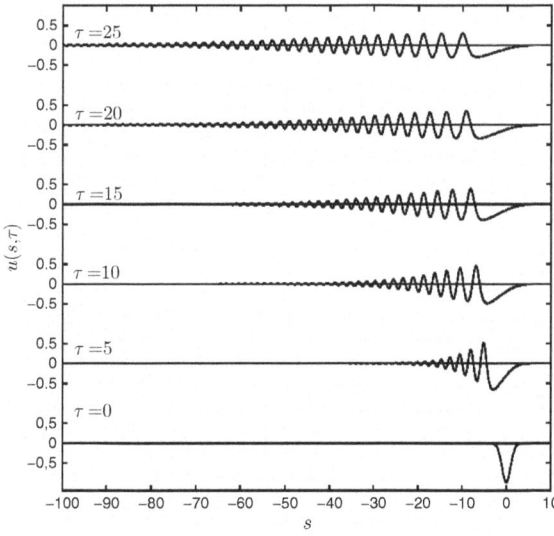

FIGURE 15.3.4b
Plots of the KdV field $u(s,\tau)$ at time instants $\tau = 0, 5, 10, 15, 20, 25$ in the case $D = 10$ ($\gamma = 0.1$) and the negative initial condition (17).

Its solution is provided by the integral (14.1.10),

$$u(s,\tau) = \int \tilde{u}_0(\kappa) \exp\left(i(\kappa s - W(\kappa)\tau)\right) d\kappa,$$

where

$$W(k) = -\gamma \kappa^3,$$

and $\tilde{u}_0$ is the Fourier image of the initial field $u_0(s)$. In the case of initial condition (17), it is equal to

$$\tilde{u}_0(\kappa) = \frac{1}{2\pi} \int u_0(s) e^{-i\kappa\tau} ds = -\frac{1}{\sqrt{2\pi}} \exp\left(-\frac{\kappa^2}{2}\right). \tag{18}$$

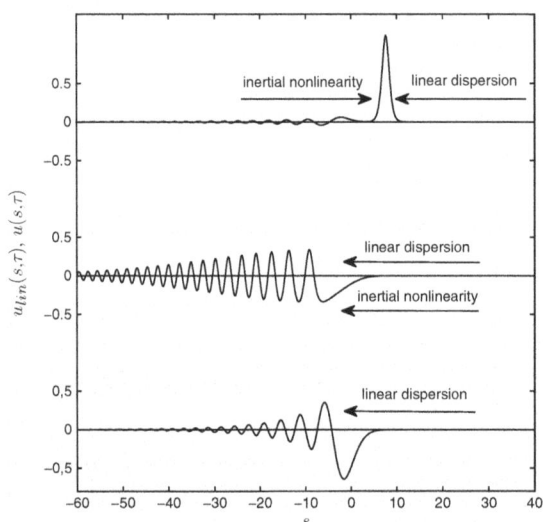

FIGURE 15.3.5
From bottom to top: Plots of (a) the linearized field $u_{\text{lin}}(s,\tau)$ (19), (b) a solution of the KdV equation (11) satisfying the negative initial condition (17), and (c) solution $u(s,\tau)$ of the KdV equation (11) satisfying the positive initial condition (10).

Accordingly, one can rewrite $u(s,\tau)$ in the following integral form:

$$u_{\text{lin}}(s,\tau) = -\sqrt{\frac{2}{\pi}} \int_0^\infty \exp\left(-\frac{\kappa^2}{2}\right) \cos\left(\kappa s - W(\kappa)\tau\right) d\kappa. \tag{19}$$

15.3. KdV Equation and Solitons

Figure 15.3.5 shows (bottom to top): (a) linearized field $u_{\text{lin}}(s,\tau)$ (19), (b) a solution of the KdV equation (11) satisfying the negative initial condition (17), and (c) solution $u(s,\tau)$ of the KdV equation (11) satisfying the positive initial condition (10). One can see that in the first two cases, the solutions are moving to the left. For the linearized field, this phenomenon may be explained by the fact that the group velocity $v(\kappa)$ (14.4.3) governing the movement of wave packets is negative. Indeed,

$$v(\kappa) = -3\gamma\kappa^2 < 0 \qquad (\kappa \neq 0).$$

The characteristic length of the linearized wave packet $u_{\text{lin}}(s,\tau)$ can be estimated using the relation

$$L(\tau) \simeq V(\kappa_0)\tau = 3\gamma\kappa_0^2\tau,$$

where κ_0 is the characteristic scale of the initial condition's Fourier image $\tilde{u}_0(\kappa)$. In the particular case (18), one can take $\kappa \simeq 2$. Thus, for $\tau = 20$ (the case depicted in Fig. 15.3.5), one has

$$L(20) \simeq 3 \cdot 0.1 \cdot 4 \cdot 20 = 24.$$

In the case of the negative initial condition (17), the inertial nonlinearity in the KdV equation tends to force the field $u(s,\tau)$ further to the left. As a result, the oscillating wave packet of the nonlinear field $u(s,\tau)$ to the left of the initial Gaussian field $u_0(s)$ becomes even longer than in the linearized case.

On the other hand, in the case of the positive initial condition (10), the linear dispersion and inertial nonlinearity in (11) act on the field $u(s,\tau)$ in the opposite directions. This competition compresses the waves to produce well-localized solitons.

In the following subsections we shall discuss analytic representations of (multi)soliton solutions of the KdV equation (11) that form the main part of solutions in the case of positive initial conditions $u_0(s)$ and small values of the constant γ.

15.3.5 Exact Soliton Solutions

There exists a comprehensive theory of the KdV equation that reduces finding its general solutions to analysis of linear scattering of quantum-mechanical particles by a quantum-mechanical potential proportional to the initial condition $u_0(s)$ of the original KdV equation. However, if we are interested only in the soliton trains solutions, which in most physical applications represent the most important part of solutions of many nonlinear dispersive equations, we can restrict ourselves to a much simpler theory of soliton-like solutions of the KdV equation, which will be explained below.

We shall solve the KdV equation in the form (13). Introducing the "potential" $h(s,\theta)$ of the solution $w(s,\theta)$,

$$w(s,\theta) = \frac{\partial h(s,\theta)}{\partial s}, \qquad (20)$$

let us rewrite (13) in a form analogous to the KPZ equation (14.3.1),

$$\frac{\partial h}{\partial t} + \frac{\sigma}{2}\left(\frac{\partial h}{\partial s}\right)^2 + \frac{\partial^3 h}{\partial x^3} = 0. \qquad (21)$$

Recall that the Hopf–Cole logarithmic transformation (14.3.9) was used to reduce the nonlinear Burgers equation to a linear diffusion equation. So, taking it as a hint, let us try to simplify equation (21) using a similar logarithmic transformation,

$$h(s,\theta) = \frac{12}{\sigma}\frac{\partial \ln \varphi(s,\theta)}{\partial s} = \frac{12}{\sigma \varphi}\frac{\partial \varphi}{\partial s}. \qquad (22)$$

Applying transformation (22) to each term of equation (21), we obtain

$$\frac{\partial h}{\partial \theta} = \frac{12}{\sigma \varphi^4}\left(\varphi^3 \frac{\partial^2 \varphi}{\partial \theta \partial s} - \varphi^2 \frac{\partial \varphi}{\partial s}\frac{\partial \varphi}{\partial \theta}\right),$$

$$\frac{\sigma}{2}\left(\frac{\partial h}{\partial s}\right)^2 = \frac{12}{\sigma \varphi^4}\left(6\varphi^2\left(\frac{\partial^2 \varphi}{\partial s^2}\right)^2 - 12\varphi \frac{\partial^2 \varphi}{\partial s^2}\left(\frac{\partial \varphi}{\partial s}\right)^2 + 6\left(\frac{\partial \varphi}{\partial s}\right)^4\right),$$

$$\frac{\partial^3 h}{\partial s^3} = \frac{12}{\sigma \varphi^4}\left(\varphi^3 \frac{\partial^4 \varphi}{\partial s^4} - \varphi^2\left(4\frac{\partial \varphi}{\partial s}\frac{\partial^3 \varphi}{\partial s^3} + 3\left(\frac{\partial^2 \varphi}{\partial s^2}\right)^2\right)\right.$$
$$\left. + 12\varphi\left(\frac{\partial \varphi}{\partial s}\right)^2 \frac{\partial^2 \varphi}{\partial s^2} - 6\left(\frac{\partial \varphi}{\partial s}\right)^4\right).$$

In the next step we shall substitute the above equalities into (21), canceling the common factor $12/\sigma\varphi^4$ and grouping terms according to the powers of the function φ:

$$a_0 + a_1 \cdot \varphi + a_2 \cdot \varphi^2 + a_3 \cdot \varphi^3 = 0. \qquad (23)$$

Due to the special structure of equation (21), the coefficients of the zeroth and first powers of φ turn out to be identically equal to zero ($a_0 = a_1 \equiv 0$), and (23) simplifies to the form

$$a_2 + a_3 \cdot \varphi = 0. \qquad (24)$$

15.3. KdV Equation and Solitons

The remaining coefficients are as follows:

$$a_2 = \frac{\partial \varphi}{\partial s}\frac{\partial \varphi}{\partial \theta} - 3\left(\frac{\partial^2 \varphi}{\partial s^2}\right)^2 + 4\frac{\partial \varphi}{\partial s}\frac{\partial^3 \varphi}{\partial s^3},$$

$$a_3 = -\frac{\partial^2 \varphi}{\partial \theta \partial s} - \frac{\partial^4 \varphi}{\partial s^4}. \tag{25}$$

This still does not look like a success story until we make an observation that the remaining coefficients a_2 and a_3 can be rearranged in the form

$$a_2 = \frac{\partial \varphi}{\partial s} \cdot \mathcal{L}\varphi + 3\left[\frac{\partial \varphi}{\partial s}\frac{\partial^3 \varphi}{\partial s^3} - \left(\frac{\partial^2 \varphi}{\partial s^2}\right)^2\right],$$

$$a_3 = -\frac{\partial}{\partial s}\mathcal{L}\varphi, \tag{26}$$

where $\mathcal{L}$ stands for the differential operator

$$\mathcal{L} = \frac{\partial}{\partial \theta} + \frac{\partial^3}{\partial s^3}.$$

Hence, if $\varphi(s, \theta)$ satisfies the linear KdV equation

$$\mathcal{L}\varphi = \frac{\partial \varphi}{\partial \theta} + \frac{\partial^3 \varphi}{\partial s^3} = 0, \tag{27}$$

then $a_3 \equiv 0$, while the coefficient a_2 reduces to

$$a_2 = 3\left[\frac{\partial \varphi}{\partial s}\frac{\partial^3 \varphi}{\partial s^3} - \left(\frac{\partial^2 \varphi}{\partial s^2}\right)^2\right]. \tag{28}$$

It is clear that if we can find a solution φ of the linear KdV equation (27) that also makes the right-hand side of (28) vanish, then substituting it into (22), we shall find a solution $h(s, \theta)$ of the nonlinear equation (21). Then the desired solution $w(s, \theta)$ of the nonlinear KdV equation (13) can be obtained by employing the formula

$$w(s, \theta) = \frac{12}{\sigma}\frac{\partial^2 \ln \varphi(s, \theta)}{\partial s^2}, \tag{29}$$

which follows directly from the relations (20) and (22).

An example of such a family of solutions of the linear equation (27), which also makes the right-hand side of (28) vanish, is given by the formulas

$$\varphi = 1 + e^{-z}, \qquad z = k(s - k^2\theta), \tag{30}$$

where k is an arbitrary constant. The constant 1 in the first equality is selected specifically to keep the logarithm of φ positive for all values of the auxiliary parameter z ($\ln\varphi > 0$, $\forall z$). Since

$$\frac{\partial}{\partial s} = k\frac{\partial}{\partial z},$$

we can rewrite (29) in the form

$$w = \frac{12k^2}{\sigma}\frac{\partial^2}{\partial z^2}\ln(1 + e^{-z}) = \frac{12k^2}{\sigma}\frac{e^z}{(1 + e^z)^2}.$$

Using the definition of the hyperbolic secant,

$$\text{sech}(z) = \frac{1}{\cosh(z)} = \frac{2}{e^z + e^{-z}},$$

the previous relation can be rewritten in a form more convenient for analysis:

$$w = \frac{3k^2}{\sigma}\text{sech}^2\left(\frac{z}{2}\right).$$

Substituting the explicit expression (30) for the auxiliary argument z gives

$$w(s,\theta) = \frac{3k^2}{\sigma}\text{sech}^2\left[\frac{k}{2}(s - k^2\theta)\right].$$

The translational invariance property (8) of the KdV equation now permits us to write a more general version of the above solution,

$$w(s,\theta) = \frac{3k^2}{\sigma}\text{sech}^2\left[\frac{k}{2}(s - d - k^2\theta)\right], \tag{31}$$

where d is an arbitrary shift of the previous special solution along the s-axis. Also, notice that in view of the evenness of the function $\text{sech}(z)$, one may, without loss of generality, select k to be positive ($k > 0$). A 3-D plot of the solution (31) corresponding to the constants $\sigma, k = 1$ and $d = 0$ is shown in Fig. 15.3.6.

We shall now provide a physical interpretation of the solution (31) using first, the relation (20) to convert it into the solution $u(s,\tau)$ of the more physically transparent KdV equation (11),

$$u(s,\tau) = \sigma\gamma w(s,\gamma\tau) = 3c\,\text{sech}^2\left[\sqrt{\frac{c}{4\gamma}}(s - c\tau - d)\right], \tag{32}$$

15.3. KdV Equation and Solitons

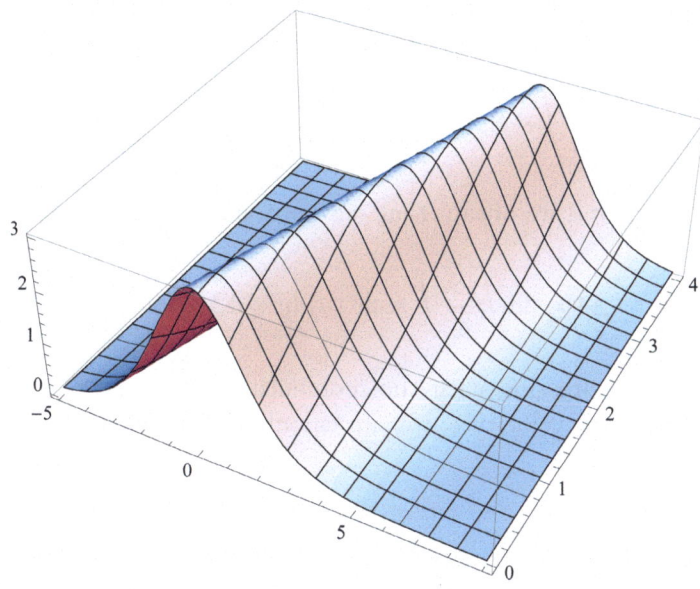

FIGURE 15.3.6
A 3-D plot of the exact soliton solution (31) of the KdV equation corresponding to constants $\sigma, k = 1$ and $d = 0$. It is shown in the time interval $0 < \theta < 4$.

where we have introduced the new notation

$$c = \gamma k^2 \quad \text{or equivalently,} \quad k = \sqrt{\frac{c}{\gamma}} \quad (k, \gamma > 0). \tag{33}$$

The solution (32) obtained above is obviously a wave, moving from left to right with velocity c without changing its shape. So it is plausible that it is the soliton that we discovered earlier in numerical experiments with solutions of the KdV equation. Equations (32) and (33) indicate that our soliton possesses the following characteristic properties: its amplitude $3c$ and its effective width ($\sim \sqrt{c/4\gamma}$) are tied uniquely to its velocity c. Qualitatively, the higher the soliton's velocity, the larger the amplitude and the narrower the width of the soliton (for a given γ).

In some physical applications, solitons can be interpreted as quasiparticles, whose momentum is equal to

$$\mathcal{M} = \int u(s, \tau) ds. \tag{34}$$

Mathematica gives us the formula

$$S(n) = \int \text{sech}^n(y)\,dy = \sqrt{\pi}\,\frac{\Gamma\left(\frac{n}{2}\right)}{\Gamma\left(\frac{n+1}{2}\right)}, \qquad n > 0, \tag{34}$$

so that $S(2) = 2$, and we obtain that our soliton's momentum is

$$\mathcal{M} = 3c\sqrt{\frac{4\gamma}{c}}\int \text{sech}^2(y)\,dy = 12\sqrt{\frac{\gamma}{c}}\,c.$$

Comparing this expression with the momentum $\mathcal{M} = mc$ of a particle of mass m moving with velocity c, we can calculate the "mass" of the soliton,

$$m = 12\sqrt{\frac{\gamma}{c}}. \tag{35}$$

In a similar fashion, the soliton's kinetic energy can be defined as the integral

$$\mathcal{E} = \frac{1}{2}\int u^2(s,\tau)\,ds. \tag{36}$$

Calculating this integral for the soliton (32) and bearing in mind that $S(4) = 4/3$, we get

$$\mathcal{E} = \frac{1}{2} 9c^2\sqrt{\frac{4\gamma}{c}}\int \text{sech}^4(y)\,dy = \frac{mc^2}{2},$$

where m is given by (35).

15.3.6 Multisoliton Solutions

Inspired by our success in finding exact analytic single soliton solutions of the KdV equation, and recalling that certain initial fields $w_0(s)$ generated several solitons numerically (Sect. 15.3.3), we now aim at finding multisoliton analytic solutions of the KdV equation.

So, let us return to the equation (24) for the function φ, defining the solution (29) of the KdV equation (13). Substituting the functions a_2 and a_3 from (26) into (24), we can rewrite (24) in the form

$$\varphi\frac{\partial^2\varphi}{\partial s\partial\theta} - \frac{\partial\varphi}{\partial s}\frac{\partial\varphi}{\partial\theta} - 4\frac{\partial\varphi}{\partial s}\frac{\partial^3\varphi}{\partial s^3} + 3\left(\frac{\partial^2\varphi}{\partial s^2}\right)^2 + \varphi\frac{\partial^4\varphi}{\partial s^4} = 0. \tag{37}$$

15.3. KdV Equation and Solitons

To search for solutions of this equation, it is convenient to introduce a symmetric bilinear operator,

$$\mathcal{P}[f,g] = \frac{1}{2}\left[\frac{\partial^2 f}{\partial s \partial \theta}g + \frac{\partial^2 g}{\partial s \partial \theta}f - \frac{\partial f}{\partial s}\frac{\partial g}{\partial \theta} - \frac{\partial g}{\partial s}\frac{\partial f}{\partial \theta} + \right.\\ \left.\frac{\partial^4 f}{\partial s^4}g + \frac{\partial^4 g}{\partial s^4}f - 4\frac{\partial^3 f}{\partial s^3}\frac{\partial g}{\partial s} - 4\frac{\partial^3 g}{\partial s^3}\frac{\partial f}{\partial s} + 6\frac{\partial^2 f}{\partial s^2}\frac{\partial^2 g}{\partial s^2}\right]. \tag{38}$$

Employing the above operator notation, one may rewrite (37) in the form

$$\mathcal{P}[\varphi,\varphi] = 0. \tag{39}$$

At this point, let us state a few useful properties of the bilinear operator $\mathcal{P}[f \cdot g]$.

Property 1. For a fixed f (or g), $\mathcal{P}[f \cdot g]$ is a linear operator in g (or f). Consequently, it possesses the following bilinear property:

$$\mathcal{P}\left[\sum_i c_i f_i, \sum_j d_j g_j\right] = \sum_i \sum_j c_i d_j \mathcal{P}[f_i, g_j], \tag{40}$$

where c_i and d_j are arbitrary constants.

Property 2. If f and g are exponential functions, then the differential operator $\mathcal{P}[f \cdot g]$ acts as multiplication by an algebraic polynomial. Indeed, consider the exponential functions

$$f = e^{\omega_1 \theta - k_1 s}, \qquad g = e^{\omega_2 \theta - k_2 s}.$$

Simple calculations give

$$\mathcal{P}\left[e^{\omega_1 \theta - k_1 s}, e^{\omega_2 \theta - k_2 s}\right] = P(k_1 - k_2, \omega_1 - \omega_2) e^{(\omega_1 + \omega_2)\theta - (k_1 + k_2)s}, \tag{41}$$

where

$$P(k,\omega) = \frac{1}{2}k\left(k^3 - \omega\right) \tag{42}$$

is a symmetric function, that is,

$$P(k,\omega) = P(-k,-\omega).$$

Property 3. It follows directly from (42) that

$$P(0,0) = 0, \qquad P(k,k^3) = 0. \tag{43}$$

First, let us demonstrate that the above operator machinery is capable of producing the familiar single solution of the KdV equation. Indeed, substituting
$$\varphi = 1 + ce^{\omega\theta - ks} \tag{44}$$
into equation (39) and using the bilinear property (40), we obtain
$$\mathcal{P}\left[1 + ce^{\omega\theta - ks}, 1 + ce^{\omega\theta - ks}\right]$$
$$= \mathcal{P}[1,1] + 2c\mathcal{P}[e^{\omega\theta - ks}, 1] + c^2\mathcal{P}[e^{\omega\theta - ks}, e^{\omega\theta - ks}] = 0. \tag{45}$$

Applying (41) and the first equality in (43), we obtain
$$\mathcal{P}[1,1] = P(0,0) = 0,$$
$$\mathcal{P}[e^{\omega\theta - ks}, e^{\omega\theta - ks}] = P(0,0)\,e^{2\omega\theta - 2ks} = 0. \tag{46}$$

Thus, the equation (45) reduces to the algebraic equation
$$\mathcal{P}[e^{\omega\theta - ks}, 1] = P(k, \omega)e^{\omega\theta - ks} = 0.$$

Due to the second equality in (43), the last equation is satisfied if $\omega = \kappa^3$. In other words, we have found a solution of equation (39) equal to
$$\varphi = 1 + c\,e^{k^3\theta - ks}. \tag{47}$$

To make the right-hand side of the relation (29) mathematically correct, for any s and θ we should take here $c > 0$, so that we can rewrite (47) in the equivalent form
$$\varphi = 1 + e^{-k(s - d - k^2\theta)}, \tag{48}$$
where $d \in (-\infty, \infty)$ is an arbitrary constant. This gives an expression—already familiar to us—for a single soliton solution (31).

Consider now the following natural generalization of the expression (47),
$$\varphi = 1 + c_1 e^{k_1^3\theta - k_1 s} + c_2 e^{k_2^3\theta - k_2 s}, \tag{49}$$
and calculate the left-hand side of (39). Using the bilinear property (40) of the operator $\mathcal{P}$, and relations (46) and (41), we obtain
$$\mathcal{P}[\varphi, \varphi] = c_1 P(k_1, k_1^3)e^{k_1^3\theta - k_1 s} + c_2 P(k_2, k_2^3)e^{k_2^3\theta - k_2 s}$$
$$+ 2c_1 c_2 P(k_1 - k_2, k_1^3 - k_2^3)e^{(k_1^3 + k_2^3)\theta - (k_1 + k_2)s},$$

15.3. KdV Equation and Solitons

or, taking into account the second equality (43),

$$\mathcal{P}[\varphi, \varphi] = 2c_1 c_2 P(k_1 - k_2, k_1^3 - k_2^3) e^{(k_1^3 + k_2^3)\theta - (k_1 + k_2)s}.$$

It is obvious that if $k_1 \neq k_2$, then the function (49) is not a solution of the equation (39). Indeed, the equation (39) is nonlinear (quadratic), so that a linear superposition of solutions is, in general, not a solution. So, to continue our search for new solutions, we have to insert into the right-hand side of the expression (49) some nonlinear combinations of the linear summands in (49). It turns out that due to the particular properties of the operator $\mathcal{P}$ expressed by (41) and (43), we can find such a solution of (39) by adding to the right-hand side of equality (49) a simple nonlinear term, namely, a product of the terms already included in (49). This means that we are now looking for a solution of the form

$$\varphi = 1 + c_1 e^{k_1^3 \theta - k_1 s} + c_2 e^{k_2^3 \theta - k_2 s} + \rho e^{(k_1^3 + k_2^3)\theta - (k_1 + k_2)s}. \tag{50}$$

Substituting it into equation (39) and following an argument that already gave us (49), we obtain

$$\mathcal{P}[\varphi \cdot \varphi] =$$

$$2\left(c_1 c_2 P(k_1 - k_2, k_1^3 - k_2^3) + \rho P(k_1 + k_2, k_1^3 + k_2^3)\right) e^{(k_1^3 + k_2^3)\theta - (k_1 + k_2)s}.$$

The right-hand side is equal to zero if

$$\rho = -c_1 c_2 \frac{P(k_1 - k_2, k_1^3 - k_2^3)}{P(k_1 + k_2, k_1^3 + k_2^3)} = c_1 c_2 \left(\frac{k_1 - k_2}{k_1 + k_2}\right)^2.$$

Replacing the constant ρ in (50) by the right-hand side of the above equality, we obtain the desired new solution of equation (39):

$$\varphi = 1 + c_1 e^{k_1^3 \theta - k_1 s} + c_2 e^{k_2^3 \theta - k_2 s} + c_1 c_2 \left(\frac{k_1 - k_2}{k_1 + k_2}\right)^2 e^{(k_1^3 + k_2^3)\theta - (k_1 + k_2)s}.$$

Finally, proceeding as in the transition from (47) to (48), we replace constants c_1 and c_2 by arbitrary space shifts $d_1, d_2 \in (-\infty, \infty)$ to produce a more geometrically transparent expression for the function $\varphi(s, \theta)$:

$$\varphi = 1 + e^{-z_1} + e^{-z_2} + \mathcal{R}\, e^{-z_1 - z_2}, \tag{51}$$

where

$$z_i = k_i(s - d_i - k^2 \theta), \qquad \mathcal{R} = \left(\frac{k_1 - k_2}{k_1 + k_2}\right)^2. \tag{52}$$

In the next subsection we shall demonstrate that the field $w(s,\theta)$ (29), where $\varphi(s,\theta)$ is given by the equality (51), describes a two-soliton collision phenomenon, which we have already observed in Fig. 15.3.3b.

Remark 2. Extension to n-soliton solutions. By similar but more sophisticated arguments, Hirota[4] obtained multisoliton solutions for the KdV equation. To better comprehend their structure, let us rewrite the formula (51) in a form involving a determinant:

$$\varphi = \begin{vmatrix} 1 + e^{-z_1} & \dfrac{2\sqrt{k_1 k_2}}{k_1 + k_2} e^{-\frac{z_1+z_2}{2}} \\ \dfrac{2\sqrt{k_1 k_2}}{k_1 + k_2} e^{-\frac{z_1+z_2}{2}} & 1 + e^{-z_2} \end{vmatrix}. \tag{53}$$

Hirota proved that the elegant generalization

$$\varphi = \big|\psi_{i,j}(s,\theta)\big|, \qquad \psi_{i,j} = \delta_{i,j} + \frac{2\sqrt{k_i k_j}}{k_i + k_j} \exp\left(-\frac{z_i + z_j}{2}\right), \tag{54}$$

$$i, j = 1, 2, \ldots, n,$$

of the relation (53) gives, in combination with the equality (29), a solution of the KdV equation (13) describing n-soliton collisions. Here $\delta_{i,j}$ is the usual Kronecker delta, equal to one if $i = j$ and zero otherwise.

15.3.7 Collision of Solitons

In discussing the results of numerical calculations in Sect. 15.3.3, we have already observed that the solitons are stable in the sense that the nonlinear interactions during their collisions do not disturb their shapes; they remain the same far after a collision. At this point, we are prepared to validate the above numerics-based observation via quantitative analytic formulas, which follow from the relations (29) and (51) describing propagation and collision of two solitons.

Our exploration of two-soliton collisions will be more geometrically transparent if we begin with a discussion not of the KdV equation's solution $w(s,\theta)$, but of its "potential" $h(s,\theta)$ (22). Substituting (51) into (22), we obtain

$$h(s,\theta) = -\frac{12}{\sigma} \cdot \frac{k_1 e^{-z_1} + e^{-z_2}\left[k_2 + \mathcal{R}(k_1 + k_2)e^{-z_1}\right]}{1 + e^{-z_1} + e^{-z_2}\left[1 + \mathcal{R}e^{-z_1}\right]}.$$

---

[4] R. Hirota. Exact solution of the Korteweg–de Vries equation for multiple collisions of solitons. *Phys. Rev. Lett.* (1971) 27, 1192–1194.

15.3. KdV Equation and Solitons

Let us introduce an auxiliary field,

$$\bar{h}(y,\theta) = h(y + d_1 + k_1^2\theta, \theta).$$

It has a transparent geometric sense. Namely, $\bar{h}(y,\theta)$ describes the time evolution of the KdV equation's solution in the coordinate system moving uniformly with the first soliton, corresponding to the parameter k_1. In other words, the new comoving coordinate y and the old fixed coordinate s are joined by the relation

$$y = s - d_1 - k_1^2\theta \quad \Longleftrightarrow \quad s = y + d_1 + k_1^2\theta.$$

The above relations imply that

$$\bar{h} = -\frac{12}{\sigma} \cdot \frac{k_1 e^{-k_1 y} + e^{-k_2(y-\chi)}\left[k_2 + (k_1+k_2)\mathcal{R}e^{-k_1 y}\right]}{1 + e^{-k_1 y} + e^{-k_2(y-\chi)}\left[1 + \mathcal{R}e^{-k_1 y}\right]}, \tag{55}$$

where

$$\chi = d_2 - d_1 + (k_2^2 - k_1^2)\theta.$$

Intuitively, χ can be thought of as a measure of the "distance" between the first and the second solitons. If $|k_1| \neq |k_2|$, then χ changes in the course of time θ. Suppose for the sake of definiteness, that

$$k_1, k_2 > 0, \qquad k_2 > k_1.$$

Then χ is an increasing function of time θ. Consequently, as χ increases from $-\infty$ to $+\infty$, the two solitons' behavior displays three distinct stages:

(i) In the first stage, when $\chi \ll -1$, the first soliton is immovable (in the coordinate system y) and the second approaches it from the left with uniform speed.

(ii) In the second stage, for $|\chi| \lesssim 1$, a collision of the two solitons occurs.

(iii) Finally, in the third stage, for $\chi \gg 1$, the first soliton becomes immovable again while the second moves away from it to the right with uniform speed.

Let us consider the above three stages in more detail. At the very beginning of the first stage, when

$$\chi \to -\infty \quad \Rightarrow \quad e^\chi \to 0,$$

the expression (55) tends to the following limit:
$$\bar{h} = -\frac{12k_1}{\sigma} \cdot \frac{e^{-k_1 y}}{1+e^{-k_1 y}} = \frac{6k_1}{\sigma}\left[\tanh\left(\frac{k_1 y}{2}\right) - 1\right],$$

since
$$\frac{e^{-x}}{1+e^{-x}} = \frac{1}{1+e^x} = \frac{1}{2} - \frac{1}{2}\tanh\left(\frac{x}{2}\right). \tag{56}$$

Consequently, the solution $\tilde{w}(y,\theta)$, expressed in the uniformly moving coordinate system, is equal to
$$\tilde{w}(y,\theta) = \frac{\partial \bar{h}(y,\theta)}{\partial y} = \frac{3k_1^2}{\sigma}\operatorname{sech}^2\left(\frac{k_1 y}{2}\right). \tag{57}$$

On the other hand, in the third stage, when
$$\chi \to \infty \quad \Rightarrow \quad e^\chi \to \infty,$$

the limit of (55) is of the form
$$\bar{h} = -\frac{12}{\sigma} \cdot \frac{k_2 + (k_1+k_2)\mathcal{R}e^{-k_1 y}}{1+\mathcal{R}e^{-k_1 y}} = -\frac{12}{\sigma}\left(k_2 + k_1\frac{\mathcal{R}e^{-k_1 y}}{1+\mathcal{R}e^{-k_1 y}}\right),$$

or equivalently,
$$\bar{h} = -\frac{12}{\sigma}\left(k_2 + k_1\frac{e^{-k_1(y-\xi)}}{1+e^{-k_1(y-\xi)}}\right), \qquad \xi = \frac{1}{k_1}\ln(\mathcal{R}).$$

Taking into account (56), we obtain
$$\bar{h} = -\frac{12}{\sigma}\left[k_2 + \frac{k_1}{2} - \frac{k_1}{2}\tanh\left(\frac{k_1}{2}(k_1 y - \xi)\right)\right].$$

The last expression means that long after the collision, the solution of the KdV equation, expressed in the coordinate system comoving with the first soliton, is equal to
$$\tilde{w} = \frac{\partial \bar{h}}{\partial y} = \frac{3k_1^2}{\sigma}\operatorname{sech}^2\left(\frac{k_1}{2}(y-\xi)\right).$$

It differs from the original (before the collision) soliton (57) only by the space shift ξ. Figure 15.3.7 shows a contour plot of the field $\tilde{w}(y,\theta)$ obtained via (57) from $\bar{h}(y,\theta)$ (55). The corresponding 3-D plot is shown in Fig. 15.3.8.

15.3. KdV Equation and Solitons

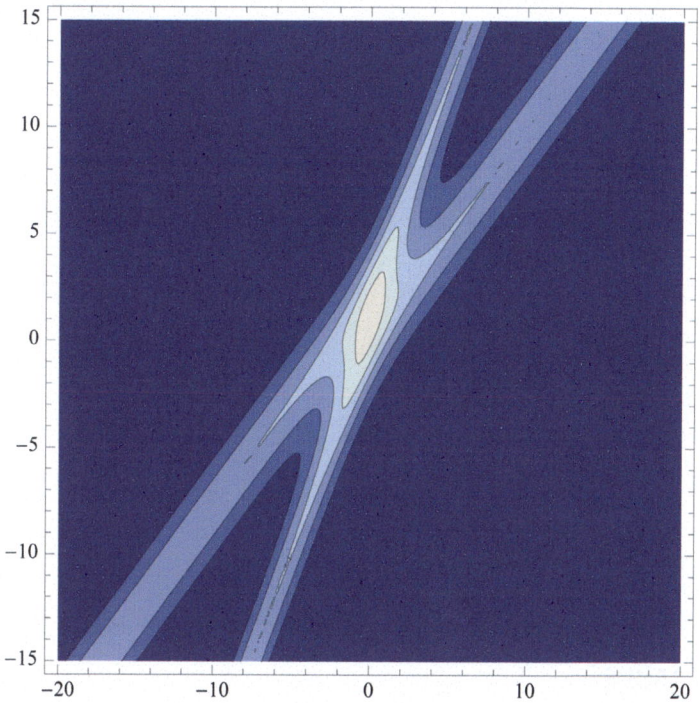

FIGURE 15.3.7
A contour plot of the two-soliton collision. Here $\sigma = 1$. The time frame is $-15 \leq t \leq 15$. The phase shifts caused by the interaction of the two solitons in a neighborhood of $x = 0, t = 0$, are clearly visible.

The picture in Fig. 15.3.7 can be interpreted in two different ways. The first is that in the course of time, the second soliton overtakes the first one, so that the ultimate result of the collision is only its finite space shift is comparison with the original motion of the solitons. However, some physicists see the collision interaction of the solitons as representing the real physical process of an inelastic collision of two quasiparticles. Each of them has a complex structure involving waves compressed into the well-localized wave packet(quasiparticle) and subject to two competing influences: linear dispersion and inertial nonlinearity (see Fig. 15.3.5). During the collision, quasiparticles do not swap places but repulse each other, exchanging their masses and, due to the energy conservation law, velocities as well. This phenomenon can be observed in Fig. 15.3.7.[5]

---
[5]Fascinating videos of KdV solitons, both simulated and photographed at ocean beaches, can be found at www.youtube.com/watch?v=ZsTe2N5_eZE.

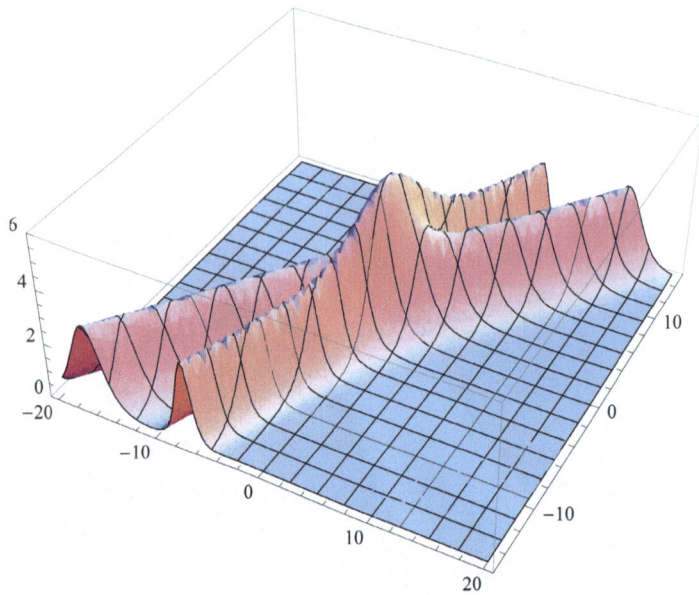

FIGURE 15.3.8
A 3-D illustration of a two-soliton collision shown as a contour plot in Fig. 15.3.7.

15.4 Flows in Porous Media

In this final section of the present volume, we provide a brief review of the celebrated problem of the flow of a compressible fluid in a homogeneous porous medium. The equation that governs the flow is qualitatively different from all the previous nonlinear equations we have considered so far, since the nonlinearity appears here in the highest-order derivative. It gives rise to a phenomenon that we have not observed previously: the existence of a solution with compact support at all times. As a matter of fact, in the case of nonzero initial data with compact support, the equation has no classically differentiable solution.[6] Thus all of our solutions must be considered weak solutions.

---

[6]See, e.g., F. Otto, The geometry of dissipative evolution equations: the porous medium equation, *Communications in Partial Differential Equations* 26 (2001), 101–174.

15.4.1 Darcy's Law and the Porous Medium Equation

The three quantities describing the flow of compressible gas in a d-dimensional porous medium are the velocity $\boldsymbol{v}(\boldsymbol{x},t)$, density $\rho = \rho(\boldsymbol{x},t)$, and pressure $P = P(\boldsymbol{x},t)$. The density and pressure are related by the usual hydrodynamic *conservation of mass law*

$$\chi \frac{\partial P}{\partial t} + \boldsymbol{\nabla} \cdot (\rho \boldsymbol{v}) = 0, \tag{1}$$

where the constant $\chi > 0$ represents the porosity of the medium, which is the fraction of the medium filled with gas. The velocity and pressure are related by *Darcy's law*,[7]

$$\nu \boldsymbol{v} = -\mu \boldsymbol{\nabla} P. \tag{2}$$

The constant $\nu > 0$ represents the viscosity of the gas, and μ is called the permeability of the medium.

To close the above system of two equations containing three unknown functions, one usually imposes an equation of state relating pressure P to density ρ. We have already encountered such an equation in Sect. 15.1, where we defined a polytropic gas. Postulating the power scaling of degree $\alpha > 0$, the closing *equation of state* is of the form

$$P = \kappa \rho^\alpha, \tag{3}$$

where the constant κ is positive.

Eliminating the functions $P(\boldsymbol{x},t)$ and $\boldsymbol{v}(\boldsymbol{x},t)$ from the above system of equations (1)–(3), introducing dimensionless time, space, and density, which we will call $u(\boldsymbol{x},t)$, we arrive at the standard form of the *porous medium equation*,

$$\frac{\partial}{\partial t} u(\boldsymbol{x},t) = \Delta(u^{1+\alpha}(\boldsymbol{x},t)), \qquad \boldsymbol{x} \in \mathbf{R}^d, \ t > 0, \ m \geq 1. \tag{4}$$

Observe that equation (4) can also be rewritten in the form

$$\frac{\partial}{\partial t} u(\boldsymbol{x},t) = \operatorname{div}\Big((1+\alpha) u^\alpha(\boldsymbol{x},t) \boldsymbol{\nabla} u(\boldsymbol{x},t)\Big), \tag{5}$$

---

[7] A comprehensive calculation showing how Darcy's law can be obtained from the general equations of flow in porous media can be found, e.g., in W.G. Grey and K. O'Neil, On the general equations of flow in porous media and their reduction to Darcy's law, *Water Resources Research* 12 (1976), 148–154, or U. Hornung, *Homogenization and Porous Media*, Springer-Verlag, New York 1997, pp. 16ff.

with the factor in front of the gradient an increasing function of u. This fact plays an important role in the determination of the behavior of solutions of the Cauchy problem for (4).

The case $\alpha = 0$, not considered here, obviously corresponds to the standard diffusion equation discussed in detail in Chap. 10. So from now on, we will always assume
$$\alpha > 0.$$
Several special cases describe specific physical phenomena: $\alpha = 1$ often is used to model thin saturated regions in porous media, $\alpha \geq 2$ is a general model of percolation of gas through porous media, $\alpha = 3$ has been used in the study of thin liquid films spreading under gravity, and $\alpha = 6$ has appeared in the investigation of radiative heat transfer by so-called Marshak waves.[8]

The fundamental property of the porous medium equation is that it possesses solutions that are compactly supported for all $t > 0$, a behavior dramatically different from the behavior of solutions of other parabolic equations we have considered thus far, in which even the solution produced by the Dirac delta initial data spread over an infinite spatial interval for every $t > 0$.

15.4.2 Barenblatt Self-Similar Solutions

The study of self-similar solutions of the porous medium equation begins with an observation that if $u(\boldsymbol{x}, t)$ is a solution of (4), then for every positive constant c, the function
$$u(\boldsymbol{x}, t; c) = c^d u(c\boldsymbol{x}, c^{d\alpha+2} t) \tag{6}$$
is also a solution of (4). In other words, the porous medium equation is invariant under substitutions
$$u \mapsto c^d u, \quad \boldsymbol{x} \mapsto c\boldsymbol{x}, \quad t \mapsto c^{d\alpha+2} t. \tag{7}$$

Barenblatt's striking observation of 1952 was that the explicit expression
$$u_B(\boldsymbol{x}, t) = \left(\max\left(0, B t^{-d\alpha/(d\alpha+2)} - \frac{\alpha}{2(\alpha+1)(d\alpha+2)} \cdot \frac{|\boldsymbol{x}|^2}{t} \right) \right)^{1/\alpha} \tag{8}$$

---

[8]See, e.g., G.I. Barenblatt, On some unsteady motions of a liquid and a gas in a porous medium, *Prikladnaya Matematika i Mekhanika* 16 (1952), 67–78; D.G. Aronson, *The Porous Medium Equation*, Lecture Notes in Mathematics 1224, Springer-Verlag 1224, 1986; M.E. Gurtin and R.C. MacCamy, On the diffusions of biological populations, *Mathematical Biosciences* 33 (1977), 35–49.

15.4. Flows in Porous Media

is a self-similar solution of (4) for every positive constant B. Obviously, it is strictly positive only in the expanding ball

$$\{\boldsymbol{x} : |\boldsymbol{x}|^2 \leq B \frac{2(\alpha+1)(d\alpha+2)}{\alpha} t^{-d\alpha/(d\alpha+2)+1}\} \subset \mathbf{R}^d. \qquad (9)$$

In the 1-D case $d = 1$ and in the classic case of the linear equation of state (3), that is, $\alpha = 1$, the nonlinearity in (4) is quadratic, and the Barenblatt solution takes a particularly simple form:

$$u_B(x,t) = \max\left(0, Bt^{-1/3} - \frac{1}{12} \cdot \frac{x^2}{t}\right). \qquad (10)$$

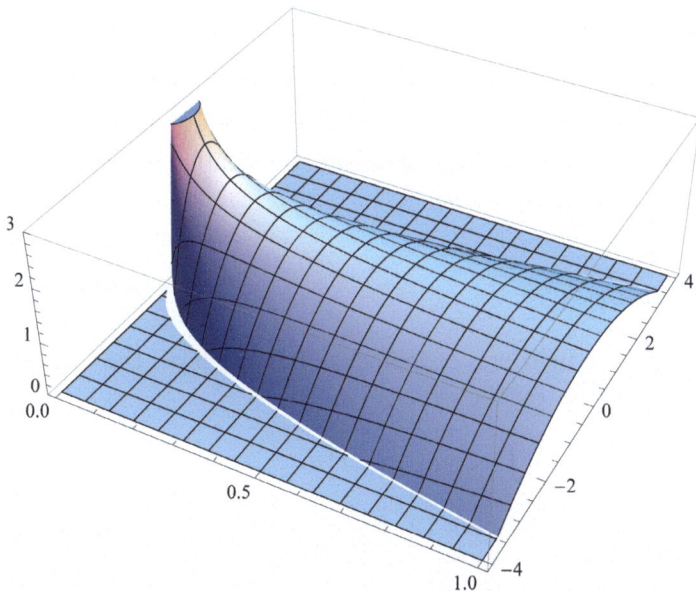

FIGURE 15.4.1
Evolution of Barenblatt's self-similar solution (8) of the porous medium equation for times $t \in (0, 1]$. We have here the constant $B = 1$.

Its support is the interval

$$-\sqrt{12B} \cdot t^{1/3} \leq x \leq \sqrt{12B} \cdot t^{1/3},$$

and Fig. 15.4.1 shows its evolution as a function of $t > 0$.

The solution $u_B(x,t)$ has a singularity at $t \to 0$, but its integral remains constant. Indeed,
$$\int_{-\sqrt{12B} \cdot t^{1/3}}^{\sqrt{12B} \cdot t^{1/3}} u_B(x,t)\, dx = C,$$
where the constant C is equal to $8B^{3/2}/\sqrt{3}$. Hence Barenblatt's solution is a source solution, and in the weak sense,
$$\lim_{t \to 0} u_B(x,t) = C \cdot \delta(x).$$

Barenblatt's solution u_B is not just a special solution of the porous medium equation. As time $t \to \infty$, other solutions to the initial value problem for (4) behave asymptotically like u_B. Indeed, we have both
$$\lim_{t \to \infty} t^{(d/(d\alpha+2))} \max_{\boldsymbol{x} \in \mathbf{R}^d} |u(\boldsymbol{x},t) - u_B(\boldsymbol{x},t)| = 0 \tag{11}$$
and
$$\lim_{t \to \infty} \int \cdots \int |u(\boldsymbol{x},t) - u_B(\boldsymbol{x},t)|\, dx_1 \cdots dx_d = 0, \tag{12}$$
for every solution u of (4) with nonnegative and integrable initial value $u(\boldsymbol{x},0) = u_0(\boldsymbol{x})$ with the same mass as u_B.[9]

15.5 Exercises

1. In the case in which the field $v(x,t)$ models a 1-D gas flow, the kinetic energy of the flow is given by the expression
$$T = \frac{1}{2} \int \rho(x,t)\, v^2(x,t)\, dx, \tag{1}$$
which is different from the formula (2) for E in the answers and solutions to Chap. 14. Show that in the case of a pressureless gas whose velocity and density are described by equations (15.1.27), there is a close relationship between the integrals E and T.

2. It is well known that the KdV equation has infinitely many invariants. Let us derive the two "most physical" of them. More precisely, show that if the solution $u(s,\tau)$ of the KdV equation (15.3.11) converges to zero fast enough as $x \to \pm\infty$, then the "momentum" $\mathcal{M}$ (15.3.34) and "energy" $\mathcal{K}$ (15.3.36) of the field $u(s,\tau)$ are constant, i.e., they do not depend on τ.

---

[9] See, e.g., J.L. Vazquez, Asymptotic behavior for the porous medium equation posed in the whole space, *Journal of Evolution Equations* 2 (2002), 1–52.

15.5. Exercises

3. Given the initial field $u_0(s)$, calculate the integral

$$S(\tau) = \int s\, u(s,\tau)\, ds\,.$$

4 Show that in the case of the n-soliton solution (15.3.29), (15.3.54) of the KdV equation (15.3.14) with

$$0 < k_1 < k_2 < \cdots < k_n,$$

we have

$$\lim_{\theta \to \pm\infty} \int \sqrt{w(s,\theta)}\, ds = \sqrt{\frac{12}{\sigma}}\, \pi\, n\,. \tag{2}$$

Obtain an analogous relation for the solution $u(s,\tau)$ of the KdV equation (15.3.11).

5. Neglecting the impact of the oscillating tail and assuming that the number of solitons in the solution $u(s,\tau)$ of the KdV equation (15.3.11) is equal to 1 ($n=1$), estimate the dependence of the soliton's velocity c on its momentum $\mathcal{M}$ and parameter γ. Produce the same estimate by relying on the value of the kinetic energy $\mathcal{K}$.

Appendix A

Answers and Solutions

Chapter 9: Potential Theory and Elliptic Equations

1.
$$G(x,y) = \frac{1}{2}\left(|x-y| - x - y + \frac{2xy}{l}\right).$$

2.
$$G(x,y) = \frac{1}{2ik}[e^{-ik(x+y)} - e^{-ik|x-y|}] = \begin{cases} e^{-iky}\sin kx, & \text{for } x < y; \\ e^{-ikx}\sin ky, & \text{for } x > y. \end{cases}$$

A standing wave forms to the left of the source. For $ky = \pi n \mapsto y = n\lambda/2$, the wave propagating to the right of the source is completely canceled out by the reflected wave.

3. A formal application of the second Green's formula indicates that the solution can be represented in the form

$$u(\boldsymbol{x}) = -\int f(y_1)\frac{\partial}{\partial y_2}G(\boldsymbol{x},\boldsymbol{y})\bigg|_{y_2=0} dy_1, \quad x_2 > 0,$$

where $G(\boldsymbol{x},\boldsymbol{y})$ is the Green's function satisfying the homogeneous boundary condition $G(\boldsymbol{x},\boldsymbol{y})|_{x_2=0} = 0$. Substituting in the above integral the Green's function,

$$G = \frac{1}{2\pi}\Big(\ln[(x_1-y_1)^2 + (x_2-y_2)^2] - \ln[(x_1-y_1)^2 + (x_2+y_2)^2]\Big),$$

which was obtained by the reflection method, we get

$$u(\boldsymbol{x}) = \int f(y_1) \frac{1}{\pi} \frac{x_2}{(x_1-y_1)^2 + x_2^2} dy_1, \quad x_2 > 0.$$

One can check by direct verification that for a sufficiently wide class of functions $f(x_1)$, the above integral provides a solution to the problem. To make the procedure rigorous, $f(x_1)$ needs to be an everywhere continuous and absolutely integrable function.

4. $\Delta u = 0$, $x_2 < 0$; $u(x_1, x_2 = 0) = -f(x_1)$.

5. The solution relies on the invariance of the Laplace operator (expressed in polar coordinates) with respect to the transformation $R = 1/\rho$, which maps the inside of the unit disk into the outside of the unit disk. First, let us write the Green's function in the unbounded plane in polar coordinates:

$$G(\rho, r, \varphi, \psi) = \frac{1}{4\pi} \ln\left[\rho^2 + r^2 - 2\rho r \cos(\varphi - \psi)\right].$$

Here (ρ, φ) are the polar coordinates of the point $\boldsymbol{x}$, and (r, ψ) are the polar coordinates of the point $\boldsymbol{y}$. Using the above Green's function, we can write the reflected (inside the unit circle) Green's function in which r is replaced by $1/r$, and obtain the following form of the desired Green's function:

$$G = \frac{1}{4\pi} \ln\left[\frac{\rho^2 + r^2 - 2\rho r \cos(\varphi - \psi)}{A(\rho^2 + 1/r^2 - 2(\rho/r)\cos(\varphi - \psi))}\right].$$

The Green's function is determined up to an arbitrary constant A, which can be found from the condition $G|_{\rho=1} = 0$. Hence $A = r^2$, so that

$$G(\rho, r, \varphi, \psi) = \frac{1}{4\pi} \ln\left[\frac{\rho^2 + r^2 - 2\rho r \cos(\varphi - \psi)}{\rho^2 r^2 + 1 - 2\rho r \cos(\varphi - \psi)}\right].$$

According to the second Green's formula, the solution of the original boundary value problem is of the form

$$u(\rho, \varphi) = \int_{-\pi}^{\pi} f(\psi) \frac{\partial}{\partial r} G \bigg|_{r=1} d\psi.$$

Finally, substituting the explicit expression for the normal derivative of the Green's function, we arrive at the celebrated *Poisson integral* formula:

$$u(\rho, \varphi) = \frac{1}{2\pi} \int_{-\pi}^{\pi} f(\psi) \frac{1 - \rho^2}{1 + \rho^2 - 2\rho \cos(\varphi - \psi)} d\psi.$$

6. The desired potential is obtained by an application of the operator $-p(\boldsymbol{n} \cdot \nabla)$ to the Green's function (9.1.15),

$$U(\boldsymbol{x}) = \frac{p}{4\pi}(\boldsymbol{n} \cdot \nabla)\frac{1}{|\boldsymbol{x}-\boldsymbol{y}|}, \quad \boldsymbol{x} \neq \boldsymbol{y},$$

of the 3-D Poisson equation. After differentiation, we finally get

$$U(\boldsymbol{x}) = -\frac{p}{4\pi}\frac{(\boldsymbol{n} \cdot \boldsymbol{m})}{|\boldsymbol{x}-\boldsymbol{y}|^2},$$

where $\boldsymbol{m} = (\boldsymbol{x}-\boldsymbol{y})/|\boldsymbol{x}-\boldsymbol{y}|$ is the unit vector directed from the dipole to the point $\boldsymbol{x}$.

7. The Fourier image of the solution

$$\tilde{u}(x,p) = \frac{1}{2\pi}\int u(x,y)e^{ipy}\,dy$$

solves the boundary value problem

$$\frac{d^2\tilde{u}}{dx^2} + (k^2 - p^2)\tilde{u} = 0, \quad \tilde{u}(x=0,p) = \tilde{f}(p),$$

where $\tilde{f}(p)$ is the Fourier image of the original boundary condition. For $|p| < k$, the solution to the problem satisfying the radiation condition has the form

$$\tilde{u}(x,p) = \tilde{f}(p)e^{-iqx}, \quad \text{where } q = \sqrt{k^2 - p^2}.$$

For $|p| > k$, the radiation condition is not applicable, and the desired solution

$$\tilde{u}(x,p) = e^{-|q|x}$$

can be obtained using an assumption about its decaying asymptotics as $x \to \infty$. Substituting the above expressions into the inverse Fourier transform formula gives

$$u(x,y) = \int_{|p|<k} \tilde{f}(p)e^{-iqx-ipy}\,dp + \int_{|p|>k} \tilde{f}(p)e^{-|q|x-ipy}\,dp,$$

which is one of the possible forms of the solution to the original problem. Physicists often interpret it as follows: The first summand represents a superposition of the planar wave that freely propagate in space. The quantity $\tilde{f}(p)$ is the complex amplitude of the planar waves propagating in the direction of the *wavevector* $\boldsymbol{k}$. Its projections on the x- and y-axes are, respectively, q and p. The norm $|\boldsymbol{k}| = k$ of the wavevector is equal to the wavenumber. The second component describes a localized field that decays exponentially with x and that is concentrated in a thin layer adjacent to the boundary plane $x = 0$.

Chapter 10: Diffusions and Parabolic Evolution Equations

1. The first step incorporates the initial condition in the equation itself to obtain

$$\frac{\partial f}{\partial t} + \alpha(t)\frac{\partial f}{\partial x} = \frac{\beta(t)}{2}\frac{\partial^2 f}{\partial x^2} + \delta(t)\delta(x), \qquad t \geq 0,\ x \in \mathbf{R}. \tag{1}$$

It turns out that the easiest way to solve equation (1) is to take the Fourier transform[10]

$$\tilde{f}(\kappa; t) = \int f(x,t)\, e^{i\kappa x}\, dx \tag{2}$$

of both sides with respect to the spatial variable x. This operation transforms the partial differential equation (2) into an ordinary differential equation,

$$\frac{d\tilde{f}}{dt} - i\kappa\alpha(t)\tilde{f} + \frac{1}{2}\beta(t)\kappa^2\tilde{f} = \delta(t). \tag{3}$$

If one takes into account the causality condition, then the solution of the above equation has the form

$$\tilde{f}(\kappa; t) = \exp\left(i\kappa h(t) - \frac{1}{2}b(t)\kappa^2\right), \qquad t > 0, \tag{4}$$

where

$$h(t) = \int_0^t \alpha(\tau)\, d\tau, \qquad b(t) = \int_0^t \beta(\tau)\, d\tau. \tag{5}$$

Applying the inverse Fourier transform

$$f(x,t) = \frac{1}{2\pi}\int_{-\infty}^{\infty} \tilde{f}(\kappa; t) e^{-i\kappa x}\, d\kappa \tag{6}$$

to the right-hand side of (4) and taking into account the formula (10.3.2), one finally obtains

$$f(x,t) = \frac{1}{\sqrt{2\pi b(t)}} \exp\left[-\frac{(x-h(t))^2}{2b(t)}\right]. \tag{7}$$

---

[10] Note that this version of the Fourier transform differs, by the minus sign in the exponent, from the Fourier transform introduced in Volume 1. This version of the Fourier transform is more convenient in the applications to random fields discussed in Volume 3.

Chapter 10. Diffusions and Parabolic Evolution Equations 331

2. Observe that the normalization condition (10.10.3) is to be expected in situations where $f(x,t)$ can be interpreted as the probability density function of a randomly diffusing particle. In any case, to check the validity of (10.10.3), let us rewrite (10.10.2) as a continuity equation

$$\frac{\partial f}{\partial t} + \frac{\partial G}{\partial x} = 0, \qquad (8)$$

where the quantity

$$G(x,t) = -a(x)f(x,t) - \frac{\beta(x)}{2}\frac{\partial f(x,t)}{\partial x} \qquad (9)$$

has an obvious physical interpretation as the *flow* of the field $f(x,t)$ along the x-axis. Integrating both sides of (8) over the whole x-axis, we obtain

$$\frac{d\mathcal{N}}{dt} + G(x=\infty,t) - G(x=-\infty,t) = 0. \qquad (10)$$

If we make a natural assumption that the flows at plus and minus infinity balance each other, that is, $G(-\infty,t) = G(\infty,t)$, then the relation (10.10.2) is reduced to the condition $\partial \mathcal{N}/\partial t \equiv 0$, which together with the initial condition $\mathcal{N}(t=0) = \int \delta(x)\,dx \equiv 1$, gives (10.10.3).

3. Applying the Fourier transform (2) to (10.10.4), and recalling that multiplication by x in the spatial domain becomes the differentiation operator in the wavenumber domain ($x \mapsto -i(\partial/\partial\kappa)$), we arrive at the first-order partial differential equation

$$\frac{\partial \tilde{f}}{\partial t} + \alpha(t)\kappa\frac{\partial \tilde{f}}{\partial \kappa} + \frac{\beta(t)}{2}\kappa^2\tilde{f} = 0, \qquad \tilde{f}(\kappa;t=0) = e^{i\kappa y} \qquad (11)$$

for the Fourier image $\tilde{f}(\kappa;t)$ of the solution $f(x,t)$ of the parabolic equation (10.10.4).

The latter equation can be solved by the method of characteristics described in Volume 1. Indeed, suppose that κ is a function of t, that is, $\kappa = \Omega(\kappa_0;t)$, where $\kappa_0 = \Omega(\kappa_0;t=0)$ is an arbitrary initial value. Then equation (11) splits into two characteristic equations,

$$\begin{aligned} d\Omega/dt &= \alpha(t)\Omega, & \Omega(\kappa_0,t=0) &= \kappa_0, \\ dF/dt &= -(\beta(t)/2)\Omega^2 F, & F(\kappa_0,t=0) &= e^{i\kappa_0 y}. \end{aligned} \qquad (12)$$

Here, the function

$$F(\kappa_0;t) = \tilde{f}\Big(\Omega(\kappa_0;t);t\Big) \qquad (13)$$

describes the behavior of the desired Fourier image $\tilde{f}(\kappa; t)$ along a chosen characteristic line.

Solutions of the equation (12) have the form

$$\Omega(t, \kappa_0) = \kappa_0 e^{h(t)}, \tag{14}$$

where $h(t)$ has been defined by the first equality in (5), and

$$F(\kappa_0, t) = \exp\left[i\kappa_0 y - \frac{\kappa_0^2}{2} E(t)\right], \tag{15}$$

with

$$E(t) = \int_0^t \beta(\tau) e^{2h(\tau)}\, d\tau. \tag{16}$$

Now to get the Fourier image $\tilde{f}(\kappa; t)$ of the solution of the initial value problem (10.10.4–5), it suffices to insert the inverse function $\kappa_0 = \Omega^{-1}(\kappa; t) = \kappa e^{-h(t)}$ into the right-hand side of (15). As a result, we obtain

$$\tilde{f}(\kappa; t) = F\left(\Omega^{-1}(\kappa; t); t\right) = \exp\left[i\kappa y e^{h(t)} - \kappa^2 E(t) e^{-2h(t)}\right], \tag{17}$$

and an application of the inverse Fourier transform yields the final result:

$$f(x, t) = \frac{e^{h(t)}}{\sqrt{2\pi E(t)}} \exp\left[-\frac{(xe^{h(t)} - y)^2}{2E(t)}\right]. \tag{18}$$

4. In this case,

$$h(t) = \alpha t, \qquad E(t) = \frac{\beta}{2\alpha}\left(e^{2\alpha t} - 1\right).$$

Substituting these expressions into the right-hand side of (18), we get

$$f(x, t) = \sqrt{\frac{\alpha}{\pi\beta(1 - e^{-2\alpha t})}} \exp\left[-\frac{\alpha}{\beta}\frac{(x - ye^{-\alpha t})^2}{1 - e^{-2\alpha t}}\right]. \tag{19}$$

As $t \to \infty$, the above solution converges to the *stationary solution*

$$\lim_{t \to \infty} f(x, t) = f_{st}(x) = \sqrt{\frac{\alpha}{\pi\beta}} \exp\left(-\frac{\alpha}{\beta}x^2\right). \tag{20}$$

5. By definition, the desired stationary solution $f_{st}(x)$ does not depend on time. So, in equation (10.10.4), we can neglect the term containing the

Chapter 10. Diffusions and Parabolic Evolution Equations

derivative with respect to the time variable, which gives an equation of the form

$$\frac{dG_{st}(x)}{dx} = 0, \tag{21}$$

where

$$G_{st}(x) = -\alpha x f_{st}(x) - \frac{\beta}{2}\frac{df_{st}(x)}{dx} \tag{22}$$

is the stationary flow. Equality (21) means that in the stationary state, the flow is constant everywhere, that is,

$$-\alpha x f_{st}(x) - \frac{\beta}{2}\frac{df_{st}(x)}{dx} = C, \tag{23}$$

where C is the magnitude of the flow. This magnitude has to be found from the normalization condition

$$\int f_{st}(x)\,dx = 1. \tag{24}$$

It is easy to check that the above integral is bounded if and only if $C = 0$, so that the corresponding general solution of equation (22) takes the form

$$f_{st}(x) = A\exp\left(-\frac{\alpha}{\beta}x^2\right),$$

where A is an integration constant, which in the case of the normalized solution, is equal to $\sqrt{\alpha/(\pi\beta)}$.

6. The stationary solution satisfies the equation

$$(\beta + \gamma x^2)\frac{df_{st}(x)}{dx} = -2\alpha x f_{st}(x). \tag{25}$$

Its general solution is of the form

$$f_{st}(x) = \frac{A}{(\beta + \gamma x^2)^\mu}, \tag{26}$$

where

$$\mu = \frac{\alpha}{\gamma}. \tag{27}$$

The normalization condition implies

$$A^{-1} = 2\int_0^\infty \frac{dx}{(\beta + \gamma x^2)^\mu}. \tag{28}$$

The above integral can be evaluated analytically (via, e.g., Mathematica or Maple), which gives

$$A = \sqrt{\frac{\gamma}{\pi\beta}} \frac{\beta^\mu \Gamma(\mu)}{\Gamma(\mu - 1/2)}, \qquad (29)$$

so that

$$f_{st}(x) = \sqrt{\frac{\gamma}{\pi\beta}} \frac{\Gamma(\mu)}{\Gamma(\mu - 1/2)} \left(1 + \frac{\gamma}{\beta}x^2\right)^{-\mu}. \qquad (30)$$

Of course, this expression is valid only if the integral converges, that is, if

$$\mu > 1/2, \quad \text{or equivalently,} \quad \gamma < 2\alpha. \qquad (31)$$

Inequality (31) is the desired condition of existence of the stationary solution (30).

7. Unfortunately, the explicit solution $f(x,t)$ of the initial value problem (10.10.7), (10.10.5) is unknown. In this situation, a study of its moments can provide some insight into its temporal behavior. Here we concentrate on a calculation of the first two moments,

$$m_1(t) = \int x f(x,t)\, dx, \qquad m_2(t) = \int x^2 f(x,t)\, dx, \qquad (32)$$

and begin by deriving equations for them.

Multiplying equation (10.10.7) by x, integrating it over the whole x-axis, and then integrating by parts, one gets a closed equation for the first moment,

$$\frac{dm_1(t)}{dt} = (\gamma - \alpha) m_1(t), \qquad m_1(0) = y. \qquad (33)$$

The initial condition obviously follows from the initial condition (10.10.5) for equation (10.10.7). The solution of the initial value problem (33) has the form

$$m_1(t) = y e^{(\gamma - \alpha)t}. \qquad (34)$$

Remark 3. Note that the improper integral

$$\int x f_{st}(x)\, dx = A \int_{-\infty}^{\infty} \frac{x\, dx}{(\beta + \gamma x^2)^\mu} \qquad (35)$$

converges if and only if $2\mu - 1 > 1$, or equivalently, if $\alpha > \gamma$. In this case, the first moment (34) decreases in time to the value of the convergent integral $\int x f_{st}(x)\, dx = 0$ (35). In the case $\gamma > \alpha$, the first moment $m_1(t)$ diverges to infinity.

Chapter 10. Diffusions and Parabolic Evolution Equations

In the remaining case $\gamma = \alpha$, the parabolic equation (10.10.7) reduces to

$$\frac{\partial f}{\partial t} = \frac{\partial^2}{\partial x^2}\left[(\alpha x^2 + \beta)f\right], \tag{36}$$

and its stationary solution turns out to be the well-known Cauchy density,

$$f_{st}(x) = \frac{1}{\pi}\frac{\varepsilon}{\varepsilon^2 + x^2}, \qquad \varepsilon = \sqrt{\frac{\beta}{\alpha}}. \tag{37}$$

The first moment of the Cauchy density f_{st} is not well defined, but one might be tempted to replace it by the principal value of the integral

$$\frac{\varepsilon}{\pi}\mathcal{PV}\int \frac{x\,dx}{\varepsilon^2 + x^2} = \frac{\varepsilon}{\pi}\lim_{M\to\infty}\int_{-M}^{M}\frac{x\,dx}{\varepsilon^2 + x^2} = 0. \tag{38}$$

However, the formula (34) gives $\lim_{t\to\infty} m_1(t) = y$. This seemingly paradoxical situation shows the perils of recklessly interchanging limit operations without a proper justification.

A similar calculation for the second moment leads us to the initial value problem

$$\frac{dm_2(t)}{dt} = (3\gamma - 2\alpha)m_2(t) + \beta, \qquad m_2(0) = y^2.$$

Its solution is

$$m_2(t) = y_0^2 e^{(3\gamma-2\alpha)t} + \frac{\beta}{3\gamma - 2\alpha}\left(e^{(3\gamma-2\alpha)t} - 1\right). \tag{39}$$

8. The stationary solution satisfies the following ordinary differential equation:

$$\frac{\beta}{2}\frac{df_{st}}{dx} + \alpha f_{st} = C, \tag{40}$$

where C is the magnitude of the stationary flow. The general solution of the above equation has the form

$$f_{st}(x) = \frac{C}{\alpha} + A\exp\left(-\frac{2\alpha}{\beta}x\right). \tag{41}$$

The boundary condition (10.10.10) forces $A = 0$, and the initial condition (10.10.5) implies that the stationary solution has to be normalized on the interval $x \in [0, \ell]$. This means that

$$f_{st} = \frac{1}{\ell}, \tag{42}$$

so that the magnitude of the stationary flow is

$$C = \frac{\alpha}{\ell}.$$

Thus the stationary flow does not depend on the parameter β and is a linear function of the parameter α.

9. Take the Fourier transform (2) of both sides of the integrodifferential equation (10.10.11), keeping in mind that the Fourier transform maps the convolution into a product. The result is an ordinary differential equation,

$$\frac{d\tilde{f}(\kappa;t)}{dt} = \nu[\tilde{w}(\kappa) - 1]\tilde{f}(\kappa;t), \qquad \tilde{f}\kappa;t=0) = 1. \qquad (43)$$

Its solution is

$$\tilde{f}(\kappa;t) = \exp\left[-\nu\left(1 - \tilde{w}(\kappa)\right)t\right], \qquad t > 0. \qquad (44)$$

Taking the inverse Fourier transform, we obtain the desired solution of the Kolmogorov–Feller equation:

$$f(x,t) = \frac{1}{2\pi}\int \exp\left[-i\kappa x - \nu\left(1 - \tilde{w}(\kappa)t\right)t\right]d\kappa. \qquad (45)$$

Let us discuss in more detail a particular case of the Gaussian kernel,

$$\tilde{w}(\kappa) = \exp\left(-\frac{\kappa^2}{2}\right), \qquad (46)$$

which converges to 0 as $\kappa \to \infty$. In this case, it makes sense to split the Fourier image of the solution of the Kolmogorov–Feller equation into a constant part (with respect to κ) and an absolutely integrable part

$$\tilde{f}(\kappa;t) = e^{-\nu t} + \tilde{f}_c(\kappa;t), \qquad (47)$$

where

$$\tilde{f}_c(\kappa;t) = e^{\nu[\tilde{w}(\kappa)-1]t} - e^{-\nu t}. \qquad (48)$$

Thus, the solution of the Kolmogorov–Feller equation can be split into the sum of singular and continuous parts,

$$f(x,t) = \delta(x)e^{-\nu t} + f_c(x,t), \qquad (49)$$

where

$$f_c(x,t) = \frac{1}{2\pi}\int\left\{\exp\left(\nu t\left[e^{-\kappa^2/2} - 1\right]\right) - e^{-\nu t}\right\}e^{-i\kappa x}d\kappa. \qquad (50)$$

Chapter 10. Diffusions and Parabolic Evolution Equations

For large times (when $\nu t \gg 1$), the asymptotic behavior of the continuous part of the solution is determined by the behavior of its Fourier image in a small vicinity of $\kappa = 0$. There, the Gaussian function (46) can be replaced by the first two terms of its power expansion,

$$\exp\left(\frac{-\kappa^2}{2}\right) = 1 - \frac{\kappa^2}{2} + \cdots, \tag{51}$$

and the small term $e^{-\nu t}$ can be neglected. As a result, we obtain the following asymptotic formula for the continuous part of the solution of the Kolmogorov–Feller equation:

$$f_c(x, t) \sim \frac{1}{2\pi} \int \exp\left(-\frac{\nu t \kappa^2}{2} - i\kappa x\right) d\kappa, \quad (\nu t \gg 1), \tag{52}$$

or, after integration,

$$f_c(x, t) \sim \frac{1}{\sqrt{2\pi\nu t}} \exp\left(-\frac{x^2}{2\nu t}\right), \quad (\nu t \gg 1). \tag{53}$$

So as time grows to infinity, the continuous part of the solution of the Kolmogorov–Feller equation with Gaussian kernel becomes more and more Gaussian itself. One can provide a rigorous proof of this fact within the framework of the celebrated *central limit theorem* of probability theory.

10. The Fourier image of the Cauchy kernel is

$$\tilde{w}(\kappa) = e^{-|\kappa|}, \tag{54}$$

and it tends to zero as $|\kappa| \to \infty$. So the solution has a structure similar to that of (49) in the preceding problem. Moreover, the singular part of the solution is the same as in the case of the Gaussian kernel, while the continuous part is given by

$$f_c(x, t) = \frac{1}{2\pi} \int \left\{ \exp\left(\nu t \left[e^{-|\kappa|} - 1\right]\right) - e^{-\nu t} \right\} e^{-i\kappa x} d\kappa. \tag{55}$$

Now the exploration of the asymptotics ($\nu t \to \infty$) of f_c follows the lines of the solution to Problem 9: the Fourier image of the kernel is approximated by the first two terms of its power expansion,

$$e^{-|\kappa|} = 1 - |\kappa| + \cdots, \tag{56}$$

and the term $e^{-\nu t}$ is dropped. As a result, after integration, we obtain

$$f_c(x,t) \sim \frac{1}{\pi} \frac{\nu t}{(\nu t)^2 + x^2}, \qquad (\nu t \gg 1). \tag{57}$$

Thus, in the case of the Cauchy kernel, the continuous part of the solution of the Kolmogorov–Feller equation converges, for large times, to the Cauchy density. This is another example of the central limit theorem effect, but in this case, the limit is not Gaussian and corresponds to the environment of so-called *Lévy flights*. We will return to this topic in Volume 3.

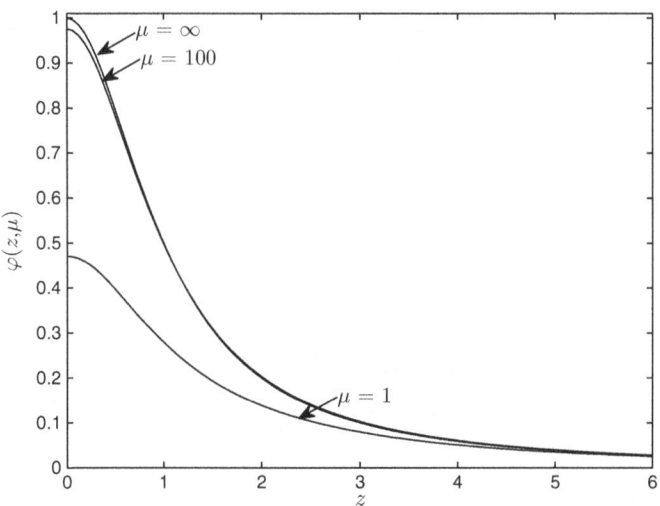

FIGURE 1
The plots of graphs of functions $\varphi(z,\mu)$ (59), for $\mu = 1, 100$, and of the function $\varphi(z,\infty)$ (60) (from bottom to top). Note that $\varphi(z,100)$ almost coincides with the Cauchy curve $\varphi(z,\infty)$. The conclusion is that for, say, $\mu = \nu t > 10$, the continuous part of the solution of the Kolmogorov–Feller equation with Cauchy kernel is well approximated by the Cauchy density itself.

11. First of all, notice that the asymptotic Cauchy density (57) is self-similar. So changing variables

$$z = \frac{x}{\nu t}, \qquad \mu = \nu t, \tag{58}$$

and introducing an auxiliary function,

$$\varphi(z,\mu) = \pi \mu f_c\left(\mu z, \frac{\mu}{z}\right), \tag{59}$$

Chapter 10. Diffusions and Parabolic Evolution Equations

we have
$$\varphi(z, \infty) = \frac{1}{1+z^2}. \tag{60}$$

Thus our numerics will have to demonstrate that $\varphi(z,\mu) \to \varphi(z,\infty)$ as $\mu \to \infty$.

Let us rewrite $\varphi(z,\mu)$ in a form more convenient for numerical calculations,
$$\varphi(z,\mu) = \int_0^\infty \psi(r,\mu) \cos(rz)\, dr, \tag{61}$$

where
$$\psi(r,\mu) = \exp\left(\mu\left[\exp\left(-\frac{r}{\mu}\right) - 1\right]\right) - \exp(-\mu). \tag{62}$$

To calculate the improper integral (61) numerically, we have to replace its upper limit by a large number, say R, and in order to avoid the Gibbs phenomenon (see Volume 1), multiply the integrand by the Cesàro factor $(1 - r/R)$. As a result, it remains to evaluate numerically the integral
$$\varphi(z,\mu) \approx \int_0^R \psi(r,\mu)\left(1 - \frac{r}{R}\right)\cos(rz)\, dr. \tag{63}$$

The selection of R and the accuracy of numerical integration of (63) will depend on your computer's power and on your common sense. Figure 1 shows the results of our computations for $R = 30$, and $\mu = 1, 100, \infty$ (from bottom to top).

12. To find the required main asymptotics, let us rewrite the asymptotic series (4.3.8) from Volume 1 in the form needed for our task,
$$f_c(x,t) \sim \frac{e^{-ix\tau}}{2\pi} \sum_{m=0}^\infty \left(\frac{1}{ix}\right)^{m+1} \lfloor \tilde{f}_c^{(m)} \rceil, \tag{64}$$

where τ is the point on the κ-axis where the Fourier image
$$\tilde{f}_c(\kappa; t) = \exp\left(\nu t\left[e^{-|\kappa|} - 1\right]\right) - e^{-\nu t} \tag{65}$$

or its derivatives have discontinuities. Recall that $\lfloor \tilde{f}_c^{(m)} \rceil$ is the size of the jump of the mth derivative (with respect to κ) of $\tilde{f}_c(\kappa; t)$ at the discontinuity point τ,
$$\lfloor \tilde{f}_c^{(m)} \rceil = \tilde{f}_c^{(m)}(\kappa = \tau + 0; t) - \tilde{f}_c^{(m)}(\kappa = \tau - 0; t). \tag{66}$$

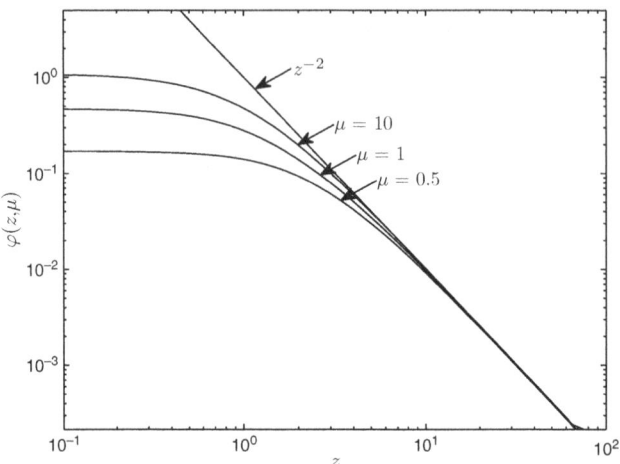

FIGURE 2
Log-log plots of functions $\varphi(z,\mu)$ (61), for different values of μ. It is evident that for every μ, the continuous part of the solution of the Kolmogorov–Feller equation with the Cauchy kernel has a slowly decaying power tail $\sim x^{-2}$.

In our case, the function (65) is continuous, but its derivative has a jump at $\kappa = 0$ ($\tau = 0$),
$$\tilde{f}'_c(+0;t) - \tilde{f}'_c(-0,t) = -2\nu t. \tag{67}$$
Thus it follows from (64) that the main asymptotics of the continuous part of the Kolmogorov–Feller equation are as follows:
$$f_c(x,t) \sim \frac{\nu t}{\pi x^2}, \qquad (x \to \infty). \tag{68}$$
Figure 2 depicts the log-log plots of the numerically calculated functions $\varphi(z,\mu)$ for different values of the dimensionless time parameter μ, and compares them with the adjusted asymptotics $\varphi(z,\mu) \sim 1/z^2$ (68).

Chapter 11: Waves and Hyperbolic Equations

1. To solve the problem, let us express the function $v(x,t)$ via Fourier images of the convolution of the functions $h(x,t)$ and $u(x,t)$. According to (11.1.1) and (11.1.2), we have
$$\tilde{v}(\omega,\kappa) = 4\pi^2 \tilde{h}(\omega,\kappa)\tilde{u}(\omega,\kappa),$$

Chapter 11. Waves and Hyperbolic Equations

so that
$$v(x,t) = \int \tilde{v}(\omega, \kappa) e^{-i\omega t + i\kappa x} d\omega \, d\kappa.$$

Using the relation (11.1.9) and the probing property of the Dirac delta with respect to ω, we obtain

$$v(x,t) = 4\pi^2 \int \tilde{f}(k)\tilde{h}(W(k), k) e^{i(kx - W(k)t)} dk.$$

Substituting $u(x,t)$ from (11.1.10) and the above expression for $v(x,t)$ into (11.12.1), utilizing the relation (11.3.3),

$$\int e^{ix(\kappa_1 + \kappa_2)} dx = 2\pi \delta(\kappa_1 + \kappa_2),$$

the probing property of the above Dirac delta, and the fact that $W(\kappa)$ is an odd function, we finally obtain

$$S = (2\pi)^3 \int |f(k)|^2 \tilde{h}(W(k), k) \, dk = const. \tag{1}$$

Some of these invariants, different for waves of different physical natures, correspond to the energy, momentum, photon number, and other conservation laws.

2. The relevant dispersion relations have the form

$$\omega^2 = a^2 k^2 + \omega_0^2.$$

So
$$\omega = \pm W(k) = \pm \sqrt{a^2 k^2 + \omega_0^2}.$$

Correspondingly,

$$c = \frac{W(k)}{k} = a\sqrt{1 + \omega_0^2/(a^2 k^2)} > a$$

and
$$v(k) = W'(k) = a/\sqrt{1 + \omega_0^2/(a^2 k^2)} < a.$$

Note that
$$c(k)v(k) = a^2 = const.$$

3. Let us represent $u(x,t)$ in the form of the spatial Fourier integral

$$u(x,t) = \int \tilde{f}(k,t) e^{ikx} dk,$$

where

$$\tilde{f}(k,t) = \tilde{f}(k) \exp\left(-ikat\sqrt{1+\omega_0^2/(a^2 k^2)}\right).$$

From the formula (4.3.3) of Volume 1, we know that if the function $f(x) = u(x, t=0)$ has a jump, then its Fourier image has the asymptotics

$$\tilde{f}(k) \sim \frac{\lfloor f \rfloor}{2\pi i k} e^{-ikx_0}, \qquad (k \to \infty).$$

Consequently, the spatial Fourier image of the wave packet $u(x,t)$ has the asymptotics

$$\tilde{f}(k,t) \sim \frac{\lfloor f \rfloor}{2\pi i k} e^{-ik(x_0+at)}, \qquad (k \to \infty). \qquad (2)$$

This means that the original jump in the wave packet does not disappear but preserves its size $\lfloor f \rfloor$, and moves with the velocity a, as if it were evolving in a nondispersive medium.

4. *Mathematical formulation of the problem:* We need to solve equations (11.11.3a) and (11.11.3b) in the domain $-\infty < x < +\infty$, $0 < t < +\infty$, taking into account the nondistortion condition $CR = LG$, and the following initial conditions:

$$v(x,0) = v_0(x), \qquad i(x,0) = i_0(x).$$

Solution: The fields $v(x,t)$ and $i(x,t)$ in an infinitely long distorsionless line are given by the expressions (11.11.14), where the functions $\Phi(x)$ and $\Psi(x)$ are determined by the initial conditions

$$\Phi(x) + \Psi(x) = v_0(x), \qquad \text{and} \qquad \sqrt{\frac{C}{L}}[\Phi(x) - \Psi(x)] = i_0(x).$$

Hence

$$\Phi(x) = \frac{1}{2}[v_0(x) + \sqrt{\frac{L}{C}} i_0(x)], \quad \text{and} \quad \Psi(x) = \frac{1}{2}[v_0(x) - \sqrt{\frac{L}{C}} i_0(x)].$$

Chapter 11. Waves and Hyperbolic Equations

Consequently,
$$v(x,t) = \frac{1}{2}e^{-\frac{R}{L}t}[v_0(x-at) + v_0(x+at)] + \frac{1}{2}\sqrt{\frac{L}{C}}e^{-\frac{R}{L}t}[i_0(x-at) - i_0(x+at)],$$
and
$$i(x,t) = \frac{1}{2}e^{-\frac{R}{L}t}[i_0(x-at) + i_0(x+at)] + \frac{1}{2}\sqrt{\frac{C}{L}}e^{-\frac{R}{L}t}[v_0(x-at) - v_0(x+at)].$$

Physical effects: In practice, one often tries to generate waves that propagate along the transmission line in a given direction. The above solution implies, for example, that if the initial voltage and current are related by the condition
$$v_0(x) = \sqrt{\frac{L}{C}}\, i_0(x),$$
then the wave will move only to the right, and in this case, the voltage is described by the following simple expression:
$$v(x,t) = e^{-\frac{R}{L}t} v_0(x-at).$$

5. *Mathematical formulation of the problem:* One has to find fields $v(x,t)$ and $i(x,t)$ satisfying equations (11.11.3a) and (11.11.3b) in the domain $0 < x < +\infty$, $0 < t < +\infty$, and the nondistortion condition $CR = LG$ in the presence of the initial conditions
$$v(x,0) = f(x), \quad i(x,0) = -\sqrt{\frac{C}{L}}\, f(x) \qquad (3)$$
and the boundary condition
$$v(0,t) = -R_0\, i(0,t). \qquad (4)$$

Solution: The general solution has the form
$$v(x,t) = e^{-\frac{R}{L}t}[\Phi(x-at) + \Psi(x+at)], \qquad (5)$$
$$i(x,t) = \sqrt{\frac{C}{L}}\, e^{-\frac{R}{L}t}[\Phi(x-at) - \Psi(x+at)]. \qquad (6)$$
From the initial conditions (3), it follows that
$$\Phi(x) = 0, \; \Psi(x) = f(x), \; 0 < x < +\infty. \qquad (7)$$

Substituting functions (6) into the boundary condition (4), after omitting $e^{-\frac{R}{L}t}$, we obtain

$$\Phi(-at) + \Psi(at) = -R_0\sqrt{\frac{C}{L}}[\Phi(-at) - \Psi(at)], \qquad 0 < t < +\infty.$$

Replacing $-at$ with x, the above equality and (4) give

$$\Phi(x) = f(-x)\,\frac{R_0\sqrt{C} - \sqrt{L}}{R_0\sqrt{C} + \sqrt{L}}, \qquad x < 0.$$

Consequently, we find that the solution of the problem is of the form (6), where

$$\Phi(x) = \begin{cases} f(-x)\,\dfrac{R_0\sqrt{C} - \sqrt{L}}{R_0\sqrt{C} + \sqrt{L}}, & \text{for } -\infty < x < 0; \\ 0, & \text{for } 0 < x < +\infty. \end{cases}$$

$$\Psi(x) = f(x), \qquad \text{for } 0 < x < +\infty.$$

Physical effects: It is natural to represent voltage in the line as a superposition of the wave running toward the endpoint $x = 0$ and the wave reflected from that endpoint:

$$v(x, t) = v_{\text{run}}(x, t) + v_{\text{refl}}(x, t),$$

where

$$v_{\text{run}}(x, t) = e^{-\frac{R}{L}t} f(x + at), \qquad v_{\text{refl}}(x, t) = K\, e^{-\frac{R}{L}t} f(at - x).$$

The reflection coefficient is given by

$$K = K(\rho) = \frac{\rho - 1}{\rho + 1}.$$

Chapter 11. Waves and Hyperbolic Equations

(see Fig. 1) depends on the nondimensional grounding resistivity

$$\rho = R_0 \sqrt{\frac{C}{L}}.$$

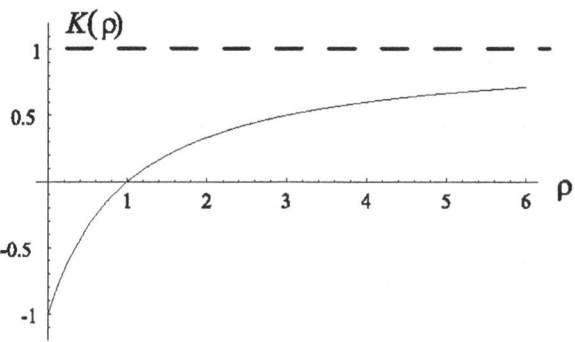

FIGURE 1
A semi-infinite transmission line: Dependence on the grounding resistivity of the reflection-from-the-grounded-end-point coefficient.

Note some peculiarities of the reflection coefficient: For small values of the grounding resistivity, the coefficient becomes negative, and in the presence of a short circuit ($R_0 = 0$), when the voltage at the left endpoint remains zero at all times, the coefficient K is equal to -1. This means that the reflected wave in the vicinity of the grounded point (i.e., for $x = +0$) is equal to the negative of the incident wave. For $R_0 \to \infty$, the reflection coefficient converges to 1, while if one chooses the grounding resistivity to be $R_0 = \sqrt{L/C}$ (the so-called *matched load*), the reflection coefficient is $K(1) = 0$, and there is no reflected wave. This effect is widely used to achieve damping of reflected waves in long transmission lines.

6. *Mathematical formulation of the problem:* The present problem differs from Problem 5 only in its boundary condition, which now becomes

$$v(0, t) = -L_0\, i_t(0, t). \tag{8}$$

Solution: As before, the functions $\Phi(x)$ and $\Psi(x)$ in (6), for $x > 0$, are given by the equalities (7), and for $x < 0$, they have to be defined so that the condition (8) is satisfied. Substituting in (8) the equalities (6) and canceling the factor $e^{-(R/L)t}$, we have

$$\Phi(-at) + \Psi(at) = -L_0\sqrt{\frac{C}{L}}[\Phi_t(-at) - \Psi_t(at)] + +L_0\frac{R}{L}\sqrt{\frac{C}{L}}[\Phi(-at) - \Psi(at)].$$

Introducing an auxiliary variable $z = -at < 0$, we obtain, for $z < 0$, an ordinary differential equation for $\Phi(z)$:

$$\Phi' - k\Phi = pf(-z) - f'(-z).$$

Here we took into account that

$$\frac{d\Phi(-at)}{dt} = -a\Phi'(z), \quad \frac{\Psi(at)}{dt} = a\Psi'(-z) = af'(-z), \qquad z < 0,$$

where the prime denotes the derivative with respect to z. We also introduced the constants

$$k = \frac{L}{L_0} - \frac{R}{La}, \quad p = \frac{L}{L_0} + \frac{R}{La},$$

which have dimensions of inverse length, and observed that $a = 1/\sqrt{LC}$.

The solution of its equations with the condition $\Phi(0) = 0$, resulting from (7), is

$$\Phi(z) = e^{kz}\int_0^z \left[pf(-y) - f'(-y)\right]e^{-ky}dy.$$

Replacing z with x, we will obtain the desired continuation of the function $\Phi(x)$ onto the domain $x < 0$.

Physical effects: Let us rewrite the expression for $\Phi(x)$ in a physically more transparent form. For this purpose, we shall use a mathematical trick. First, notice that the integrand in the obtained solution can be written as

$$(pf - f')e^{-ky} = -\frac{d}{dy}\left(fe^{-ky}\right) + \frac{2R}{La}fe^{-ky}.$$

This means that the function $\Phi(x)$ in which we are interested, and which describes the behavior of the reflected wave, can be split into two components,

$$\Phi(x) = -f(-x) + \frac{2R}{La}e^{kx}\int_0^x f(-y)e^{-ky}dy,$$

the first corresponding to the wave reflected from the short-circuited point of the line, and the second corresponding to the inertial process of reflection. For $L_0 \to 0$, as $k \to \infty$, the second summand disappears.

Chapter 11. Waves and Hyperbolic Equations

7. *Mathematical formulation of the problem:* One has to solve the equation (11.11.3a), with $RC = LG$, in the domain $0 < x < +\infty$, $0 < t < +\infty$, with the boundary condition

$$v(0, t) = E(t).$$

Solution: In the general case, the voltage in a long distortionless transmission line is described by the equation (5),

$$v(x, t) = e^{-\frac{R}{L}t} \left[\Phi(x - at) + \Psi(x + at) \right].$$

From the radiation condition, which in our case means that the voltage source applied at the left endpoint of the line can only generate a wave propagating to the right, we obtain that $\Psi \equiv 0$. The function $\Phi(x)$ in (6) can be found by checking the boundary condition

$$E(t) = e^{-\frac{R}{L}t} \Phi(-at).$$

Setting $-at = x$, we have

$$\Phi(x) = E\left(-\frac{x}{a}\right) e^{-\frac{R}{L}\frac{x}{a}}.$$

Substituting this equality into the expression for voltage, we finally get

$$\Phi(x) = e^{-\frac{R}{L}\frac{x}{a}} E\left(t - \frac{x}{a}\right).$$

Physical effects: In a distortionless transmission line, for every point $x > 0$, the voltage depends on the time variable in exactly the same way the voltage source depends on the time variable, but with the added attenuation factor $\exp(-Rx/La)$ and time delay of magnitude $t = x/a$.

8. *Mathematical formulation of the problem:* One needs to solve the equations

$$v_{xx} = LC\, v_{tt} + (CR + LG)\, v_t + RG\, v, \qquad (9)$$

$$i_{xx} = LC\, i_{tt} + (CR + LG)\, i_t + RG\, i, \qquad (10)$$

for $t > 0$, with the initial conditions

$$v(x, 0) = v_0(x), \quad i(x, 0) = i_0(x), \qquad 0 < x < +\infty,$$

and the boundary condition

$$v(0, t) = -R_0\, i(0, t), \qquad 0 < t < +\infty.$$

The functions $v_0(x)$ and $i_0(x)$ describe a stationary state in the line at time $t = 0$, which can be found from the telegrapher's equation (9) with the proviso that $v_0(x)$ does not depend on t:

$$\frac{d^2 v_0}{dx^2} - GR\, v_0 = 0, \quad v(0) = E \quad (0 < x < +\infty).$$

The general solution of this equation is

$$v_0(x) = A\, e^{-\sqrt{RG}\, x} + B\, e^{\sqrt{RG}\, x}.$$

Since as $x \to +\infty$, the solution remains bounded, we have $B = 0$. Using the condition $v_0(0) = E$, we obtain

$$v_0(x) = E\, e^{-\sqrt{RG}\, x}.$$

From equation (11.12.1), which in this instance reduces to

$$\frac{dv_0(x)}{dx} + R\, i_0(x) = 0,$$

we find that

$$i_0(x) = E\sqrt{\frac{G}{R}}\, e^{-\sqrt{RG}\, x}.$$

Consequently, the initial conditions are of the form

$$v(x, 0) = E\, e^{-\sqrt{RG}\, x}, \quad i(x, 0) = E\sqrt{\frac{G}{R}}\, e^{-\sqrt{RG}\, x} \quad (x > 0).$$

Solution: The solution of the problem has the form

$$v(x, t) = e^{-\frac{R}{L}t}[\Phi(x - at) + \Psi(x + at)],$$

$$i(x, t) = \sqrt{\frac{C}{L}}\, e^{-\frac{R}{L}t}[\Phi(x - at) - \Psi(x + at)].$$

Taking into account the initial conditions and the equality $CR = GL$, we have

$$\Phi(x) + \Psi(x) = E\, e^{-\sqrt{RG}\, x}, \quad \Phi(x) - \Psi(x) = E\, e^{-\sqrt{RG}\, x}.$$

As a result,

$$\Phi(x) = E\, e^{-\sqrt{GR}\, x}, \quad \Psi(x) \equiv 0 \quad (x > 0).$$

Now this solution must be substituted into the boundary condition

$$e^{-\frac{R}{L}t}\Phi(-at) = -R_0 \sqrt{\frac{C}{L}}\, e^{-\frac{R}{L}t}\Phi(-at).$$

It follows that $\Phi(-at) = 0$ for $t > 0$. In other words, we have defined the function $\Phi(x)$ for negative values of x. For other values of x,

$$\Phi(x) = E\chi(x)e^{-\sqrt{GR}x},$$

where $\chi(z)$ is the Heaviside unit step function.

Now substitute these functions into (6). This gives

$$v(x,t) = e^{-\frac{R}{L}t} E\chi(x-at)e^{-\sqrt{GR}(x-at)}.$$

Hence, after simple transformations, and taking into account the equalities $CR = GL$ and $a = 1/\sqrt{LC}$, we get

$$v(x,t) = E\chi(x-at)e^{-\sqrt{GR}x}.$$

Similarly,

$$i(x,t) = E\sqrt{\frac{C}{L}}\,\chi(x-at)e^{-\sqrt{GR}x}.$$

Physical effects: As we could have guessed, the radiation condition implies that a sudden change of the regime at $x = 0$ creates a wave propagating to the right. One could think about it as a rigid stationary profile $v_0(x)$, $i_0(x)$, preserved for $x > at$, when the information about the changed conditions at the line's left endpoint has not yet been received. The profile is "evaporating" inside the segment $0 < x < at$ attached to the endpoint.

9. *Mathematical formulation of the problem:* We have to find the voltage distribution satisfying the boundary conditions

$$v_x(0,t) = 0, \quad v(l,t) = E_0 \sin\omega t, \tag{11}$$

and the initial conditions

$$v(x,0) = 0, \quad v_t(x,0) = 0. \tag{12}$$

As in Section 11.11, instead of voltage $v(x,t)$, we will work with a more suitable auxiliary function $u(x,t)$ satisfying equation (11.11.6) and the boundary and initial conditions obtained via (11)–(12), (11.11.4), and (11.11.17):

$$\frac{\partial u(0,t)}{\partial x} = 0, \quad u(l,t) = E_0 e^{\mu t}\sin\omega t,$$

$$u(x,0) = 0, \quad \frac{\partial u(x,0)}{\partial t} = 0.$$

Solution: It seems reasonable to approach the problem via the Fourier method. However, the latter is directly applicable only in the case of homogeneous boundary conditions; in our case, the condition on the right boundary is not homogeneous. Nevertheless, our problem can be reduced to a more convenient problem with homogeneous boundary conditions by splitting the solution into two parts and seeking $u(x,t)$ of the form

$$u(x,t) = \tilde{u}(x,t) + w(x,t), \tag{13}$$

where $w(x,t)$ is an arbitrary particular solution of the problem

$$\frac{\partial^2 w}{\partial t^2} = a^2 \frac{\partial^2 w}{\partial x^2} + b^2 w, \tag{14}$$

$$\frac{\partial w(0,t)}{\partial x} = 0, \quad w(l,t) = E_0 \, e^{\mu t} \sin \omega t, \tag{15}$$

and $\tilde{u}(x,t)$ is a solution of the problem

$$\frac{\partial^2 \tilde{u}}{\partial t^2} = a^2 \frac{\partial^2 \tilde{u}}{\partial x^2} + b^2 \, \tilde{u}, \tag{16}$$

$$\frac{\partial \tilde{u}(0,t)}{\partial x} = \tilde{u}(l,t) = 0,$$

$$\tilde{u}(x,0) = -w(x,0), \quad \frac{\partial \tilde{u}(x,0)}{\partial t} = -\frac{\partial w(x,0)}{\partial x}, \tag{17}$$

with homogeneous boundary conditions.

To find a solution of (14)–(15), notice that equation (14) allows a particular solution of the form

$$\Phi(x) \, e^{(\mu+i\omega)t},$$

where $\Phi(x)$ satisfies the ordinary differential equation

$$a^2 \, \Phi'' = [(\mu + i\omega)^2 - b^2] \, \Phi. \tag{18}$$

Additionally, if conditions

$$\Phi'(0) = 0, \quad \Phi(l) = E_0, \tag{19}$$

are satisfied, then

$$w(x,t) = \operatorname{Im}\left[\Phi(\mathrm{x}) \, e^{(\mu+i\omega)t}\right]. \tag{20}$$

The general solution of equation (18) has the following form:

$$\Phi(x) = c_1 \, e^{(\alpha+i\beta)x} + c_2 \, e^{-(\alpha+i\beta)x},$$

Chapter 11. Waves and Hyperbolic Equations

where
$$\alpha + i\beta = \frac{1}{a}\sqrt{(\mu + i\omega)^2 - b^2}\,. \tag{21}$$

Substituting (21) into (19), we obtain
$$\Phi'(0) = c_1(\alpha + i\beta) - c_2(\alpha + i\beta) = 0$$

and
$$\Phi(l) = c_1 e^{(\alpha+i\beta)l} + c_2 e^{-(\alpha+i\beta)l} = E_0\,.$$

Therefore
$$c_1 = c_2 = \frac{E_0}{2\cosh(\alpha+i\beta)l}\,.$$

Consequently, according to equality (20), we have
$$w(x,t) = E_0 e^{\mu t} \operatorname{Im}\left[\frac{\cosh(\alpha+i\beta)x}{\cosh(\alpha+i\beta)l} e^{i\omega t}\right]. \tag{22}$$

A solution of the problem (16)–(17) can be found by the method of separation of variables, which gives
$$\tilde{u}(x,t) = \sum_{n=0}^{\infty}(a_n \cos\omega_n t + b_n \sin\omega_n t)\tilde{X}_n(x)\,, \tag{23}$$

where
$$\tilde{X}_n(x) = \sqrt{\frac{2}{l}}\cos\sqrt{\lambda_n}x\,,\quad \lambda_n = \left(\frac{\pi(2n+1)}{2l}\right)^2,$$

$$a_n = (-w(x,0), \tilde{X}_n(x)) = E_0\sqrt{\frac{2}{l}}\operatorname{Im}\int_0^l \frac{\cosh(\alpha+i\beta)x \cos\sqrt{\lambda_n}x}{\cosh(\alpha+i\beta)l}dx\,,$$

$$b_n = \frac{1}{\omega_n}(-w_t(x,0), \tilde{X}_n(x))$$

$$= \frac{\mu}{\omega_n}a_n - \frac{\omega}{\omega_n}E_0\sqrt{\frac{2}{l}}\operatorname{Im}\int_0^l \frac{\operatorname{ch}(\alpha+i\beta)x \cos\sqrt{\lambda_n}x}{\operatorname{ch}(\alpha+i\beta)l}dx\,.$$

Substituting (22) and (23) into (13) and returning, via equalities (4), from the auxiliary function $u(x,t)$ to the voltage function $v(x,t)$, we finally obtain

$$v(x,t) = E_0 \operatorname{Im} \left[\frac{\cosh(\alpha + i\beta)x}{\cosh(\alpha + i\beta)l} e^{i\omega t} \right] +$$

$$+ e^{-\mu t} \sum_{n=0}^{\infty} (a_n \cos \omega_n t + b_n \sin \omega_n t) \tilde{X}_n(x).$$

For large t, the voltage distribution is described by the formula

$$v(x,t) = E_0 \operatorname{Im} \left[\frac{\cosh(\alpha + i\beta)x}{\cosh(\alpha + i\beta)l} e^{i\omega t} \right].$$

Chapter 12: First-Order Nonlinear PDEs

1. Let us begin by writing out the solution of the Cauchy problem (12.1.2), (12.2.4). It has its simplest form in the Lagrangian coordinate system, where the velocity field is given by

$$V(y,t) = u_0(y) + Hy. \tag{1}$$

The Eulerian coordinates are connected with the Lagrangian ones by the formula

$$x = y(1 + Ht) + u_0(y)t. \tag{2}$$

To find the Eulerian velocity field, we have to substitute the inverse function $y(x,t)$ in the right-hand side of (1). The Eulerian density field is described by the formula (12.2.6),

$$\rho(x,t) = \varrho_0(y(x,t)) \frac{\partial y(x,t)}{\partial x}. \tag{3}$$

Furthermore, observe that the relationship (2) between Lagrangian and Eulerian coordinates can be written in a more familiar form if we introduce new Eulerian coordinates and time,

$$x' = \frac{x}{1 + Ht}, \qquad t' = \frac{t}{1 + Ht}.$$

Then the equality (2) can be rewritten in the form

$$x' = y + u_0(y) t'. \tag{4}$$

Substituting on the right-hand side of (1) the inverse function

$$y(x,t) = \tilde{y}(x', t'), \tag{5}$$

Chapter 12. First-Order Nonlinear PDEs

where $\tilde{y}(x', t')$ is the function inverse to (4), we obtain

$$v(x,t) = u_0(\tilde{y}(x',t')) + H\tilde{y}(x',t'). \tag{6}$$

Obviously, the first term on the right-hand side is nothing but $u(x',t')$, the solution of the Cauchy problem

$$\frac{\partial u}{\partial t'} + u\frac{\partial u}{\partial x'} = 0, \quad u(x', t'=0) = u_0(x').$$

The second summand on the right-hand side of (6) can also be expressed via $u(x', t')$ if we recall the relation (12.1.10), which in our notation has the form

$$\tilde{y}(x', t') = x' - u(x', t')t'.$$

Substituting it into (6), regrouping the terms on the right-hand side, and expressing x' and t' through the "natural" coordinate x and time t, we obtain our final expression for the velocity field in the "expanding" universe:

$$v(x,t) = \frac{1}{1+Ht} u\left(\frac{x}{1+Ht}, \frac{t}{1+Ht}\right) + \frac{Hx}{1+Ht}.$$

Physical effects: Note that the first term in the above formula indicates that the expansion of the universe slows down the nonlinear evolution of the initial velocity fluctuations, and the spatial dependence of the fluctuations is obtained from the velocity field $u(x,t)$ of the "nonexpanding" universe via the simple operations of dilation by the factor of $1 + Ht$ of both space and time, and compression by the same factor of the field $u(x,t)$ itself.

Substituting expression (6) into (3), we arrive at the analogous expression for the density field

$$\rho(x,t) = \frac{1}{1+Ht} \varrho\left(\frac{x}{1+Ht}, \frac{t}{1+Ht}\right),$$

where $\varrho(x', t')$ solves the auxiliary continuity equation

$$\frac{\partial \varrho}{\partial t'} + \frac{\partial}{\partial x'}(u\varrho) = 0, \quad \varrho(x', t'=0) = \varrho_0(x').$$

2. Let us solve the given equation by passing to the corresponding characteristic equations

$$\frac{dX}{dt} = V, \quad \frac{dV}{dt} = -\frac{1}{\tau}V,$$
$$X(y, t=0) = y, \quad V(y, t=0) = v_0(y).$$

Their solutions are

$$V(y,t) = v_0(y)\, e^{-t/\tau} \quad \text{and} \quad X(y,t) = y + v_0(y)\,\theta\,,$$

where the auxiliary time is given by

$$\theta = \tau(1 - e^{-t/\tau})\,.$$

Comparing this solution with the solution of the standard Riemann equation, it is easy to see that

$$v(x,t) = e^{-t/\tau}\, u(x,\theta)\,,$$

where $u(x,\theta)$ satisfies the Riemann equation

$$\frac{\partial u}{\partial \theta} + u\frac{\partial u}{\partial x} = 0\,, \quad u(x, \theta = 0) = v_0(x)\,. \tag{8}$$

Physical effects: For $t \to \infty$, the auxiliary time is $\theta \to \tau$. Physically, this means that because of the velocity "dissipation," the nonlinear effects become weaker with the passage of time, and the shape of the velocity field "freezes," that is, it remains the same as if the time elapsed never exceeded τ.

3. Substituting the solution of the Riemann equation (12.1.9) into the Fourier integral, we have

$$\tilde{v}(\kappa, t) = \frac{1}{2\pi} \int v_0(y(x,t))\, e^{-i\kappa x}\, dx\,.$$

Changing the variable of integration to the Lagrangian coordinate, we obtain

$$\tilde{v}(\kappa, t) = \frac{1}{2\pi} \int v_0(y)\, e^{-i\kappa X(y,t)}\, \frac{\partial X}{\partial y}\, dy\,.$$

Integration by parts now gives

$$\tilde{v}(\kappa, t) = -\frac{i}{2\pi\kappa} \int e^{-i\kappa X(y,t)}\, v_0'(y)\, dy\,.$$

Substituting the explicit expression (12.1.3) for the function $X(y,t)$, we find that

$$\tilde{v}(\kappa, t) = -\frac{i}{2\pi\kappa} \int e^{-i\kappa y}\, v_0'(y)\, e^{-i\kappa v_0(y)}\, dy\,,$$

which can be rewritten in the form

$$\tilde{v}(\kappa, t) = \frac{1}{2\pi\kappa^2 t} \int e^{-i\kappa y}\, d\left[e^{-i\kappa v_0(y)t} - 1\right]\,.$$

Chapter 12. First-Order Nonlinear PDEs

Integrating by parts again, we arrive at the desired formula (12.2.14).

4. In mathematical terms, we need to find the Fourier series expansion of the solution $v(x,t)$ of the Riemann equation. As in the case of the Fourier series expansion for the density field discussed earlier in the chapter, we shall initially find the generalized Fourier image of the field $v(x,t)$. It is described by the formula (12.2.14) derived above in the solution to Exercise 3, which, in the case under consideration, takes the form

$$\tilde{v}(\kappa,t) = \frac{i}{2\pi\kappa kt} \int e^{-i\mu z} \left(e^{-i\mu\tau \sin z} - 1\right) dz,$$

where we used nondimensional variables

$$\mu = \kappa/k, \quad z = ky, \quad \tau = kat.$$

Using formula (12.2.16), and replacing the integrals by the corresponding Dirac deltas, the Fourier image can be written as the Fourier series

$$\tilde{v}(\kappa,t) = \frac{ia}{k}\left\{\left[J_0\left(\frac{\kappa\tau}{k}\right) - 1\right]\delta(\kappa) + \frac{k}{\kappa\tau}\sum_{n=1}^{\infty}\left[J_n\left(-\frac{\kappa\tau}{k}\right)\delta(\kappa - kn) + J_{-n}\left(-\frac{\kappa\tau}{k}\right)\delta(\kappa + kn)\right]\right\}.$$

Using the probing property of the Dirac delta, the fact that $J_0(0) = 1$, as well as the symmetry properties of the Bessel functions, we obtain

$$\tilde{v}(\kappa,t) = ia\sum_{n=1}^{\infty}(-1)^n \frac{J_n(n\tau)}{n\tau}\left[\delta(\kappa - kn) - \delta(\kappa + kn)\right].$$

Substituting this expression into the inverse Fourier integral, we obtain the following expansion of the solution of the Riemann equation:

$$v(x,t) = 2a\sum_{n=1}^{\infty}(-1)^{n+1}\frac{J_n(n\tau)}{n\tau}\sin(nkx), \tag{9}$$

which starts out as a purely harmonic wave.

Remark 4. A savvy student could answer the above question much faster by recalling that in the case of initially homogeneous density, $v(x,t)$

and $\rho(x,t)$ are related by the equality (12.2.9), which can be interpreted as an ordinary differential equation for $v(x,t)$,

$$\frac{dv(x,t)}{dx} = \frac{1}{t}\left(1 - \frac{\rho(x,t)}{\rho_0}\right),$$

and which contains the time t as a parameter. Substituting here the series (12.2.19) obtained earlier for the density field and taking into account the obvious boundary condition $v(x=0,t) = 0$, we obtain again the desired solution.

5. The continuity equation implies that the cumulative mass function satisfies the equation

$$\frac{\partial m}{\partial t} + v\frac{\partial m}{\partial t} + \rho_0 v = 0.$$

Multiplying it by $-4\pi e$, we obtain the electric field equation

$$\frac{\partial E}{\partial t} + v\frac{\partial E}{\partial t} = 4\pi e\rho_0 v.$$

The above equation, together with the first equation in (12.3.2), can be solved by the method of characteristics. The corresponding equations are

$$\frac{dX}{dt} = V, \qquad \frac{dV}{dt} = -\frac{e}{m}E, \qquad \frac{dE}{dt} = 4\pi e\rho_0 V.$$

Assume that the initial electric field is $E_0(x)$. Then the solutions of the characteristic equations are of the form

$$V(y,t) = v_0(y)\cos\omega t - E_0(y)\frac{1}{\gamma}\sin\omega t,$$
$$E(y,t) = E_0(y)\cos\omega t + v_0(y)\gamma\sin\omega t, \qquad (10)$$
$$X(y,t) = y + \frac{E_0(y)}{\gamma\omega} + \frac{1}{\gamma\omega}[v_0(y)\gamma\sin\omega t + E_0(y)\cos\omega t],$$

where the plasma frequency ω and the parameter γ are defined as follows:

$$\omega = \sqrt{\frac{4\pi e^2 \rho_0}{m}}, \qquad \gamma = \frac{m}{e}\omega.$$

The expression (10) for the Lagrangian fields indicates that the electrons in cold plasma oscillate harmonically, and the electric field oscillates with them harmonically. Nevertheless, the corresponding Eulerian fields are not

Chapter 12. First-Order Nonlinear PDEs

harmonic. Let us demonstrate this fact by analyzing in detail the case in which at the initial time, the electron and ion densities are balanced. In view of the relationship established above between the density and electric field fluctuations, the above statement means that $E_0(y) \equiv 0$, and the Lagrangian fields (10) simplify drastically:

$$V(y,t) = v_0(y)\cos\omega t, \quad E(y,t) = \gamma v_0(y)\sin\omega t,$$

$$X(y,t) = y + \frac{v_0(y)}{\omega}\sin\omega t.$$

A comparison of these Lagrangian fields with the Lagrangian velocity and coordinate fields for the Riemann equation shows that the Eulerian velocity and electric fields for cold plasma can be written in the form

$$v(x,t) = u(x,\theta)\cos\omega t, \quad E(x,t) = u(x,\theta)\gamma\sin\omega t,$$

where

$$\theta = \frac{\sin\omega t}{\omega},$$

and $u(x,\theta)$ is the solution of the Riemann equation (8).

6. In our case, the directional diagram of the snowfall is described by

$$D(\theta') = c\delta(\theta' - \theta_0).$$

The condition (12.3.3) guarantees that for every point of the initial profile $h_0(x)$, the Dirac delta is not concentrated outside the interval of integration in (12.2.16). Consequently, the velocity of the snow accretion on the interface segment that has the normal inclined at the angle θ to the z-axis is

$$c(\theta) = c\cos(\theta - \theta_0) = \cos\theta\cos\theta_0 + \sin\theta\sin\theta_0.$$

Substituting this expression into (12.2.11), (12.2.12), changing from θ to u, and introducing the new notation

$$c_\perp = c\cos\theta_0, \quad c_\| = c\sin\theta_0,$$

we arrive at the following equation for $h(x,t)$:

$$\frac{\partial h}{\partial t} = c_\perp + c_\| u,$$

or

$$\frac{\partial h}{\partial t} - c_\| \frac{\partial h}{\partial x} = c_\perp, \quad h(x, t=0) = h_0(x).$$

This equation is linear and has the obvious solution

$$h(x,t) = h_0(x - c_\| t) + c_\perp t. \tag{11}$$

The form of the above solution indicates that in the course of time, the bumps on the snow surface do not change shape, but just move upwind.

7. It is not difficult to show that growth of the interface $h(x,t)$ also has to satisfy the linear equation developed in the solution for Exercise 6, where in the present case,

$$c_\perp = \int_{-\pi/2}^{\pi/2} D(\theta') \cos\theta' \, d\theta', \qquad c_\| = \int_{-\pi/2}^{\pi/2} D(\theta') \sin\theta' \, d\theta'.$$

8. Utilizing formula (12.2.15), we can calculate the rate of interfacial growth in the direction θ:

$$c(\theta) = c \int_{-\theta-\pi/2}^{\pi/2} \cos^2\theta' \cos(\theta' - \theta) \, d\theta' = \frac{1}{3}(1 + \cos\theta)^2.$$

Substituting this expression into (12.2.12) and changing from θ to u, we obtain

$$\Phi(u) = \frac{1}{3}\left(1 + \frac{1}{\sqrt{1+u^2}}\right)^2 \sqrt{1+u^2}.$$

Since we intend to solve the problem in the small angle approximation, let us expand $\Phi(u)$ in a Taylor series and restrict our attention to the first nonzero term containing a power of u:

$$\Phi(u) \approx \frac{4}{3}c + \frac{c}{12}u^4.$$

Substituting this expression into the right-hand side of equation (12.2.11) and dropping the trivial constant term (which physically indicates the transition to the coordinate system comoving with the interface at its average speed), we find that in the small angle approximation, the interface $h(x,t)$ satisfies the following nonlinear partial differential equation:

$$\frac{\partial h}{\partial t} = \frac{c}{12}\left(\frac{\partial h}{\partial x}\right)^4.$$

The auxiliary functions entering the solution of the characteristic equations (12.2.19) are as follows:

$$\Phi(u) = \frac{c}{12}u^4, \qquad C(u) = \frac{c}{3}u^3, \qquad \Lambda(u) = -\frac{c}{4}u^4.$$

Chapter 12. First-Order Nonlinear PDEs

Substituting them into formulas (12.3.22), which describe the evolution of the interface $h(x, t)$, we arrive at its parametric description

$$x = y + \frac{c}{3}(u_0(y))^3 t, \qquad h = h_0(y) - \frac{c}{4}(u_0(y))^4 t.$$

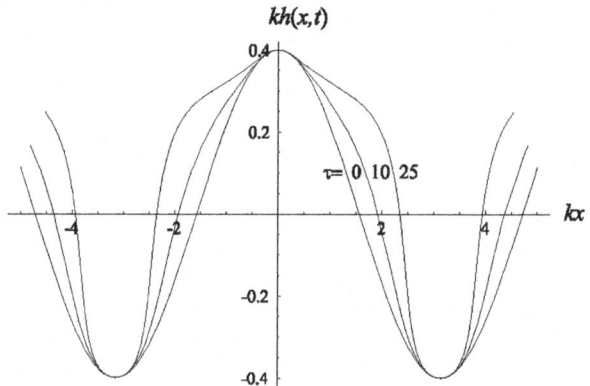

FIGURE 1
Time evolution of the interface $h(x, t)$ from Exercise 8, for $\varepsilon = 0.4$ and $\tau = 0$, 10, and 25.

In the special case of the sine initial profile, these equations take the form

$$kx = \mu + \frac{\tau}{3}\varepsilon^3 \sin^3 \mu, \qquad kh = \varepsilon \cos \mu - \frac{\tau}{4}\varepsilon^4 \sin^4 \mu,$$

where the nondimensional variables

$$\mu = ky, \qquad \varepsilon = kh, \qquad \tau = ckt$$

have been introduced.

Graphs illustrating the evolution of the interface $h(x, t)$ for various τ are shown in Fig. 1 above.

9. In this case, the evolution equation (12.2.11) is of the form

$$\frac{\partial h}{\partial t} = \frac{c}{\sqrt{1 + u^2}}.$$

The corresponding small angle approximation equation (after changing to a coordinate system comoving with the average speed ct) is

$$\frac{\partial h}{\partial t} + \frac{c}{2}\left(\frac{\partial h}{\partial x}\right)^2 = 0.$$

The auxiliary functions entering the right-hand side of equalities (12.3.22) are
$$\Phi(u) = -\frac{c}{2}u^2, \qquad C(u) = -cu, \qquad \Lambda(u) = \frac{c}{2}u^2.$$

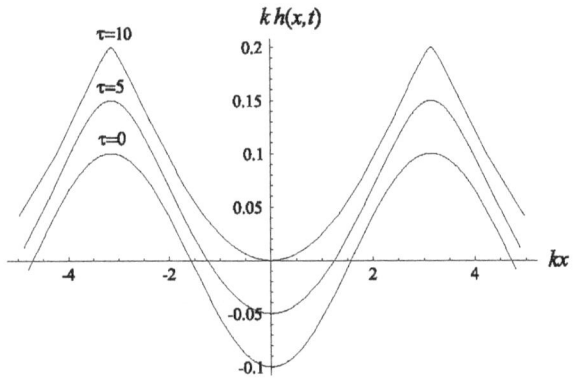

FIGURE 2
Time evolution of the interface $h(x,t)$ from Exercise 9, for $\varepsilon = 0.4$ and times $\tau = 0$, 5, and 10.

Consequently, the shape of the interface is described by the parametric equations
$$x = y - cu_0(y)\, t, \qquad h = h_0(y) + \frac{c}{2}u_0^2(y)\, t.$$
For the sine initial profile (12.3.4), this yields
$$kx = \mu + \varepsilon\tau \sin\mu, \qquad kh = -\varepsilon\cos\mu + \tfrac{\tau}{2}\varepsilon^2 \sin^2\mu,$$
$$\varepsilon = kh, \qquad \mu = ky, \qquad \tau = kct.$$

The corresponding graphs are shown in Fig. 2. To better show the evolution of the profile, the graphs are shifted slightly upward along the vertical axis.

If we think of the scenario of the wave front evolution in an isotropic medium shown in Fig. 12.3.4 as a standard deposition, then in the current problem, the interface evolves in the reverse direction, following a "corrosion" scenario as shown in Fig. 12.3.5. Its characteristic features are sharpening of the crests and smoothing out of the troughs.

10. Note that the gradient function $u(x,t)$ (12.2.10) of the interface $h(x,t)$ is described by the parametric equations
$$x = y + khc\sin(ky)t, \qquad u = -khc\sin(ky)\, t,$$

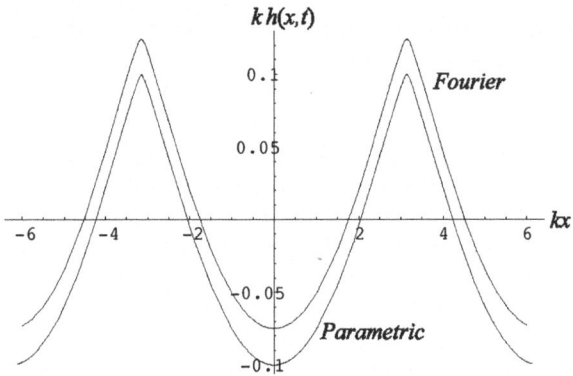

FIGURE 3
The interface $h(x,t)$ from Exercise 10, for $\tau = 10$ and $\varepsilon = 0.1$. The *lower curve* was constructed from the parametric equations, and the *upper curve* by taking the first ten terms of the Fourier series.

which coincide with the solution of the Riemann equation with the initial condition $v_0(x) = -khc\sin(kx)$ and time running backward. We have already obtained the Fourier series for the solution of the Riemann equation with the sine initial condition. Hence, we can write immediately

$$\frac{dh(x,t)}{dx} = 2\sum_{1}^{\infty}(-1)^n \frac{J_n(n\varepsilon\tau)}{n\tau}\sin(nkx).$$

The ordinary (not partial!) derivative notation with respect to x emphasizes the fact that time t appears here only as a parameter. Integrating both parts of the above equality, we obtain

$$k\,h(x,t) = 2\sum_{1}^{\infty}(-1)^n \frac{J_n(n\varepsilon\tau)}{n^2\tau}\cos(nkx) + C(t).$$

Here $C(t)$ is constant. If we are just interested in the shape of the interface $h(x,t)$ rather than its actual position, we can disregard $C(t)$ altogether. Figure 3 shows a graph of the interface $h(x,t)$ for $\tau = 10$ and $\varepsilon = 0.1$. The lower curve was constructed from the parametric equations, and the upper curve by taking the first ten terms of the Fourier series. One can see that the shapes are almost identical.

Chapter 13: Generalized Solutions of First-Order Nonlinear PDEs

1. The simplest way to prove the inequality relies on the form (13.2.5) of the solution of the Riemann equation. Differentiating both sides of that relation with respect to x, we get

$$q(x,t) = \frac{1}{t}\left(1 - \frac{\partial y(x,t)}{\partial x}\right).$$

Since in the single-stream regime function, $y(x,t)$ is monotonically increasing, its derivative is nonnegative and the expression in the parentheses is less than or equal to 1, which proves the inequality (13.7.1).

Let us illustrate the validity of the inequality (13.7.1) by drawing a graph of $tq(x,t)$ as a function of the argument x. It will be convenient to produce this graph by relying on the parametric form of the function $q(x,t)$, using as a parameter the Lagrangian coordinate y:

$$x = X(y,t) = y + v_0(y)\,t, \qquad q = Q(y,t).$$

Here we employed our standard convention and denoted by $Q(y,t)$ the Lagrangian field corresponding to the Eulerian field $q(x,t)$. Let us find $Q(y,t)$ by noticing that the following relationship holds:

$$\left.\frac{\partial y(x,t)}{\partial x}\right|_{x=X(y,t)} = \left(\frac{\partial X(y,t)}{\partial y}\right)^{-1} = \frac{1}{J(y,t)}.$$

Here $J(y,t)$ is the Jacobian of the Lagrangian-to-Eulerian mapping (13.2.11). Taking into account the explicit formula for the Jacobian, we arrive at the following relations defining $q(x,t)$:

$$x = y + v_0(y)\,t, \qquad q = \frac{v_0'(y)}{1 + v_0'(y)\,t}. \qquad (1)$$

Substituting the sinusoidal initial condition (13.3.8), we finally get

$$\eta = \zeta + \tau \sin\zeta, \qquad \theta = \frac{\tau \cos\zeta}{1 + \tau \cos\zeta},$$

where

$$\eta = kx, \qquad \zeta = ky, \qquad \tau = kat \quad \text{and} \quad \theta = tq$$

Chapter 13. Generalized Solutions of Nonlinear PDEs

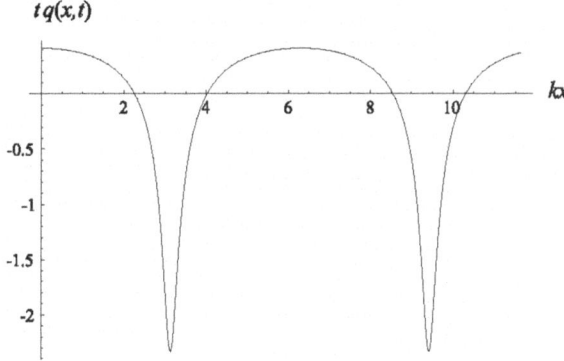

FIGURE 1
The field $tq(x,t)$, where $q(x,t)$ is the spatial derivative of the solution of the Riemann equation with sinusoidal initial condition. All the variables have been nondimensionalized. The graph has been constructed for the value $\tau = 0.7$ of the nondimensional time. Inequality (13.7.1) is obviously satisfied. Additionally, one can see the typical deep troughs of the field $q(x,t)$ corresponding to the steepening fronts of the velocity field $v(x,t)$ (the areas of the profile $v(x,t)$, where $\partial v/\partial x$ is negative).

are nondimensional variables. A graph of $tq(x,t)$, constructed by utilizing these equalities, is shown in Fig. 1.

2. To solve the problem, it is not necessary to analyze the Eulerian fields (the same approach was taken in Problem 1.) Indeed, in the time interval $t < t_n$ of the single-stream motion, when the Eulerian-to-Lagrangian mapping $y = y(x,t)$ is strictly monotone and maps $\mathbf{R}_x$ onto $\mathbf{R}_y$, the following pairs of inequalities are equivalent:

$$a(t) \leq q(x,t) \leq b(t), \quad x \in \mathbf{R}, \quad \Longleftrightarrow \quad a(t) \leq Q(y,t) \leq b(t), \quad y \in \mathbf{R}.$$

Hence, it suffices to study the Lagrangian field $q = Q(y,t)$, which is more convenient for analysis. It follows from (1) that

$$tQ < c(z), \qquad c(z) = \frac{z}{1+z}, \qquad z = tv_0'. \tag{2}$$

Furthermore, note that $c(z)$ increases monotonically in the interval $z \in (-1, \infty)$. Consequently, substituting the maximal value of $z = \mu t$ in (2), we obtain the maximum value of q,

$$tq \leq \frac{\mu t}{1 + \mu t} < 1. \tag{3}$$

Similarly, as long as the inequality

$$\nu t < 1 \tag{4}$$

is valid, we find the lower bound for q,

$$qt \geq -\frac{\nu t}{1 - \nu t}. \tag{5}$$

To complete our solution, we should observe that the inequality (4) is equivalent to the single-streamness condition $t < t_n$. Consequently, as long as the solution $v(x, t)$ of the Riemann equation is single-streamed, both inequalities (3) and (5) hold.

Remark 1. In the case $\nu = \mu = ka$, considered in the previous problem, the spatial derivative $v(x, t)$ satisfies the inequality

$$-\frac{\tau}{1 - \tau} \leq t\, q(x, t) \leq \frac{\tau}{1 + \tau} \qquad (\tau = kat).$$

Remark 2. Observe that the dependence of the supremum of the values of the field $q(x, t)$ on the initial maximal value μ weakens as $t \to \infty$, and asymptotically converges to the universal upper bound $1/t$, established in the previous problem. A similar weakening with the growth of t of the dependence on the initial conditions is fairly typical for nonlinear fields and waves.

Remark 3. By contrast, the infimum of the values of $q(x, t)$ as $t \to t_n$ diverges to $-\infty$ and leads to the gradient catastrophe at $t = t_n$.

3. First, let us find the center of mass of the particle cloud. For this purpose, let us multiply both sides of the equality (13.3.4) by x and integrate them over the whole x-axis. As a result, after changing the order of integration in the iterated integral on the right-hand side, we get

$$\int x\, \rho(x, t)\, dx = \int d\kappa\, \tilde{\rho}(\kappa, t) \int dx\, x\, e^{ikx}.$$

Chapter 13. Generalized Solutions of Nonlinear PDEs

The last integral, see Volume 1, is the derivative of the Dirac delta,

$$\int x\, e^{ikx}\, dx = -2\pi i \frac{\partial}{\partial \kappa} \delta(\kappa).$$

Thus we have

$$\int x\, \rho(x,t)\, dx = 2\pi i \frac{\partial}{\partial \kappa} \tilde{\rho}(\kappa, t)\big|_{\kappa=0}. \tag{6}$$

Substituting here the derivative with respect to κ of the integral $\int x\rho(x,t)\,dx$, see (13.3.3), we obtain

$$\int x\rho(x,t)\, dx = \int X(y,t)\, \rho_0(y)\, dy.$$

Recalling the Lagrangian-to-Eulerian mapping (13.2.1) for uniformly moving particles, we finally get that

$$\overline{x}(t) = x_c + \overline{v}_0\, t, \tag{7}$$

where

$$\overline{v}_0 = \frac{1}{M} \int v_0(y)\, \rho_0(y)\, dx.$$

Now let us calculate the particle cloud's dispersion. Some simple algebra yields

$$D(t) = \overline{x^2}(t) - \overline{x}^2(t), \tag{8}$$

where the density's second moment is given by

$$\overline{x^2}(t) = \frac{1}{M} \int x^2\, \rho(x,t)\, dx.$$

By analogy with (7),

$$\int x^2\, \rho(x,t)\, dx = -2\pi \frac{\partial^2}{\partial \kappa^2} \tilde{\rho}(\kappa,t)\big|_{\kappa=0},$$

so that in view of (13.3.3),

$$\overline{x^2}(t) = \frac{1}{M} \int X^2(y,t)\, \rho_0(y)\, dy.$$

Substituting now the explicit expression (13.2.1) for $X(y,t)$ and utilizing equalities (7) and (8), we obtain

$$D(t) = D_y + t^2 D_v + 2t \left[\overline{y\, v_0(y)} - \overline{y}\, \overline{v_0(y)}\right],$$

where
$$D_y = \overline{y^2} - \overline{y}^2, \qquad D_v = \overline{v_0^2(y)} - \overline{v_0(y)}^2,$$
and where we have also introduced the operator
$$\overline{f(y)} = \frac{1}{M} \int f(y) \rho_0(y)\, dy$$
of spatial averaging with respect to the initial density $\rho_0(y)$ of the particle cloud.

4. Let us write the equation of the original interface shape in the form
$$\boldsymbol{r} = \boldsymbol{r}_0(s),$$
where
$$\boldsymbol{r}_0(s) = \boldsymbol{l}\,\zeta(s) + \boldsymbol{m}\,\eta(s),$$
and $\boldsymbol{l}$ and $\boldsymbol{m}$ are unit vectors on the coordinate axes x and z, respectively. The normal vector to the initial interface can be written in the form
$$\boldsymbol{n} = \frac{\boldsymbol{l}\,\dot{\eta} - \boldsymbol{m}\,\dot{\zeta}}{\sqrt{\dot{\zeta}^2 + \dot{\eta}^2}},$$
where the dot denotes differentiation with respect to the parameter s. Obviously, the equation of the growing interface is
$$\boldsymbol{r} = \boldsymbol{r}_0(s) + ct\,\boldsymbol{n}. \tag{9}$$
Substituting the above formula for the normal vector and passing to the coordinate notation, we obtain the desired parametric description of the growing interface,
$$\begin{cases} x = \zeta(s) + \dfrac{ct\dot{\eta}(s)}{\sqrt{\dot{\zeta}^2 + \dot{\eta}^2}}, \\[2mm] z = \eta(s) - \dfrac{ct\dot{\zeta}(s)}{\sqrt{\dot{\zeta}^2 + \dot{\eta}^2}}. \end{cases}$$
Now suppose that the initial condition is given explicitly in the form $z = h_0(x)$. Then the initial parametric conditions are
$$x = s, \qquad z = h_0(s).$$

Chapter 13. Generalized Solutions of Nonlinear PDEs

Consequently, in this case, the parametric solution can be obtained by substituting the expressions $\zeta(s) = s$ and $\eta(s) = h_0(s)$, and we have

$$x = s + \frac{ct\dot{h}_0(s)}{\sqrt{1+\dot{h}_0^2(s)}}, \qquad z = h_0(s) - \frac{ct}{\sqrt{1+\dot{h}_0^2(s)}}. \tag{10}$$

Here $\dot{h}_0(x)$ has a clear geometric meaning: it is the tangent of the angle between the normal to $h_0(x)$ and the z-axis. If it is small, that is, if

$$|\dot{h}_0(x)| \ll 1,$$

then with good accuracy, we can expand the right-hand side of the equalities (10) in a Taylor series in powers of $\dot{h}_0$ and retain only the first nonzero term of the expansion. This gives

$$x = s + ct\,\dot{h}_0(s), \qquad z = h_0(s) + \frac{ct}{2}\dot{h}_0^2(s).$$

It is not difficult to verify that the above formulas give a parametric representation of the exact solutions of the approximate equation (13.1.5).

5. Let us introduce the local unit vectors $(\mathbf{e}_r, \mathbf{e}_\varphi)$ of the polar coordinate system and write the equation of the initial contour in the vector form

$$\mathbf{r} = \varrho_0(\varphi)\,\mathbf{e}_r. \tag{11}$$

To write a similar vector equation

$$\mathbf{r} = \varrho(\varphi, t)\,\mathbf{e}_r$$

of the growing contour $\mathcal{L}_t$, it is necessary to complement the right-hand side of (11) by the summand $ct\,\mathbf{n}$, where $\mathbf{n}$ is the unit vector normal to the contour. The normal vector itself can be found by rotating by 90° the vector tangent to the contour of the form

$$\boldsymbol{\tau} = \frac{\mathbf{r}'}{|\mathbf{r}'|},$$

where the prime indicates differentiation with respect to the angle φ. Differentiating (11) with respect to φ, and taking into account the fact that $\mathbf{e}'_r = \mathbf{e}_\varphi$, we have

$$\boldsymbol{\tau} = \frac{\varrho'_0\,\mathbf{e}_r + \varrho_0\,\mathbf{e}_\varphi}{\sqrt{\varrho_0^2 + \varrho_0'^2}} \quad\Rightarrow\quad \mathbf{n} = \frac{\varrho_0\,\mathbf{e}_r - \varrho'_0\,\mathbf{e}_\varphi}{\sqrt{\varrho_0^2 + \varrho_0'^2}}.$$

The relationship between the vectors discussed above is shown in Fig. 2 (left). It is clear from Fig. 2 that mapping of the points of the original contour into points of the evolving contour changes the angular coordinate; the points of the evolving contour not only escape from the origin of the coordinate system but are subject to a rotating motion as well. For that reason, in analogy with the Lagrangian and Eulerian coordinates of particles in a hydrodynamic flow, it is convenient to introduce the "Lagrangian" angle ψ, and the current "Eulerian" angle φ; see, Fig. 2 (right). The picture makes it clear that these two angles are related by the equation

$$\varphi = \psi + \beta(\psi),$$

where β is the angle between the angular coordinates of the point on the original contour and the corresponding point on the evolving contour. It can be expressed via the angle α between the normal vector $\boldsymbol{n}$ and the unit vector $\boldsymbol{e}_r$ of the polar coordinate system. The mutual position of the vectors $\boldsymbol{n}$ and $\boldsymbol{e}_r$ indicates that

$$\sin\beta = \frac{ct}{\varrho}\sin\alpha \quad \Rightarrow \quad \beta = \arcsin\left(\frac{ct}{\varrho}\sin\alpha\right),$$

where

$$\varrho = \varrho(\varphi, t) = \sqrt{(\boldsymbol{r}\cdot\boldsymbol{r})} = \sqrt{c^2 t^2 + 2ct\varrho_0 \cos\alpha + \varrho_0^2}$$

is the distance from the origin to the point of the evolving contour.

Since

$$\tan\alpha = -\frac{\varrho_0'}{\varrho_0}, \quad \sin\alpha = -\frac{\varrho_0'}{\sqrt{\varrho_0^2 + \varrho_0'^2}}, \quad \text{and} \quad \cos\alpha = \frac{\varrho_0}{\sqrt{\varrho_0^2 + \varrho_0'^2}},$$

we finally get

$$\begin{cases} \varrho = \varrho(\psi, t) = \sqrt{c^2 t^2 + 2ct\varrho_0(\psi)\cos\alpha(\psi) + \varrho_0^2(\psi)}, \\ \varphi = \psi(\psi, t) + \arcsin\left(\dfrac{ct \sin\alpha(\psi)}{\varrho(\psi, t)}\right), \\ \alpha(\psi) = -\arctan\left(\dfrac{\varrho_0'(\psi)}{\varrho_0(\psi)}\right). \end{cases} \quad (12)$$

The above system of equations provides a parametric description of the evolving contour in the polar coordinate system.

Chapter 13. Generalized Solutions of Nonlinear PDEs

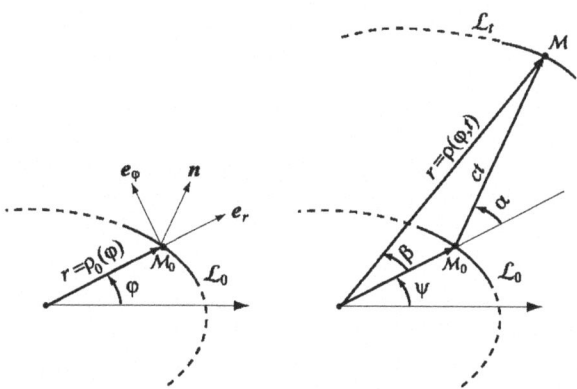

FIGURE 2
A geometric construction of the evolving contour in polar coordinates. The picture on the left shows a fragment of the initial contour, its radius vector, the local unit vectors of the polar coordinate system, and the normal vector at a selected point $\mathcal{M}_0$ of the contour $\mathcal{L}_0$. The picture on the right shows fragments of the contour at the initial time $t=0$ and some $t>0$. Shown are the radius vector of the selected point $\mathcal{M}_0$ on the original contour and its image $\mathcal{M}$ on the evolved contour $\mathcal{L}_t$ at time $t>0$. The vector $ct\boldsymbol{n}$ connecting the above two points is also pictured.

6. Expressions (12) are rather bulky and opaque. These shortcomings can be partly removed by an asymptotic analysis for $ct \to \infty$, which suggests introduction of a small parameter

$$\mu = \frac{\varrho_0}{ct}.$$

Expanding the right-hand sides in (12) into powers of this parameter and retaining only the terms involving the first and zeroth powers in μ, we obtain

$$\varrho = ct + \varrho_0(\psi)\cos\alpha(\psi), \qquad \varphi = \psi + \alpha(\psi). \tag{13}$$

The geometric meaning of these equalities is clear: for $ct \gg \varrho_0$, every point $\mathcal{M}_0$ of the original contour is mapped into a point $\mathcal{M}$ of the evolving contour located in the normal direction to the original contour, and the distance of $\mathcal{M}$ from the origin is equal to the sum of the magnitude of the vector $ct\boldsymbol{n}$ and that of the projection of $\varrho_0(\psi)\boldsymbol{e}_r$ onto the vector $ct\boldsymbol{n}$.

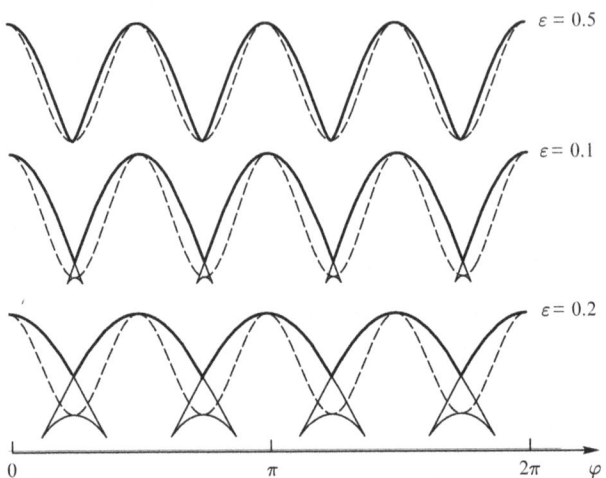

FIGURE 3
Asymptotic shapes of the evolving contour as described by their dependence on the angle φ in the polar coordinate system. The case of three values of ε are explored. The vertical scales on different pictures are different and adjusted to the amplitude of the original condition (*dashed lines*). The two bottom graphs display multivalued fragments. In these cases, the actual shape of the growing contour corresponds to the upper branch of the function $\varrho_\infty(\varphi)$. In the two bottom graphs, the actual growing contour is marked by a *thick line*.

Notice that if the first term on the right-hand side of the first equation in (13), which does not affect the shape of the contour, is dropped, then we obtain a parametric equation of the contour that is time-independent:

$$\varrho = \varrho_\infty(\varrho) = \frac{\varrho_0^2(\psi)}{\sqrt{\varrho_0^2(\psi) + \varrho_0'^2(\psi)}}, \quad \varphi = \psi - \arcsin\left(\frac{\varrho_0'(\psi)}{\sqrt{\varrho_0^2(\psi) + \varrho_0'^2(\psi)}}\right). \tag{14}$$

In other words, in the course of time, the shape of the growing contour does not change. Figure 3 shows the "frozen" shape of the evolving contour in the case that the original shape is given by the equality (13.7.2).

Solution 7. Let us recall how to determine the shape of the ice floe at $t = T$. The procedure is to roll a disk $\mathcal{C}_R$ of radius $R = cT$ around the exterior

Chapter 13. Generalized Solutions of Nonlinear PDEs

of the original contour $\mathcal{L}_0$. The locus of the centers of the rolling disk is the desired boundary $\mathcal{L}_T$ of the freezing ice floe at $t = T$; see Fig. 4.

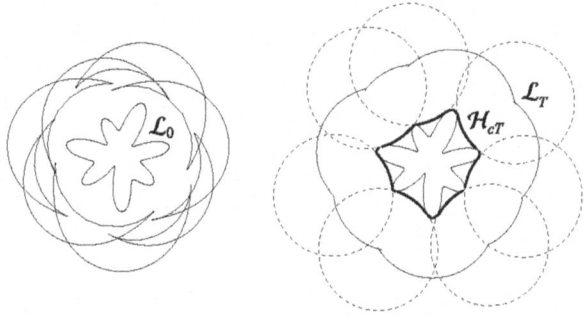

FIGURE 4
Determination of the shape of an ice floe subject to a freezing and thawing cycle. The original shape $\mathcal{L}_0$ and its image given by the equations (9) and (12) are shown on the left. The picture on the right shows only the segments of the mapping that correspond to the actual shape of the contour $\mathcal{L}_T$. The *thick line* represents the envelope $\mathcal{H}_{cT}$ of the original contour $\mathcal{L}_0$. All the disks of radius $R = cT$ that touch the original contour at exactly two points are shown. Their arcs between the points of tangency form pieces of the envelope $\mathcal{H}_{cT}$.

Let us take a closer look at what happens when the disk $\mathcal{C}_R$ is rolled around the contour $\mathcal{L}_0$. Sometimes, it will not be able to touch all the points of the original contour $\mathcal{L}_0$, just as happened in the case pictured in Fig. 4. In this case, the form of the contour $\mathcal{L}_T$ will not change if the original contour $\mathcal{L}_0$ is replaced by its (nonconvex) *envelope* $\mathcal{H}_{cT}$. The envelope $\mathcal{H}_R$ (see Fig. 4) of the contour $\mathcal{L}_0$ is here defined as a contour consisting of all the points of tangency of the external circles of radius R with $\mathcal{L}_0$, complemented, in the case of separated adjacent points of tangency, by a circular arc of radius R connecting those two points.

Let us now take a look at the process of thawing. To find the floe's shape after the thawing process is complete, one has to roll the same disk of radius cT, but this time around the interior side of the contour $\mathcal{L}_T$. The locus $\mathcal{L}_{2T}$ of the centers of such internally osculating disks will provide the shape of the boundary of the thawed floe after time $2T$. It is clear that the final shape of the floe at $t = 2T$ will coincide with the shape of the envelope $\mathcal{H}_{cT}$ shown in Fig. 4.

Consequently, if the initial contour $\mathcal{L}_0$ coincides with its envelope $\mathcal{H}_{cT}$, then after the cycle of freezing and thawing, the floe returns to its original shape. If this condition is not satisfied, then the final shape is the envelope $\mathcal{H}_{cT}$ of the original shape. A geometric illustration of the above conclusion is shown in Fig. 4.

Remark 1. As the freezing time T goes to ∞, then after the thawing cycle, the floe's shape will approach the shape of the *convex hull* of the original shape. One can visualize the boundary of the latter by stretching an elastic band around the original shape.

Remark 2. Physicists often like to see what happens if one reverses the course of time and to check whether the events retrace themselves. In this spirit, we can think about the process of thawing as a process of freezing in reverse time. The above analysis shows that the above process of the contour growth is time-reversible only if the original contour coincides with its envelope. In the general case, the process of contour growth is not reversible. Indeed, in the process of contour growth, the information contained between the original contour and its envelope is irretrievably lost. On the other hand, the growth of a convex contour is completely reversible, since it coincides with its convex envelope.

Remark 3. From the mathematicians' viewpoint, the mappings (9) and (12), of $\mathcal{L}_0$ onto $\mathcal{L}_T$ can be interpreted as solutions of the interface evolution equation (12.2.7) applied to the closed contour $\mathcal{L}_0$. As long as the initial contour $\mathcal{L}_0$ coincides with its envelope $\mathcal{H}_{cT}$, the mappings determine a classical solution of (12.2.7). Otherwise, the rolling of the disk around the contour $\mathcal{L}_0$ gives the weak solution.

8. Let us denote the 1-D matter density along the r-axis by

$$\rho_0(r) = \varrho \, \Omega \, r^2.$$

According to the ERS principle, to solve the problem we need to find the coordinate $q(r, t)$ of the global minimum of the function

$$\phi(q, t) - r\, M(q) \qquad (q > 0), \tag{15}$$

where

$$\phi(q, t) = P(q)t + N(q), \quad \text{and} \quad M(q) = \int_0^q \rho_0(r)\, dr = \frac{1}{3} \varrho \, \Omega \, q^3, \tag{16}$$

Chapter 13. Generalized Solutions of Nonlinear PDEs

and $P(q)$ and $N(q)$ are given by an expression similar to (13.5.28):

$$P(q) = \begin{cases} \dfrac{p_0}{\varepsilon} q, & 0 < q < \varepsilon, \\ p_0, & \varepsilon < q, \end{cases} \qquad N(q) = \frac{1}{4} \varrho \Omega q^4.$$

Here q plays the role of the Lagrangian coordinate.

Following the geometric construction based on the ERS principle, to determine the law of motion of the detonation wave it is necessary to find the line $M(q)\,r+h$, touching the curve (15) at exactly two points, q^- and q^+. The position of the left point is obvious: $q^- = 0$. Consequently, $h = 0$. We shall find the position of the right point by making sure that the functions $\phi(q,t)$ and $M(q)r$ and their derivatives with respect to q are equal. As a result, we obtain two equations with respect to q and r:

$$\gamma t + 3q^4 = 4q^3 r \quad \text{and} \quad q = r,$$

wheremthe auxiliary parameter γ is given by

$$\gamma = \frac{12\,p_0}{\sigma \Omega}.$$

Thus

$$r = q^+ = \sqrt[4]{\gamma t}.$$

This is the desired law of motion of the detonation wave. If we substitute the above expression for q in the formula for the cumulative mass function (16), then we find the law of growth of the mass accumulated by the detonation wave:

$$M(t) = (4p_0 t)^{3/4} \left(\frac{\sigma \Omega}{3} \right)^{1/4}.$$

9. Recall that the weak solution of the Riemann equation is of the form

$$v_w(x,t) = \frac{x - y_w(x,t)}{t}, \tag{17}$$

where $y_w(x,t)$ is the coordinate of the point of tangency of the initial potential

$$s_0(y) = s\,\chi(y)$$

and the parabola (13.4.13). Here, as before, $\chi(y)$ is the Heaviside unit step function equal 0 for $y < 0$ and 1 otherwise. The graph of the mapping $y = y_w(x,t)$ and the accompanying geometric construction are shown in

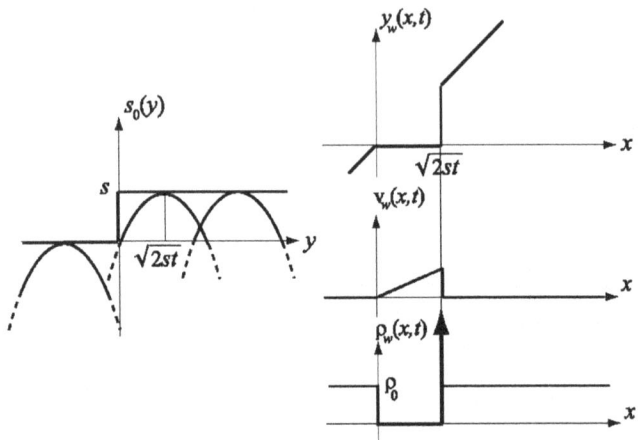

FIGURE 5
Left: The initial potential function and the osculating (from below) parabolas. Three typical positions of the parabolas are shown that correspond to different portions of the mapping $y = y_w(x,t)$. *Right (top to bottom)*: Graphs of the Eulerian-to-Lagrangian mapping, the solution velocity field, and the density field of the flow of sticky particles. The *thick arrow* in the bottom picture indicates the singularity of the generalized density field.

Fig. 5. The graph and equation (17) demonstrate that in our case, the weak solution of the Riemann equation is a sawtooth-shaped function of the form

$$v(x,t) = \begin{cases} 0, & x < 0, \\ \dfrac{x}{t}, & 0 < x < x^*(t), \\ 0, & x > x^*(t), \end{cases} \quad (18)$$

where

$$x^*(t) = \sqrt{2st} \quad (19)$$

is the coordinate of the discontinuity (shock front) of the weak solution.

A physical explanation of the shape of the weak solution (18) is as follows: A comparison of the ERS and the global minimum principles shows that the latter describes the motion of inelastically colliding particles given uniform initial density ρ_0. The assumed Dirac delta initial condition means that for $t = 0$, the matter does not move, with the exception of the particle located at $x = 0$, which is given momentum $\rho_0 s$. Its running coordinate is given by the expression (19), and its velocity is

$$v^*(t) = \frac{d}{dt}\sqrt{2st} = \sqrt{\frac{s}{2t}}.$$

This moving particle is transformed into a macroparticle, collecting matter that was initially located to its right, inside the interval $x \in [0, x^*]$. The mass of the thus formed macroparticle is obviously

$$m^*(t) = \rho_0\, x^*(t) = \rho_0 \sqrt{2st}.$$

By multiplying it by the velocity of the macroparticle, we ensure that the momentum conservation law is satisfied:

$$p^* = m^*(t)\, v^*(t) = \rho_0 \sqrt{2st}\, \sqrt{\frac{s}{2t}} = s = const.$$

Now we are ready to find the matter density $\rho_w(x, t)$ at an arbitrary point x. Recall that the cumulative mass function of the flow of sticky particles is described by the expression (13.5.38),

$$m(x, t) = M(y_w(x, t)).$$

In our case, $M(y) = \rho_0\, y + C$, where C is an inessential constant. In the case $C = 0$, we have

$$m(x, t) = \rho_0\, y_w(x, t).$$

To find the current flow density it remains to differentiate both sides of the above equality with respect to x. The simplest way to do it is by inspection of the graph of the mapping $y = y_w(x, t)$ in Fig. 5. As a result, we get

$$\rho_w(x, t) = m^*(t)\delta(x^*(t) - x) + \rho_c(x, t),$$

where

$$\rho_c(x, t) = \begin{cases} \rho_0, & x < 0, \\ 0, & 0 < x < x^*(t), \\ \rho_0, & x^*(t) < x. \end{cases}$$

Remark 1. Observe that for x contained in the interval $[0, x^*]$, the flow density is zero (see also Fig. 5). The reason for this is obvious: the macroparticle sweeps all the matter located in its path, compressing it into an infinitely dense cluster with a singular density. Moreover, the sawtooth field $v_w(x, t)$ (18), if interpreted as a velocity field of the flow of sticky particles, turns out to be vacuous—it describes the velocity of matter of zero density and mass.

Because of this phenomenon, the above weak solution of the Riemann equation may seem to be devoid of physical meaning. But this is not the case. In nonlinear acoustics, where $v_w(x, t)$ describes the pressure field, such a solution is fully realizable, and the corresponding pressure field can be measured experimentally.

10. According to the ERS principle, the coordinate $x^*(t)$ of the macroparticle can be determined from the fact that the function (13.5.26) has two identical global minima at two different points with coordinates y^- and y^+. In the context of our problem, it is easy to establish the relationship between those two coordinates. Indeed, since the particles to the right of the macroparticle do not move, their Lagrangian coordinates coincide with their Eulerian coordinates. This observation also applies to the particle immediately adjacent to the macroparticle. This means that

$$y^+(t) = x^*(t) > 0.$$

Similar considerations apply to the particles on the left that move with the identical velocity V. Thus we get another useful formula,

$$y^-(t) = x^*(t) - Vt < 0.$$

To obtain an equation for $x^*(t)$, it remains to equate the values of the function (13.5.26) at specified points, taking advantage of the equality $x = x^*$. Hence, the desired equation is of the form

$$x^* \left[M(x^*) - M(x^* - Vt) \right] = [N(x^*) - N(x^* - Vt)] - VtM(x^* - Vt), \quad (20)$$

where we used the relationships (13.5.28) and took into account the fact that in our case, the momentum field is

$$P(y) = V M(y) \chi(-x).$$

Let us take a look at the limit cases of equation (20). If, for example, the initial density is the same everywhere, then

$$M(y) = y \quad \text{and} \quad N(y) = \frac{y^2}{2}.$$

We assumed here, without loss of generality, that the initial uniform density was equal to 1. Substituting the above expression into (20), we arrive at the formula

$$x^*(t) = \frac{V}{2} t.$$

Chapter 13. Generalized Solutions of Nonlinear PDEs 377

This formula, together with the initial condition, implies that the weak solution of the Riemann equation found via the global minimum principle is

$$v_w(x,t) = V \chi \left(\frac{V}{2}t - x \right).$$

In other words, in view of the global minimum principle, the shock front of the field $v(x,t)$ propagates with velocity $V/2$, equal to one-half the magnitude of the discontinuity.

Now let us find the asymptotic solution of the equation (20) for $t \to \infty$, assuming, as demanded by the problem, that the total mass of the flow is finite. Physically, it is evident that as $t \to \infty$, all the material in the flow becomes glued into one macroparticle moving with a constant speed. In this case, physical intuition is in agreement with the mathematical derivations. Namely, one can show rigorously that the following asymptotics are valid:

$$M(x^*) - M(x^* - Vt) \to M,$$
$$N(x^*) - N(x^* - Vt) \to x_c M, \qquad (t \to \infty),$$
$$M(x^* - Vt) \to -M_-,$$

where M is the total mass in the flow, x_c is its initial center of mass, and

$$M_- = \int_{-\infty}^{0} \rho_0(y)\, dy$$

is the mass of particles moving at $t = 0$. Substituting the above asymptotics into equation (20), we obtain a physically obvious expression for the limit law of motion of the macroparticle that contains the whole mass of the flow:

$$x^*(t) = x_c + \frac{M_-}{M} Vt.$$

11. Before we attempt to find the velocity and density fields, it is useful to study the motion of individual particles. Consider a particle in the flow with an arbitrary Lagrangian coordinate y. As always, denote its Eulerian coordinate by $x = X(y,t)$. Motion of the particles is subject to Newton's second law, which in the case of a one-dimensional UGL takes the form

$$\frac{d^2 X}{dt^2} = \gamma \left[M_r(y,t) - M_l(y,t) \right],$$

where $M_r(y,t)$ is the cumulative mass of the flow to the right of the selected particle, and $M_l(y,t)$ is the cumulative mass to its left. As long as the particles maintain their original order, the above cumulative masses are independent of time and equal to

$$M_l(y) = \int_{-\infty}^{y} \rho_0(y')\,dy \quad \text{and} \quad M_r(y) = \int_{y}^{\infty} \rho_0(y')\,dy,$$

and Newton's equation becomes extremely simple:

$$\frac{dX(y,t)}{dt} = V(y,t), \qquad \frac{dV(y,t)}{dt} = \gamma[M_r(y) - M_l(y)]. \tag{21}$$

To obtain the above crisp display of the Lagrangian velocity field $V(y,t)$, we have split the second-order equation into two equations of first order.

Equations (21) have to be complemented by the initial conditions

$$X(y, t=0) = y, \qquad V(y, t=0) = v_0(y).$$

Their solutions provide the desired Lagrangian velocity field

$$v = V(y,t) = v_0(y) + \gamma\left[M - 2M(y)\right]t \tag{22}$$

and the Lagrangian-to-Eulerian mapping

$$x = X(y,t) = y + v_0(y)t + \frac{\gamma}{2}\left[M - 2M(y)\right]t^2. \tag{23}$$

In the above formulas, $M(y) := M_l(y)$, and we took advantage of the fact that the cumulative masses on the left and the right are related by the identity

$$M_l(y) + M_r(y) \equiv M.$$

As a result, we have

$$M_r(y) - M_l(y) = M - 2M_l(y) = M - 2M(y).$$

On the other hand, we know that the Lagrangian density field is always described by the expression

$$R(y,t) = \frac{\rho_0(y)}{J(y,t)}, \tag{24}$$

where in our case, the Lagrangian-to-Eulerian Jacobian is given by

$$J(y,t) = \frac{\partial X}{\partial y} = 1 + v_0'(y)t - \frac{\gamma}{2}\rho_0(y)t^2. \tag{25}$$

Chapter 13. Generalized Solutions of Nonlinear PDEs

Remark 1. Notice a particular feature of the density field (24). Whereas the velocity field (24) depends on the total mass of the matter to the left and to the right of a point with given Lagrangian coordinate y, the density field is determined by the *local* properties of the flow: its behavior is fully determined by the values of the initial velocity and density at one point only.

12. First of all, let us calculate the cumulative mass located to the left of a point with Lagrangian coordinate y:

$$M(y) = \rho_0 \ell^2 \int_{-\infty}^{y} \frac{dz}{z^2 + \ell^2} = \frac{M}{\pi} \left[\arctan\left(\frac{y}{\ell}\right) + \frac{\pi}{2} \right],$$

where

$$M = \pi \rho_0 \ell$$

is the total mass of the matter in the flow (Fig. 6).

Separately, let us write the combination of the cumulative masses on the left and on the right that enter into the expressions for the Lagrangian velocity field (22) and the Lagrangian-to-Eulerian mapping (23):

$$M - 2M(y) = -2\rho_0 \ell \arctan\left(\frac{y}{\ell}\right).$$

Substituting this expression into (22) and (23), we obtain a parametric description of the Eulerian velocity and density fields:

$$\eta = \zeta - \tau^2 \arctan(\zeta), \quad u = -2\tau \arctan(\zeta), \quad r = \frac{1}{1 + \zeta^2 - \tau^2},$$

where we have introduced the nondimensional time, velocity, and density

$$\eta = \frac{x}{\ell}, \quad \zeta = \frac{y}{\ell}, \quad \tau = \sqrt{\gamma \rho_0}, \quad u = \frac{v}{\sqrt{\gamma \rho_0 \ell}}, \quad r = \frac{\rho}{\rho_0}.$$

13. In this case, utilizing the fact that $g(z)$ is even, it is possible to write the combination of cumulative masses entering into equations (22) and (23) in the form

$$M - 2M(y) = \int_{-y}^{y} \rho_0(y') \, dy' = \rho_0 \ell \int_{-y/\ell}^{y/\ell} g(z) \, dz,$$

which demonstrates that the behavior of the velocity field depends essentially only on the density of particles in the interval $[-y, y]$, and is independent of

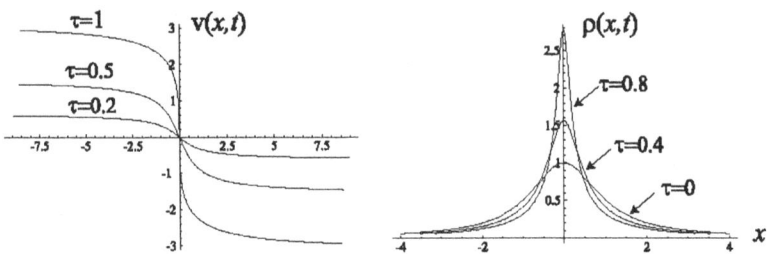

FIGURE 6
Graphs of the nondimensional Eulerian velocity and density fields for the gravitationally interacting particles shown at a progression of time instants. It is clear that the gravitational interaction leads to an ever accelerating convergence of the particles to the origin, and to the growth of the density curve in the neighborhood of the center of the 1-D universe.

the behavior of the initial density field outside this interval. This opens up a possibility of easy transition to the limit of uniform initial density. Indeed, letting ℓ to infinity and observing that in view of the continuity of the function $g(z)$, the last integral converges to $2y/\ell$, we get

$$\lim_{\ell \to \infty} [M - 2M(y)] = 2\rho_0 \, y \, .$$

Substituting the limit value into the right-hand sides of (23) and (24), we can find the velocity field and the Lagrangian-to-Eulerian mapping for the initially uniformly distributed flow of gravitationally interacting particles:

$$x = y\left(1 - \gamma\rho_0 \, t^2\right) + v_0(y) \, t \, , \qquad v = v_0(y) + 2\gamma\rho_0 \, y \, t \, .$$

Consequently, in view of the expressions (24) and (25), the Lagrangian density field is given by

$$R(y,t) = \frac{\rho_0}{1 + v_0'(y) \, t - \gamma\rho_0 \, t^2} \, .$$

For a uniformly expanding universe, when $v_0(y) = Hy$, the density remains uniform but diverges to infinity as $\tau \to \tau_n$, where the nondimensional collapse time is given by

$$\tau_n = \delta + \sqrt{1 + \delta^2} \, , \qquad \tau = \sqrt{\gamma\rho_0} \, t \, , \qquad \delta = \sqrt{\frac{H}{4\gamma\rho_0}} \, .$$

Chapter 14: Nonlinear Waves and Growing Interfaces: 1-D Burgers–KPZ Models

1. Multiplying equation (14.1.11) by $u(x,t)$ and then integrating it term by term over the whole x-axis, one gets

$$\frac{1}{2}\frac{d}{dt}\int u^2(x,t)\,dx + \frac{1}{3}\int \frac{\partial u^3(x,t)}{\partial x}\,dx = \mu \int u(x,t)\frac{\partial^3 u(x,t)}{\partial x^3}\,dx.$$

The second integral on the left-hand side is obviously equal to zero. Thus, to prove the invariance of the above functional, it is sufficient to show that the integral on the right-hand side is zero. Integration by parts shows that this indeed is the case:

$$\int u \frac{\partial^3 u}{\partial x^3}\,dx = -\frac{1}{2}\int \frac{\partial}{\partial x}\left(\frac{\partial u}{\partial x}\right)^2 dx = \left(\frac{\partial u}{\partial x}\right)^2\bigg|_{-\infty}^{\infty} = 0.$$

So, for a solution of the KdV equation, we have

$$\frac{1}{2}\frac{d}{dt}\int u^2\,dx \equiv 0 \quad \Rightarrow \quad \int u^2\,dx = \text{const}.$$

Remark 1. A similar calculation for the Burgers equation (14.1.8) results in the relation

$$\frac{1}{2}\frac{d}{dt}\int v^2(x,t)\,dx = -\mu \int \left(\frac{\partial v}{\partial x}\right)^2 dx < 0, \tag{1}$$

which reflects the fact that the viscosity ($\mu > 0$) dissipates the "energy"

$$E = \frac{1}{2}\int v^2(x,t)\,dx \tag{2}$$

of the wave.

2. Let us calculate the derivative

$$\frac{dv(x)}{dx} = -\frac{U^2}{2\mu \cosh^2\left(\frac{Ux}{2\mu}\right)},$$

of the stationary solution appearing under the integral sign in (14.5.1). Substituting this expression into the integral (14.5.1), we obtain

$$\Gamma = \frac{U^4}{4\mu} \int \frac{dx}{\cosh^4\left(\frac{Ux}{2\mu}\right)} = \frac{U^3}{2} \int \frac{dz}{\cosh^4(z)}.$$

Calculating the remaining integral, we have

$$\Gamma = \frac{2}{3} U^3. \tag{3}$$

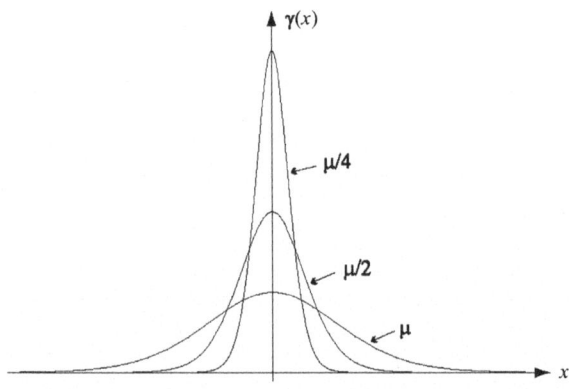

FIGURE 1
Plots of the dissipation of the density function $\gamma(x)$ of a stationary wave, shown here for identical U's and different μ: $\mu_1 = \mu$, $\mu_2 = \mu/2$, $\mu_3 = \mu/4$. It is clear that as μ decreases, the densities concentrate around their center point while keeping the area underneath constant.

Chapter 14. Nonlinear Waves and Growing Interfaces

Remark 1. Observe that the energy dissipation rate depends only on the amplitude of the wave U and is independent of the viscosity coefficient μ. Here, we see an echo of the Cheshire cat effect of Sect. 14.1: the viscosity disappears ($\mu \to 0_+$), but the dissipation of the field remains. For a detailed analysis of the mechanism underlying the Cheshire cat effect, it is useful to explore the *density of dissipation* field

$$\gamma(x,t) = \mu \left(\frac{\partial v(x,t)}{\partial x} \right)^2, \tag{4}$$

whose the integral over x gives the dissipation rate of the field (14.5.1). The graphs of dissipation density for the stationary wave

$$\gamma(x) = \frac{U^4}{4\mu \cosh^4 \left(\frac{Ux}{2\mu} \right)} \tag{5}$$

for identical U and different values of μ are shown in Fig. 1.

It is easy to prove that for $\mu \to 0_+$, the dissipation density (5) weakly converges to the Dirac delta,

$$\lim_{\mu \to \mu_+} \gamma(x) = \frac{2}{3} U^3 \delta(x).$$

This weak limit is yet another mathematical manifestation of the familiar fact that in the inviscid limit ($\mu \to 0_+$), all dissipation processes take place in the infinitesimal neighborhoods of jumps of the corresponding weak solutions of the nonlinear equations.

3. Let us first recover the initial conditions of the linear diffusion equation (14.3.11) corresponding to the given initial conditions of the Burgers equation. The initial potential is

$$s_0(x) = \int_0^x v_0(y)\, dy = -2\mu \ln \left[\cosh\left(\frac{U_1}{2\mu}(x-\ell_1)\right) \cosh\left(\frac{U_2}{2\mu}(x-\ell_2)\right) \right].$$

This and (14.3.11) give

$$\varphi_0(x) = \cosh\left(\frac{U_1}{2\mu}(x-\ell_1)\right) \cosh\left(\frac{U_2}{2\mu}(x-\ell_2)\right),$$

or

$$\varphi_0(x) = \cosh\left(\frac{U_+}{2\mu}(x-\ell_+)\right) + \cosh\left(\frac{U_-}{2\mu}(x-\ell_-)\right).$$

Here we have taken into account the fact that the initial potential $s_0(x)$ is defined only up to an additive constant, and the initial field $\varphi_0(x)$ is defined only up to a multiplicative constant. The notation is as follows:

$$U_\pm = U_2 \pm U_1, \qquad \ell_\pm = \frac{U_2 \ell_2 \pm U_1 \ell_1}{U_2 \pm U_1}.$$

The solution $\varphi(x,t)$ of the linear diffusion equation is obtained by noticing that the diffusion equation

$$\frac{\partial \varphi}{\partial t} = \mu \frac{\partial^2 \varphi}{\partial x^2},$$

with the initial condition $\cosh(ax)$, has a solution with separable variables,

$$e^{\mu a^2 t} \cosh(ax).$$

Thus

$$\varphi(x,t) = \exp\left(\frac{U_+^2}{4\mu} t\right) \cosh\left(\frac{U_+}{2\mu}(x - \ell_+)\right)$$

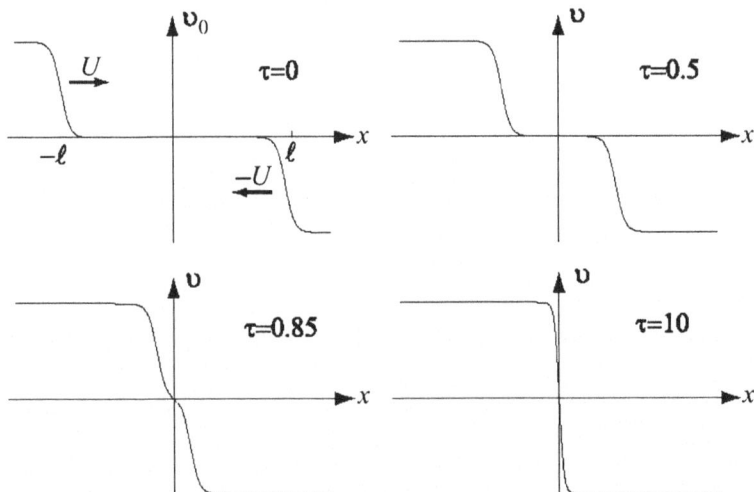

FIGURE 2
Merging stationary wave solutions, shown as functions of x, for several values of the dimensionless time $\tau = Ut/\ell$ and the dimensionless parameter $U\ell/\mu = 25$. Observe that the effective width of the transition area of the merged wave is roughly only one-half the analogous areas for the initial stationary wave solutions.

Chapter 14. Nonlinear Waves and Growing Interfaces

$$+ \exp\left(\frac{U_-^2}{4\mu}t\right)\cosh\left(\frac{U_-}{2\mu}(x-\ell_-)\right).$$

Substituting this expression in (14.3.9a), we finally get

$$v(x,t) = -\frac{U_+\sinh\left(\frac{U_+}{2\mu}(x-\ell_+)\right) + \exp\left(-\frac{U_1 U_2}{\mu}t\right)U_-\sinh\left(\frac{U_-}{2\mu}(x-\ell_-)\right)}{\cosh\left(\frac{U_+}{2\mu}(x-\ell_+)\right) + \exp\left(-\frac{U_1 U_2}{\mu}t\right)\cosh\left(\frac{U_-}{2\mu}(x-\ell_-)\right)}. \tag{6}$$

For concreteness, assume that $U_1 U_2 > 0$. Then, as t grows, the first terms in the numerator and denominator of (8) dominate the second terms, and the solution of the Burgers equation tends asymptotically to the stationary solution

$$\lim_{t\to\infty} v(x,t) = -U_+\tanh\left(\frac{U_+}{2\mu}(x-\ell_+)\right). \tag{7}$$

In other words, for large times, two initially separate stationary waves form a single stationary wave with amplitude $U_+ = U_1 + U_2$. The wave is centered on the point

$$\ell_+ = \frac{U_2\ell_2 + U_1\ell_1}{U_2 + U_1}. \tag{8}$$

In the special case $U_1 = U_2 = U$, $\ell_2 = -\ell_1 = \ell$, the solutions (6) of the Burgers equation are shown in Fig. 2. The plots illustrate the process of jump merging.

4. The skeleton (in the inviscid limit $\mu \to 0_+$) of the initial field $v_0(x)$, see (14.5.2), is shown in Fig. 3.

The theory of weak solutions of the Riemann equation, see Sect. 13.5.5, tells us that the jump's (shock) velocity is the average of the velocities of the jumps immediately to the left and to the right. In the case under consideration, this means that the left jump moves to the right with velocity U_2, and the right jump moves to the left with velocity U_1.

Thus, the equations of jumps' motion before their collision are

$$\ell_1(t) = \ell_1 + U_2 t, \quad \text{and} \quad \ell_2(t) = \ell_2 - U_1 t.$$

At the moment of collision,

$$\ell_1(t_*) = \ell_2(t_*) \quad \Rightarrow \quad t_* = \frac{\ell_2 - \ell_1}{U_2 + U_1},$$

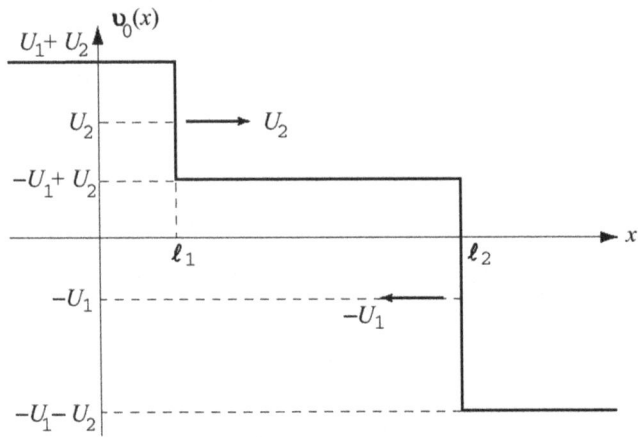

FIGURE 3
The skeleton of the initial field $v_0(x)$, see (14.5.2). The jumps move in the directions indicated by the *arrows*.

the jumps merge, creating a standing stationary wave, whose jump coordinate is
$$\ell_1(t_*) = \ell_2(t_*) = \ell_+.$$

5. In this case, the initial potential and the initial condition of the corresponding linear diffusion equation (14.3.11) are
$$s_0(x) = S\left[\chi(x) + \chi(x - \ell)\right]$$
and
$$\varphi_0(x) = 1 + \left(e^{-R} - 1\right)\chi(x) + \left(e^{-2R} - e^{-R}\right)\chi(x - \ell).$$

Substituting the above expression in (14.3.13), we obtain
$$\varphi(x,t) = 1 + \left(e^{-R} - 1\right)\Phi\left(\frac{x}{2\sqrt{\mu t}}\right) + \left(e^{-2R} - e^{-R}\right)\Phi\left(\frac{x-\ell}{2\sqrt{\mu t}}\right).$$

Thus, the solution of the Burgers equation has the form
$$v(x,t) = -\sqrt{\frac{\mu}{\pi t}}\,\frac{\left(e^{-R} - 1\right)\exp\left(-\frac{x^2}{4\mu t}\right) + \left(e^{-2R} - e^{-R}\right)\exp\left(-\frac{(x-\ell)^2}{4\mu t}\right)}{1 + \left(e^{-R} - 1\right)\Phi\left(\frac{x}{2\sqrt{\mu t}}\right) + \left(e^{-2R} - e^{-R}\right)\Phi\left(\frac{x-\ell}{2\sqrt{\mu t}}\right)}. \quad (9)$$

Chapter 14. Nonlinear Waves and Growing Interfaces

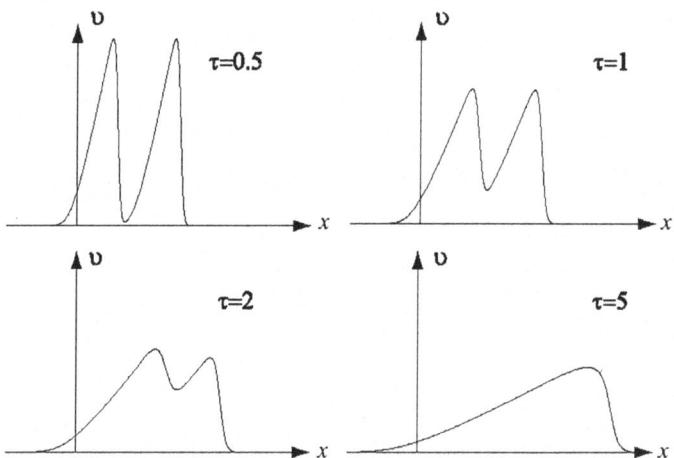

FIGURE 4
Plots of the solution (12) of the Burgers equation illustrating merging of triangular waves.

The plots of this solutions, for different values of τ, are shown in Fig. 4. They illustrate the process of merging of two triangular waves corresponding to the initial condition in the form of a sum of two Dirac deltas (14.5.3).

6. At the initial stage, the skeleton of the field consists of two separate triangles with the same area S. The triangles are bounded on the right by the jumps. Let us denote the coordinates of the jumps by $x_1(t)$ and $x_2(t)$, respectively. As long as the triangular waves do not overlap (see the upper graph in Fig. 5), the jumps' coordinates are easy to find using the fact that the areas of the triangles are preserved:

$$x_1(t) = \ell\sqrt{\tau}, \qquad x_2(t) = \ell(1 + \sqrt{\tau}), \qquad \tau = \frac{t}{t^*} = \frac{2St}{\ell^2}. \qquad (10)$$

At the time

$$t^* = \frac{\ell^2}{2S} \qquad (\tau = 1),$$

the left jump begins to overlap the right triangle (see the lower graph in Fig. 5), which changes the law of motion of the left jump. Let us find that law by solving the left jump's equation of motion,

$$\frac{dx_1(t)}{dt} = \frac{1}{t}\left(x_1(t) - \frac{\ell}{2}\right), \qquad x_1(t^*) = \ell.$$

This equation can be derived by recalling that the jump's velocity is equal to the average of the jumps' velocities to the left $(v_-(t) = x_1(t)/t)$ and to the right $(v_+(t) = (x_1(t) - \ell)/t)$.

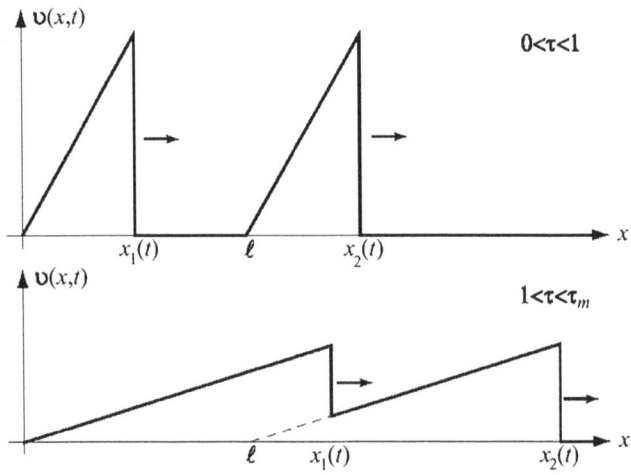

FIGURE 5
The skeleton of the solution (9) of the Burgers equation.

As a result, we get
$$x_1(t) = \frac{\ell}{2}(\tau + 1). \tag{11}$$
Comparing the coordinates of the right (10) and left (11) jumps, we obtain the desired merger time:
$$\tau_m = (1 + \sqrt{2})^2 \approx 5.83$$

Remark 1. Observe that at the beginning (for $\tau < 1$), the field $v(x,t)$ consists of two nonoverlapping triangular waves, for each of which the Reynolds number is $R = S/2\mu$. In the course of time, they merge into a single triangular wave with doubled area and doubled Reynolds number. Thus the merger of jumps leads to growth in the current Reynolds number and a loss of information about the fine structure of the initial field $v_0(x)$.

7. Suppose that at the jump point of the function $\{y\}(x,t)$, the parabolas $\Pi_1(x)$ and $\Pi_2(x)$ intersect. Retaining the two corresponding summands in (14.4.19), we shall rewrite $\{y\}(x,t)$ as

Chapter 14. Nonlinear Waves and Growing Interfaces

$$\{y\}(x,t) \approx \frac{y_1 \exp\left(-\frac{1}{2\mu t}\left[s_1 t + \frac{(x-y_1)^2}{2}\right]\right) + y_2 \exp\left(-\frac{1}{2\mu t}\left[s_2 t + \frac{(x-y_2)^2}{2}\right]\right)}{\exp\left(-\frac{1}{2\mu t}\left[s_1 t + \frac{(x-y_1)^2}{2}\right]\right) + \exp\left(-\frac{1}{2\mu t}\left[s_2 t + \frac{(x-y_2)^2}{2}\right]\right)}. \quad (12)$$

Define
$$x_1 = \frac{y_1 + y_2}{2}, \qquad \ell = \frac{y_2 - y_1}{2}, \qquad z = x - x_1.$$

Now, we can rewrite the expression (12) as

$$\{y\} \approx x_1 + \ell \, \frac{\exp\left(-\frac{s_2 t - z\ell}{2\mu t}\right) - \exp\left(-\frac{s_1 t + z\ell}{2\mu t}\right)}{\exp\left(-\frac{s_2 t - z\ell}{2\mu t}\right) + \exp\left(-\frac{s_1 t + z\ell}{2\mu t}\right)}.$$

Thus, finally,

$$\{y\} \approx \frac{y_1 + y_2}{2} + \frac{y_2 - y_1}{2} \tanh\left(\frac{\ell}{2\mu t}(x - x_1 - Ut)\right), \qquad U = \frac{s_2 - s_1}{y_2 - y_1}.$$

8. The corresponding solution of the linear diffusion equation (14.3.11) is

$$\varphi(x,t) = I_0(\mathrm{R}) + 2 \sum_{n=1}^{\infty} I_n(\mathrm{R}) \exp\left(-\frac{n^2 \tau}{2\mathrm{R}}\right) \cos(nz), \quad (13)$$

where
$$z = \kappa x, \qquad \tau = a\kappa t, \qquad \text{and} \quad \mathrm{R} = \frac{a}{2\mu\kappa}.$$

Substituting (13) in (14.3.9a) and retaining only the first summands in the numerator and denominator, we get

$$v \approx \frac{2a}{\mathrm{R}} \frac{I_1(\mathrm{R})}{I_0(\mathrm{R})} e^{-\frac{\tau}{2\mathrm{R}}} \sin(z) \qquad (\tau \gg \mathrm{R}). \quad (14)$$

In this case, it is useful to give the Reynolds number R an interpretation more appropriate from the viewpoint of acoustic applications: the number is equal to the ratio of the initial amplitude a of the field to the characteristic amplitude of the nonlinear effects,

$$a_n = \frac{1}{2\mu\kappa} \quad \Rightarrow \quad \mathrm{R} = \frac{a}{a_n}.$$

Moreover, it is natural to consider the dimensionless field

$$u = \frac{v}{a_n} = f(R) \, e^{-\frac{\tau}{2R}} \sin(z),$$

where

$$f(R) = 2 \frac{I_1(R)}{I_0(R)}.$$

When $f(R) \simeq R$, the field (14) coincides with the solution of the linear diffusion equation with initial condition (14.5.5), and the nonlinear effects can be neglected. A plot of the function $f(R)$ is shown in Fig. 6. It is clear that for large values of R, the function $f(R)$ stabilizes at level 2. This expresses the familiar fact that in the course of time, the nonlinear field "forgets" about the initial conditions, and in particular about its amplitude.

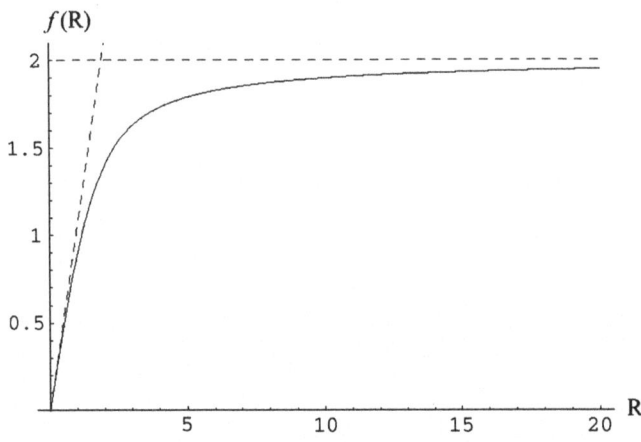

FIGURE 6
A plot of the dimensionless amplitude for an initially harmonic field at the linear evolution stage. As long as it is well approximated by the graph of the function $g(R) = R$, the nonlinear effects in the evolution of the field $v(x,t)$ can be neglected.

9. Obviously, in this case, the solution of the corresponding linear diffusion equation is

$$\varphi(x,t) = 2\mu t + x^2.$$

Thus the solution to the Burgers equation has the form

$$v(x,t) = -\frac{x}{t} \frac{2}{1+z^2}, \qquad z = \frac{x}{\sqrt{2\mu t}}. \qquad (15)$$

Plots of the equation at three consecutive time instants are shown in Fig. 7. At first sight, the above solutions are counterintuitive. Indeed, our mental picture would indicate that for $x < 0$, where $v(x,t) > 0$, the field should move to the right, and the negative field, for $x > 0$, should be displaced to the left. However, a more careful analysis of the solution and the graphics below shows that just the opposite is true. This phenomenon is due to the fact that the effective Reynolds number of the field (15) is close to 1, so that the anticipated inertial effects are overwhelmed by a diffusive washout.

10. The function $f(y;x,t)$, see (14.3.17), is uniquely determined by the form of the function $G(y;x,t)$. In the presence of V, it is

$$G(y;x,t|V) = s_0(y)t + Vyt + \frac{1}{2}(y-x)^2.$$

Multiply the numerator and denominator of the right side of the equality (14.3.17) by

$$\exp\left(\frac{Vx}{2\mu} - \frac{V^2 t}{4\mu}\right).$$

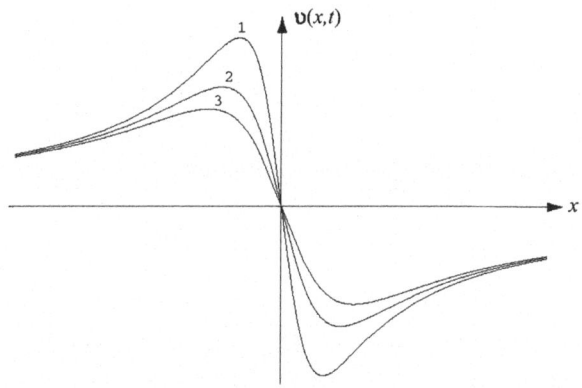

FIGURE 7
Plots of the solution (15) of the Burgers equation at three successive times, $t_1 = 1$, $t_2 = 2$, $t_1, t_3 = 3t_1$. The "anomalous" behavior of the solution is explained by the dominance of the diffusive effects over the nonlinear inertial effects.

As a result, the function $G(y;x,t)$ is transformed into

$$G(y;x,t|V) = s_0(y)t + \frac{1}{2}(y - x + Vt)^2,$$

which proves the validity of the equality (14.5.6).

Chapter 15: Other Standard Nonlinear Models

1. Indeed, it follows from (15.1.28) that

$$T = \rho_0 E - \frac{1}{3}\rho_0 t \int \frac{\partial v^3(x,t)}{\partial x}\,dx\,.$$

If the last integral is equal to zero, then

$$T = \rho_0 E\,.$$

2. Initially, let us prove that $\mathcal{M} = \text{const}$. To accomplish this, notice that one can rewrite equation (15.3.11) in the divergence form

$$\frac{\partial u}{\partial \tau} + \frac{\partial}{\partial s}\left(\frac{u^2}{2} + \gamma \frac{\partial^2 u}{\partial s^2}\right) = 0. \qquad (1)$$

Integrating this equality term by term over all of s values and assuming that the integral of u is finite while its first derivatives tend to zero as $s \to \pm\infty$, we obtain

$$\frac{\partial}{\partial \tau}\int u(s,\tau)ds = 0 \quad \Rightarrow \quad \mathcal{M} = \int u(s,\tau)ds = \text{const}\,. \qquad (2)$$

Analogously, one can obtain the integral equality

$$\frac{\partial}{\partial \tau}\left(\frac{u^2}{2}\right) + \frac{\partial}{\partial s}\left[\left(\frac{u^3}{3}\right) + \gamma\left(u\frac{\partial^2}{\partial s^2} - \frac{1}{2}\left(\frac{\partial u}{\partial s}\right)^2\right)\right] = 0\,.$$

Integrating the last relation over the whole s-axis, we obtain

$$\frac{1}{2}\frac{\partial}{\partial \tau}\int u^2(s,\tau)ds = 0 \quad \Rightarrow \quad \mathcal{K} = \frac{1}{2}\int u^2(s,\tau)ds = \text{const}\,. \qquad (3)$$

3. Using the relation (1), we obtain

$$\frac{d\mathcal{S}(\tau)}{d\tau} = \frac{1}{2}\int u^2(s,\tau)d\tau\,.$$

Taking into account the fact that the last integral does not depend on τ, one gets

$$\mathcal{S}(\tau) = \int s\,u_0(s)\,ds + \frac{\tau}{2}\int u_0^2(s)\,ds\,.$$

4. It is easy to check that the n-soliton solution (15.3.29), (15.3.53), is positive everywhere, and for any θ, it exponentially tends to zero as $s \to \pm\infty$. Thus the integral on the left-hand side of equality (15.5.2) exists. Since the solitons are moving with different velocities, they scatter over the course of time. As a result, as $\theta \to \infty$, the solitons are not overlapping, and one may rewrite the n-soliton solution in the form

$$w(s,\theta) \sim \frac{3k^2}{\sigma} \sum_{i}^{n} \operatorname{sech}^2 \left[\frac{k}{2}(s - d_i - k_i^2\theta)\right], \qquad \theta \to \pm\infty$$

$$\operatorname{sech}^2 \left[\frac{k}{2}(s - d_i - k_i^2\theta)\right] \cdot \operatorname{sech}^2 \left[\frac{k}{2}(s - d_j - k_j^2\theta)\right] \simeq 0, \qquad i \neq j.$$

In turn, as $\theta \to \infty$, the integral on the left-hand side of equality (15.5.2) becomes equal to the sum of the integrals over all solitons

$$\int \sqrt{w(s,\theta)}\, ds \sim \sqrt{\frac{12}{\sigma}} \sum_{i=1}^{n} k_i \int \operatorname{sech}\left[\frac{k}{2}(s - d_i - k_i^2\theta)\right] ds, \qquad \theta \to \pm\infty.$$

After a change of the variables of integration, we obtains

$$\int \sqrt{w(s,\theta)}\, ds \sim \sqrt{\frac{12}{\sigma}} \sqrt{\frac{12}{\sigma}}\, n \int \operatorname{sech}(z)\, dz = \sqrt{\frac{12}{\sigma}}\, \pi n.$$

We have used here the integral formula (15.3.34), taking into account that $S(1) = \pi$.

Using relation (15.3.15), we can rewrite (15.5.2) in the form

$$n = \frac{1}{\pi\sqrt{12\gamma}} \lim_{\tau \to \pm\infty} \int \sqrt{u(s,\tau)}\, ds.$$

Remark 2. The general theory of the KdV equation shows that if $w(s,0) > 0$ and

$$\int \sqrt{w(s,0)}\, ds \gg 1,$$

then the following approximate relations are true:

$$n \simeq \frac{1}{\pi\sqrt{12\gamma}} \int \sqrt{u_0(s)}\, ds \qquad (n \gg 1). \tag{4}$$

It is not unusual for asymptotic relations that are rigorously accurate for large values of some parameter (in our case, the number n of solitons) to also work satisfactorily for the intermediate values of the parameters. Let us check whether this observation applies to the relation (4). Substituting the initial condition (15.3.10) into (4), we obtain

$$n \simeq \frac{1}{\sqrt{3\pi\gamma}}.$$

In particular, in the case $\gamma = 0.1$ corresponding to Fig. 15.3.1, we obtain $n \simeq 1.03$, while for the case $\gamma = 0.04$ corresponding to Fig. 15.3.2, we have $n \simeq 1.63$.

5. To begin, assume, for the sake of generality, that the initial field $u_0(s)$ triggers an arbitrary number n of solitons. Neglecting the impact of oscillating tails and taking into account that for τ large enough, solitons don't overlap and are described by the relation (15.3.32), we obtain

$$\int u(s,\tau) \simeq 6\sqrt{\gamma} \sum_{n=1}^{n} \sqrt{c_i} \int \text{sech}^2(z)dz = 12\sqrt{\gamma} \sum_{i=1}^{n} \sqrt{c_i}$$

and

$$\int u^2(s,\tau) \simeq 18\sqrt{\gamma} \sum_{n=1}^{n} c_i \sqrt{c_i} \int \text{sech}^4(z)dz = 24\sqrt{\gamma} \sum_{i=1}^{n} c_i \sqrt{c_i}.$$

Here, we have used the relation (15.3.32) and the integral formula (15.3.34). Assuming that the soliton is unique and employing the momentum and energy invariants (2) and (3), we obtain two approximate relations for the soliton's velocity,

$$\sqrt{c} \simeq \frac{M}{12\sqrt{\gamma}}, \qquad c\sqrt{c} \simeq \frac{K}{12\sqrt{\gamma}}.$$

Let us check the accuracy of the above approximate relations in the case of the initial field (15.3.10) and $\gamma = 0.1$. The corresponding numerically calculated plots of the field $u(s,\tau)$ are pictured in Fig. 15.3.1. It is easy to calculate that in the case of the initial field (15.3.10), the momentum and energy are equal to

$$M = \sqrt{2\pi}, \qquad K = \frac{\sqrt{\pi}}{2}.$$

Accordingly, the values of a soliton's velocity, depending on the momentum and energy, are

$$c \simeq 0.437 \qquad \text{and} \qquad c \simeq 0.379.$$

Let us check the above estimates, relying on the results of numerical calculations for the solution (15.3.11) of the KdV equation depicted in Fig. 15.3.1. It follows from results of numerical calculations that the maximal value of the field $u(s, \tau)$, for $\tau = 50$, is equal to

$$\max u(s, \tau = 50) = 1.1289.$$

Assuming that this maximal value is equal to the soliton's maximal value, which is equal to three times the soliton's velocity, we obtain

$$3c = 1.1289 \quad \Rightarrow \quad c = 0.376.$$

This last result agrees well with the above velocity estimate, which relied on the energy, and is in satisfactory agreement with the estimate relying on the momentum.

Appendix B
Bibliographical Notes

The **history of distribution theory** and its applications in physics and engineering goes back to

[1] O. HEAVISIDE, On operators in mathematical physics, *Proc. Royal Soc. London*, **52**(1893), 504–529, and **54** (1894), 105–143, and

[2] P. DIRAC, The physical interpretation of the quantum dynamics, *Proc. Royal Soc. A, London*, **113**(1926-7), 621–641.

A major step toward the **rigorous theory** and its application to weak solutions of partial differential equations was made in the 1930s by

[3] J. LERAY, Sur le mouvement d'un liquide visquex emplissant l'espace, *Acta Math.* **63** (1934), 193–248,

[4] R. COURANT, D. HILBERT, *Methoden der mathematischen Physik*, Springer, Berlin 1937, and

[5] S. SOBOLEV, Sur une théorème de l'analyse fonctionelle, *Mat. Sbornik* **4** (1938), 471–496.

The theory obtained its **definitive** elegant **mathematical form** (including the locally convex linear topological spaces formalism) in a classic treatise of

[6] L. SCHWARTZ, *Théorie des distributions*, vol. I (1950), vol. II (1951), Hermann, Paris,

which reads well even today.

In its **modern mathematical depth** and richness, the theory of distributions and its application to Fourier analysis, differential equations, and other areas of mathematics can be studied from many sources, starting with the massive multivolume works

[7] I.M. GELFAND et al. *Generalized Functions*, six volumes, Moscow, Nauka 1959–1966 (English translation: Academic Press, New York and London, 1967), and

[8] L. HÖRMANDER, *The Analysis of Linear Partial Differential Operators*, four volumes, Springer, 1983–1985,

to smaller, one-volume research-oriented monographs

[9] E.M. STEIN, G. WEISS, *Introduction to Fourier Analysis on Euclidean Spaces*, Princeton University Press 1971,

[10] L.R. VOLEVICH, S.G. GINDIKIN, *Generalized Functions and Convolution Equations*, Moscow, Nauka 1994, (English translation: Gordon and Breach, 1990) and

[11] A. FRIEDMAN, *Generalized Functions and Partial Differential Equations*, Prentice Hall, Englewood Cliffs, N.J., 1963.

to textbook-style volumes less dependent on the locally convex topological vectors space technology,

[12] R. STRICHARTZ, *A Guide to Distribution Theory and Fourier Transforms*, CRC Press, Boca Raton 1994,

[13] V.S. VLADIMIROV, *Equations of Mathematical Physics*, Moscow, Nauka 1981 (English translation available on: http://www.sps.org.sa/BooksandMagazinesLibraryFiles/Book_19_152.pdf),

[14] G. GRUBB, *Distributions and Operators*, Springer-Verlag 2009,

[16] J.J. DUISTERMAAT, J.A.C. KOLK, *Distributions: Theory and Applications*, Birkhäuser-Boston, 2010.

An **elementary**, but rigorous, **construction** of distributions based on the notion of equivalent sequences was developed by

[14] J. MIKUSIŃSKI, R. SIKORSKI, *The Elementary Theory of Distributions*, I (1957), II (1961), PWN, Warsaw, and

[17] P. ANTOSIK, J. MIKUSIŃSKI, R. SIKORSKI, *Generalized Functions, the Sequential Approach*, Elsevier Scientific, Amsterdam 1973.

Appendix B Bibliographical Notes 399

Applications of the theory of distributions have appeared in uncountable physical and engineering books and papers. As far as more recent, applied-oriented textbooks are concerned, which have some affinity to our book, we would like to quote

[18] F. CONSTANTINESCU, *Distributions and Their Applications in Physics*, Pergamon Press, Oxford 1980, and

[19] T. SCHÜCKER, *Distributions, Fourier Transforms and Some of Their Applications to Physics*, World Scientific, Singapore 1991,

which however, have a different spirit and do not cover some of the modern areas covered by our book.

The classics on **Fourier integrals** are

[20] S. BOCHNER, *Vorlesungen über Fouriersche Integrale*, Akademische Verlag, Leipzig 1932, and

[21] E.C. TITCHMARSH, *Introduction to the Theory of Fourier Integrals*, Clarendon Press, Oxford 1937,

with numerous modern books on the subject, including the above-mentioned monograph [9] and elegant expositions by

[22] H. BREMERMANN, *Complex Variables and Fourier Transforms*, Addison-Wesley, Reading, Mass. 1965,

[23] H. DYM, H.P. MCKEAN, *Fourier Series and Integrals*, Academic Press, New York 1972, and

[24] T.W. KÖRNER, *Fourier Analysis*, Cambridge University Press 1988.

The **asymptotic problems** (including the method of stationary phase) discussed in this book are mostly classical. A well-known reference is

[25] N.G. DE BRUIJN, *Asymptotic Methods in Analysis*, North-Holland, Amsterdam 1958,

with a newer source being

[26] M.B. FEDORYUK, *Asymptotics, Integrals, Series*, Nauka, Moscow 1987.

The classic text on **divergent series** is

[27] G.H. HARDY, *Divergent Series*, Clarendon Press, Oxford 1949,

but the problem has broader implications and connections with asymptotic expansions and functional-analytic questions concerning infinite matrix operators; see, e.g.,

[28] R.B. DINGLE, *Asymptotic Expansions*, Academic Press, New York 1973, and

[29] I.J. MADDOX , *Infinite Matrices of Operators*, Springer-Verlag, Berlin 1973.

A modern viewpoint is presented in

[30] B. SHAWYER, B. WATSON, *Borel's Methods of Summability: Theory and Applications*, Clarendon Press, Oxford 1994.

Finally, an exhaustive discussion of **Shannon's sampling theorem** and related interpolation problems can be found in

[31] R.J. MARKS II, *Introduction to Shannon's Sampling and Interpolation Theory*, Springer-Verlag, Berlin 1991.

The linear problems considered in Part III are classical, and there is an enormous literature on the subject going back to

[32] J.-B. D'ALEMBERT, Recherches sur la courbe que forme une corde tendu mise en vibration, *Histoire de l'Académie royale des sciences et belles lettres de Berlin*, 3 (1747), 214–219,

for the **hyperbolic wave equations**, and

[33] J.-B. FOURIER, *Théorie analytique de la chaleur*, Paris 1822,

for the **parabolic diffusion (heat) equations**. A modern mathematical treatment can be found in the above-mentioned four-volume treatise by Lars Hörmander. The monumental, multivolume classic

[34] L. LANDAU AND E. LIFSCHITZ, *Course of Theoretical Physics*, Moscow 1948–1957 (English translation: The first eight volumes were translated into English by the late 1950s. The last two volumes were written in the early 1980s. Vladimir Berestetskii and Lev Pitaevskii also contributed to the series. They were published by various publishers: Pergamon press, Addison-Wesley and Butterworth–Heinemann),

Appendix B Bibliographical Notes

provides a complete classical picture of the **equations of mathematical physics**. Two volumes,

[35] R. COURANT AND D. HILBERT, *Methods of Mathematical Physics*, Berlin 1924, (English translation: Interscience Publishers Inc. 1953), and

[36] R. COURANT, *Partial Differential Equations*, New York 1962,

were also historically important.

The theory of **harmonic functions, elliptic partial differential equations**, and related problems of **potential theory** and **Markov processes** can be found in the magisterial, almost 900-page,

[37] J.L. DOOB, *Classical Potential Theory and Its Probabilistic Counterpart*, Springer-Verlag, New York 1984.

A more accessible textbook in the field is the recent

[38] L.L. HELMS, *Potential Theory*, Springer-Verlag, 2009.

Parabolic equations are lucidly presented in

[39] A. FRIEDMAN, *Partial Differential Equations of Parabolic Type*, Prentice Hall, Englewood Cliffs, N.J., 1964.

Two well-known pathbreaking monographs on connections between parabolic equations and diffusion stochastic processes and stochastic differential equations are

[40] D.W. STROOCK AND S.R.S. VARADHAN, *Multidimensional Diffusion Processes*, Springer-Verlag, New York, 1979,

and

[41] N. IKEDA AND S. WATANABE, *Stochastic Differential Equations and Diffusion Processes*, North-Holland, Amsterdam 1981.

Every textbook on partial differential equations contains chapters on **linear hyperbolic equations**. Excellent sources here are

[42] F. JOHN, *Partial Differential Equations*, third edition, Springer-Verlag, New York 1978,

[43] L.C. EVANS, *Partial Differential Equations*, American Mathematical Society, Providence, R.I., 1998.

The latter is a popular text for PDE courses for American graduate students in pure mathematics. On the other hand,

[44] L. GARDING, Linear hyperbolic partial differential equations with constant coefficients, *Acta Math.*, 85 (1950), 1–62.

provides an example of an in-depth modern research article.

Nonlinear problems discussed in Part IV have a vast literature both mathematical and physical, and we just provide a small sample of some our favorites:

[45] L. HÖRMANDER, *Lectures on Nonlinear Hyperbolic Differential Equations*, Springer-Verlag, New York 1997,

[46] R. COURANT AND K.O. FRIEDRICHS, *Supersonic Flow and Shock Waves*, Springer-Verlag, New York 1976,

[47] G.B. WHITHAM, *Linear and Nonlinear Waves*, J. Wiley and Sons 1974,

[48] C.M. DAFERMOS, *Hyperbolic Conservation Laws in Continuum Physics*, Springer-Verlag, New York 2000, and

[49] J. SMOLLER, *Shock Waves and Reaction–Diffusion Equations*, Springer-Verlag, New York 1994.

[50] M.E. TAYLOR, *Partial Differential Equations III, Nonlinear Equations*, Springer-Verlag, New York 1996.

The **scaling philosophy** and its implementation, especially to the **porous medium equation**, is explained in

[51] G.I. BARENBLATT, *Scaling, Self-Similarity, and Intermediate Asymptotics*, Cambridge University Press, 1996.

Finally, we mention two titles by the authors of the present volume:

[52] S. GURBATOV, A. MALAKHOV, AND A. SAICHEV, *Nonlinear Random Waves and Turbulence in Non-dispersive Media: Waves, Rays and Particles*, Manchester University Press, 1991, and

[53] W.A. WOYCZYŃSKI, *Burgers–KPZ Turbulence, Göttingen Lectures*, Springer-Verlag, New York 1998.

These go beyond the material discussed in Chaps. 14–15 by considering the dynamics of random fields governed by nonlinear hyperbolic and parabolic equations, the subject matter of the forthcoming Volume 3 of the present series, *Distributions in the Physical and Engineering Sciences*.

Index

3-D Burgers equation, 289
3-D KPZ equation, 289

isocline trajectory , 165
linear waves, 93

absorbing barrier, 75
absorbing boundary, 76
absorption rule, 204
acoustic time, 285
admissibility criterion, 203
Airy's function, 118
Alice in Wonderland, 78
amplitude, 96
anisotropic surface growth, 162
antenna radiation, 20
anti-invariance, 237
antisymmetry, 237
averaged Lagrangian, 257

Barenblatt solution, 324
Bessel equation, 15
Bessel function, 16
Bessel functions, 15, 157, 355
Bessel inequality, 50
Bessel transform, 23
boundary condition, homogeneous, 30
boundary conditions, inhomogeneous, 31
boundary conditions, third kind, 37

boundary problem, interior, 30
Burgers equation, 233, 255
Burgers equation, linear regime, 260
Burgers equation, N waves, 262
Burgers equation, U waves, 262

canonical form of KdV, 298
Cauchy kernel, 91
caustic, 115
Cesàro method, 139
characteristic regimes of Burgers equation, 259
Cheshire cat, 234, 383
circular antenna, 25
combustion region, 220
complex parabolic equations, 65
compressible gas, 282
concentration field, 291
conservation law, 233
conservation of mass, 321
conserved quantities, 100
continuity equation, 109, 150, 282
convex envelope, 194
critical parabola, 192
cumulative density of particles, 61
cumulative mass field, 151
cylindrical functions, 16

Darcy's law, 321
density of 3-D flow, 218

density of particles, 60
detonating fuse, 81
detonation wave, 226
diffraction problem, 67
diffusion coefficient, 59
diffusion equation, 59, 255
dimension reduction, 13
dimensional analysis, 64
dimensionless variables, 64
dimensionless KdV, 298
Dirac delta, Fourier series, 52
Dirac delta, integral representation, 25
directional antenna, 23
directional diagram, 23, 27
Dirichlet problem, exterior, 18
Dirichlet condition, 76
Dirichlet conditions, homogeneous, 30
disk antenna, 25
dispersion, 234
dispersion curves, 98
dispersion relation, 93, 95
dissipation of energy, 234
distortionless line, 122
divergence field, 150
Doppler effect, 131

E–Rykov–Sinai principle, 197
eigenfunctions, 135
eigenvalue, multiple, 41
eigenvalue, simple, 41
eigenvalues, 135
eigenvalues, comparison theorems, 47
eigenvalues, extremal properties, 45
elliptic equations, 3
end of the world, 259
energy conservation law, 101
entropy conditions, 204
equation of state, 321

error function, 62
Euler's formula, 16
Eulerian coordinates, 147
Eulerian field, 147
explosion, 107
external scales, 240

fire front, 219
floor function, 56
forest fire, 161
Fourier series, 49
Fraunhofer zone, 22
Fresnel approximation, 68
Fresnel zones, 68
fundamental solution, 3, 59

Galilean invariance, 236, 296
Gaussian kernel, 91
generalized solutions, 171
geometric dispersion, 94
geometric optics, 113
global maximum principle, 221
gradient catastrophe, 177, 241, 259
gradient field, 162
gravitational instability, 167
gravitational water waves, 99
Green's function, 3, 8, 59
Green's function, localized, 56
Green's identity, 5
group velocity, 102

Hamilton–Jacobi equation, 162
Hankel function, 14, 16
Hankel transform, 23
heat equation, 59
Helmholtz equation, 9
Helmholtz equation, 2-D, 13
Helmholtz equation, 3-D, 12
Hopf–Cole formula, 256
Hopf–Cole substitution, 255
Hubble constant, 167

Hubble expansion solution, 241
Huygens–Fresnel principle, 68
hydrodynamic flow, 106
hyperbolic equations, 93

inelastic collision, 198
inhomogeneous diffusion equation, 63
inhomogeneous media, 29
inner product, 135
integral conservation laws, 194
interface growth, 172
interface growth equation, 158, 219
internal scales, 240
invariance, 236
invariance property, 79
invariants, 100
inverse Bessel transform, 24
inviscid equation, 233
inviscid limit, 234
irreversible process, 60

Jacobian, 217

KdV equation, 294, 295
Khokhlov's solution, 246, 286
Kolmogorov–Feller equation, 91, 336
Korteweg–de Vries equation, 99, 235
KPZ equation, 230, 255

Lagrangian coordinates, 146
Lagrangian field, 147
Laplace operator, 4
Laplacian, 4
Lego blocks, 253
linear dependence, 39
local sound velocity, 282
localized potential, 261

Mach number, 285
macroparticle, 198
mass conservation law, 154

mass distribution in universe, 167
master equations, 172
Maxwell density, 184
Maxwell's rule, 194
melting ice cube, 161
melting surface, 167
mixed boundary problem, 76
modes, 56
momentum conservation law, 154, 284
monochromatic source, 10
monochromatic wave, 17
moving boundary, 81, 129
multidimensional flows, 215
multiray regime, 115
multistream solutions, 174
multivalued function, 176, 178

nabla operator, 214
narrow-band wave packet, 110
Navier–Stokes equation, 233
near-field zone, 21
Neumann conditions, homogeneous, 30
Neumann function, 16
nondispersive medium, 97
nonlinear acoustics, 173
nonlinear diffusion equation, 240
nonlinear PDEs, 229
nonlinear waves, 215, 233

ocean as waveguide, 33
Ohm's law, 127
Oleinik–Lax minimum principle, 190
operator, positive definite, 42
optical rays, 161
orthogonality, w.r.t. weight function, 39
osculating parabola, 192

parabolic equation, 59

Parseval's equality, 50, 101
particle density, 150
particle flow, 106
peculiar velocity, 242
phase, 96
phase screen, 111
phase space, 106
phase velocity, 96
Planck's constant, 65
Poisson equation, 3, 4
Poisson integral, 73, 328
polytropic gas, 282, 321
porous medium, 321
potential field, 214
potential initial field, 291
potential theory, 3
potential well, 87
pressureless gas, 287
propagator, 67

quasilinear equations, 281
quasioptics equation, 68
quasiparticle, 108
quasiperiodic waves, 272

radiation condition, 11, 128
radiation pattern, 23
rarefaction, 108
ray, 159
reciprocity theorem, 29, 30
recurrence relations, 25
reflection invariance, 236
reflection method, 75
regularization of PDEs, 229
Reynolds number, 238, 247, 258
Reynolds number, power law, 277
Riemann equation, 145

sawtooth waves, 266
scale invariance, 238
Schrödinger equation, 65, 88

self-adjoint operator, 38
self-similar solutions, 61, 248, 259
separation of variables, 33, 134
shallow water approximation, 99
shock creation, 192
single stream motion, 174, 175
skeleton of self-similar solution, 253
slope field, 162
smoothing heat bath, 183
smoothing out of jumps, 246
snowfall in absence of wind, 162
snowfall in presence of wind, 169
solitary waves, 294
soliton, 294, 299
soliton collisions, 316
soliton's "mass", 312
soliton's kinetic energy, 312
source solution, 259
spatial dispersion, 97
stationary phase, 93
stationary phase method, 112
stationary wave, 245
steepest descent, 35
Steklov's theorem, 50
sticky particles, 198, 271
strongly nonlinear equation, 281
Sturm–Liouville expansions, 48
Sturm–Liouville operator, 37
Sturm–Liouville problem, 37
Sturm–Liouville problem, eigenfunctions, 38
Sturm–Liouville problem, eigenvalues, 38
summing over streams, 180
supersingular distributions, 212
support hyperplanes, 194
symmetries, 236
symmetries of the Burgers equation, 236
symmetries of the KPZ equation, 236

Index

telegrapher's equation, 119
time-reversible medium, 95
total particle density, 180
transfer function, 67
translation invariance, 236
traveling wave, 245

velocity potential, 172
vibrations of a rod, 99
viscous medium, 260

warm particle flow, 183
water waves, 99
wave as quasiparticle, 106
wave diffraction, 17
wave function, 66
wave intensity, 74, 100

wave packet, 96
wave turnover time, 176
wave zone, 21, 67
wave's energy, 100
waveguide, 33, 53, 97
waveguide mode, 98
wavelength, 11
wavenumber, 11, 98
waves, 93
wavevector, 97
weak solutions, 190
Weierstrass theorem, 6
well-posed problem, 60
Wronskian determinant, 40

zero-viscosity limit, 234

Applied and Numerical Harmonic Analysis

A.I. Saichev and W.A. Woyczyński: *Distributions in the Physical and Engineering Sciences* (ISBN 978-0-8176-3942-6)

R. Tolimieri and M. An: *Time-Frequency Representations* (ISBN 978-0-8176-3918-1)

G.T. Herman: *Geometry of Digital Spaces* (ISBN 978-0-8176-3897-9)

A. Procházka, J. Uhlíř, P.J.W. Rayner, and N.G. Kingsbury: *Signal Analysis and Prediction* (ISBN 978-0-8176-4042-2)

J. Ramanathan: *Methods of Applied Fourier Analysis* (ISBN 978-0-8176-3963-1)

A. Teolis: *Computational Signal Processing with Wavelets* (ISBN 978-0-8176-3909-9)

W.O. Bray and C.V. Stanojević: *Analysis of Divergence* (ISBN 978-0-8176-4058-3)

G.T Herman and A. Kuba: *Discrete Tomography* (ISBN 978-0-8176-4101-6)

J.J. Benedetto and P.J.S.G. Ferreira: *Modern Sampling Theory* (ISBN 978-0-8176-4023-1)

A. Abbate, C.M. DeCusatis, and P.K. Das: *Wavelets and Subbands* (ISBN 978-0-8176-4136-8)

L. Debnath: *Wavelet Transforms and Time-Frequency Signal Analysis* (ISBN 978-0-8176-4104-7)

K. Gröchenig: *Foundations of Time-Frequency Analysis* (ISBN 978-0-8176-4022-4)

D.F. Walnut: *An Introduction to Wavelet Analysis* (ISBN 978-0-8176-3962-4)

O. Bratteli and P. Jorgensen: *Wavelets through a Looking Glass* (ISBN 978-0-8176-4280-8)

H.G. Feichtinger and T. Strohmer: *Advances in Gabor Analysis* (ISBN 978-0-8176-4239-6)

O. Christensen: *An Introduction to Frames and Riesz Bases* (ISBN 978-0-8176-4295-2)

L. Debnath: *Wavelets and Signal Processing* (ISBN 978-0-8176-4235-8)

J. Davis: *Methods of Applied Mathematics with a MATLAB Overview* (ISBN 978-0-8176-4331-7)

G. Bi and Y. Zeng: *Transforms and Fast Algorithms for Signal Analysis and Representations* (ISBN 978-0-8176-4279-2)

J.J. Benedetto and A. Zayed: *Sampling, Wavelets, and Tomography* (ISBN 978-0-8176-4304-1)

E. Prestini: *The Evolution of Applied Harmonic Analysis* (ISBN 978-0-8176-4125-2)

O. Christensen and K.L. Christensen: *Approximation Theory* (ISBN 978-0-8176-3600-5)

L. Brandolini, L. Colzani, A. Iosevich, and G. Travaglini: *Fourier Analysis and Convexity* (ISBN 978-0-8176-3263-2)

W. Freeden and V. Michel: *Multiscale Potential Theory* (ISBN 978-0-8176-4105-4)

O. Calin and D.-C. Chang: *Geometric Mechanics on Riemannian Manifolds* (ISBN 978-0-8176-4354-6)

J.A. Hogan and J.D. Lakey: *Time-Frequency and Time-Scale Methods* (ISBN 978-0-8176-4276-1)

C. Heil: *Harmonic Analysis and Applications* (ISBN 978-0-8176-3778-1)

K. Borre, D.M. Akos, N. Bertelsen, P. Rinder, and S.H. Jensen: *A Software-Defined GPS and Galileo Receiver* (ISBN 978-0-8176-4390-4)

Applied and Numerical Harmonic Analysis (Cont'd)

T. Qian, V. Mang I, and Y. Xu: *Wavelet Analysis and Applications* (ISBN 978-3-7643-7777-9)

G.T. Herman and A. Kuba: *Advances in Discrete Tomography and Its Applications* (ISBN 978-0-8176-3614-2)

M.C. Fu, R.A. Jarrow, J.-Y. J. Yen, and R.J. Elliott: *Advances in Mathematical Finance* (ISBN 978-0-8176-4544-1)

O. Christensen: *Frames and Bases* (ISBN 978-0-8176-4677-6)

P.E.T. Jorgensen, K.D. Merrill, and J.A. Packer: *Representations, Wavelets, and Frames* (ISBN 978-0-8176-4682-0)

M. An, A.K. Brodzik, and R. Tolimieri: *Ideal Sequence Design in Time-Frequency Space* (ISBN 978-0-8176-4737-7)

B. Luong: *Fourier Analysis on Finite Abelian Groups* (ISBN 978-0-8176-4915-9)

S.G. Krantz: *Explorations in Harmonic Analysis* (ISBN 978-0-8176-4668-4)

G.S. Chirikjian: *Stochastic Models, Information Theory, and Lie Groups, Volume 1* (ISBN 978-0-8176-4802-2)

C. Cabrelli and J.L. Torrea: *Recent Developments in Real and Harmonic Analysis* (ISBN 978-0-8176-4531-1)

M.V. Wickerhauser: *Mathematics for Multimedia* (ISBN 978-0-8176-4879-4)

P. Massopust and B. Forster: *Four Short Courses on Harmonic Analysis* (ISBN 978-0-8176-4890-9)

O. Christensen: *Functions, Spaces, and Expansions* (ISBN 978-0-8176-4979-1)

J. Barral and S. Seuret: *Recent Developments in Fractals and Related Fields* (ISBN 978-0-8176-4887-9)

O. Calin, D. Chang, K. Furutani, and C. Iwasaki: *Heat Kernels for Elliptic and Sub-elliptic Operators* (ISBN 978-0-8176-4994-4)

C. Heil: *A Basis Theory Primer* (ISBN 978-0-8176-4686-8)

J.R. Klauder: *A Modern Approach to Functional Integration* (ISBN 978-0-8176-4790-2)

J. Cohen and A. Zayed: *Wavelets and Multiscale Analysis* (ISBN 978-0-8176-8094-7)

D. Joyner and J.-L. Kim: *Selected Unsolved Problems in Coding Theory* (ISBN 978-0-8176-8255-2)

J.A. Hogan and J.D. Lakey: *Duration and Bandwidth Limiting* (ISBN 978-0-8176-8306-1)

G. Chirikjian: *Stochastic Models, Information Theory, and Lie Groups, Volume 2* (ISBN 978-0-8176-4943-2)

G. Kutyniok and D. Labate: *Shearlets* (ISBN 978-0-8176-8315-3)

For a fully up-to-date list of ANHA titles, visit http://www.springer.com/series/4968?detailsPage=titles or http://www.springerlink.com/content/t7k8lm/.

The manufacturer's authorised representative in the EU is Springer Nature Customer Service Centre GmbH, Europaplatz 3, 69115 Heidelberg, Germany. If you have any concerns regarding our products, please contact ProductSafety@springernature.com

Printed and bound by CPI Group (UK) Ltd, Croydon, CR0 4YY

23/03/2026

02076380-0007